America's Founding Heritage

3RD EDITION

Frank W. Fox Clayne L. Pope

BYU ACADEMIC PUBLISHING

America's Founding Heritage

3RD EDITION

Frank W. Fox *Clayne L. Pope*

BYU ACADEMIC PUBLISHING

Publisher
Roger Reynolds

Managing Editor
Jennifer Berry

Assistant Managing Editor
Kent Minson

Authors' Assistant
Karen Bryce

Text Editors
Jennifer Berry, Matthew Haslam, Karen Bryce

Typesetting
Kent Minson, Jennifer Berry

Illustrations
Devin LuBean, Kent Minson, Jennifer Berry

Primary Photographer
Devin LuBean

Proofreaders
Matthew Haslam, Kent Minson, Samantha Jones, Jae Hong

Technical Support
Justin Caldwell

Printer
BYU Print and Mail Production Center
Doug Maxwell, Ben Bean, Ed Godinez

For more information contact:
BYU Academic Publishing
3995 WSC, Provo, UT 84602
(801) 422–6231
academicpublishing@byu.edu
http://academicpublishing.byu.edu

To report ideas or text corrections email us at:
textideas@byu.edu

ISBN 10: 0-8425-2663-3
ISBN 13: 978-0-8425-2663-0

Acknowledgements

This book is an outgrowth of more than two decades teaching American Heritage at Brigham Young University. Over the years, we have had innumerable discussions with colleagues in economics, history and political science on the core concepts of this book. We would like to thank the many colleagues who have helped us refine our thinking. For this book, we note particularly Brent Gilchrist, Ralph Hancock, Matt Holland, Chris Karpowitz, James Kearl, Richard Kimball, Matthew Mason, Jeremy Pope, Rulon Pope, Jenny Pulsipher, David Spencer and Larry Wimmer. Additionally, Brett Latimer, Matt Heimburger, and Rick Griffin, helped us teach this course for a number of years and we learned significantly from them.

We are grateful for the direction and support of the College of Family, Home and Social Sciences led by Dean David B. Magleby and Associate deans Renata Forste and Elaine Walton. They have given both financial and moral support.

The staff of American Heritage has given us never-ending support in our efforts to write this text and to develop the course. Cindy Ingersoll has translated her passion for the course into countless excellent examples and applications in teaching, many of which find their way into this book in various forms. Karen Bryce has managed the crucial, but thankless work of moving a manuscript from draft to finished text while keeping the teaching of the course going. Both Cindy and Karen have demonstrated the kind of civic commitment we hope to engender in our students.

Many teaching assistants have done double-duty teaching discussion sections while conducting research for this text. In particular, Trevor Higbee, Tyler Johnson, Tally Payne and Sarah Rapp made numerous contributions to the writing and research. Also Mike Allen, Hugh Cherry, Emily DeMaso, April Hallam, Scott Howell, Kiersten Kariya, Jake Naylor, Dallas Perkins, David Stott, Talia Strong, Elizabeth Wickham, and Trevor Winn helped in the formulation of study guides and other details of the text.

Kay Darowski and Virginia Rush have worked on several editions of the textbook in both copyediting and image choices. Kay has faithfully tracked down obscure facts and elusive dates, while Virginia has spent hours on the internet examining thousands of available photographs and illustrations.

Professor Jack Welch and Heather Seferovich of BYU Studies graciously helped us publish preliminary versions of this text for student use. We deeply appreciate their help and support.

Jennifer Berry and staff at BYU Academic Publishing have worked diligently to bring the book to production in a timely manner. They have tolerated the naïve and, sometimes, time-consuming suggestions of the authors.

Our wives, Elaine Fox and Carolyn Pope, and children, David and Michael Fox, Jeremy Pope and Hilary Pope Erickson, have had to listen to interminable discussions of American heritage for three decades. Their patience and good humor has been unfailing.

But this book owes most to the tens of thousands of students who have taken our classes in American heritage over the years. They have consistently pressed us to clarify our argument, to substitute simplicity for unneeded complexity and to never gloss over the tough issues. This book is dedicated to them and to the tens of thousands who will follow.

Contents

INTRODUCTION

Preamble

We the people of the United States, in Order to form a more perfect Union, establish Justice, insure domestic Tranquility, provide for the common defense, promote the general welfare, and secure the Blessings of Liberty, to ourselves and our Posterity, do ordain and establish this Constitution for the United States of America.

As far back as we can peer through history, humankind has always lived under some form of government. Families, clans, tribes, city-states, nation-states, and various other groupings have always devised some means by which a family or particular group live together in a more or less orderly way. The need for government, a social structure that allows participating individuals some sense of belonging, safety and fair treatment, seems to lie in the socializing nature of the human species. This need is coupled with our instinctive desire to seek pleasure, avoid pain, and express ourselves. Through the chosen (or reluctantly accepted) government structure under which we live, we must constantly wrestle with the financial and social dilemmas that historically have always pushed us together or pulled us apart.

Founding

A conscious, deliberate act of creating a system of government that benefits the people.

Patria

A sense of homeland.

In an ideal world, government would resolve such dilemmas unfailingly, and the rational majority would recognize the outcome as fair, reasonable, compassionate, helpful and wise. In the real world, alas, government often becomes an instrument of tyranny. The great horrors of history—from Pharaoh oppressing the Israelites to Hitler incinerating the Jews to Cambodia's Pol Pot murdering practically every citizen who had an education or wore glasses—are essentially the work of government.

The question then is this: how can we enjoy the benefits of government without incurring its dreadful costs? Great thinkers and ambitious leaders throughout recorded history have proposed many solutions—but the majority of those solutions have failed, often dismally, and frequently with long-lasting ripples of misery for everyone involved. We have come to expect progress in many areas of life, such as science, technology and standards of living, but progress is elusive in the area of government. Years after technology helped land man on the moon, we have still witnessed such catastrophic government failures as Mao's China, Charles Taylor's Liberia, and Idi Amin's Uganda, to name but a few sad examples.

Most solutions to the problem of government have resulted abruptly and dramatically from war and conquest, or else have emerged slowly over time from historical experience. Once in a while circumstances have made it possible for the problem to be solved by conscious intention. A single individual, or more likely a small group of them, deliberates about what a better form of government might look like and how its elements might fit together. If the group is truly anxious to form a better government, it also tries to determine what the long-term result might possibly be. It happened just like that in ancient Greece, in the fifth century BC, when an Athenian tyrant named Cleisthenes, for reasons still unclear, sat down with the leading men of Athens and sketched out a democracy with various assemblies, law courts, and voting districts, together with all the rules and procedures necessary to make it work—an extraordinary feat at any time in history. We speak of such a conscious, deliberate act of creation as a **founding**.

While foundings are difficult to define precisely, they typically include a variety of elements, such as an outline of government, a body of law, definitions of citizenship, modes of participation and some evocation of divine approval. They may create an identity for some new group, together with a sense of homeland, or **patria**. A successful founding might result in a new culture blossoming. A new player might emerge in the game of nations. New political ideas may appear. And all of this likely transpires in an atmosphere of upheaval, for foundings rarely arise in quiet times.

While the founding of a new government *could* result from foreign conquest or a palace coup, we tend to insist otherwise. A true founding, as we now define such an event, must address the problem of government posed above, and it must also create reciprocal benefits, good things for both governed and government. We don't think of William the Conqueror as founding England because he merely invaded and subdued it. It was a century and a half later, when one of William's successors reluctantly signed the Magna Carta, that England experienced a *founding*.

Accordingly, foundings rest on ideas, beliefs, and values that are widely shared. The nearly 700-year-old Magna Carta includes strong reflections of right and wrong as generally accepted in Norman England:

> 38. No bailiff for the future shall, upon his own unsupported complaint, put anyone to his "law," without credible witnesses brought for this purpose.
>
> 39. No freeman shall be taken or imprisoned or disseised [dispossessed] or exiled or in any way destroyed, nor will we go upon him nor send upon him, except by the lawful judgment of his peers or by the law of the land.
>
> 40. To no one will we sell, to no one will we refuse or delay, right or justice.

The principles set down by the English magnates confronting King John are pretty clear. No one gets accused without witnesses. No one gets tried except by a jury of peers. No one can buy justice. No one can be denied it. Such principles are unifying elements. All who accept those of the Magna Carta become English in a sense and buy into the English founding. Foundings, in other words, are not solely about organizing governments—they are about building nationhood.

But organizing a government still holds center stage. Government alone can keep the strong from preying on the weak, can resolve disputes and settle controversies, can mobilize the community in time of peril, and can lay down rules to facilitate cooperation. Only when principles are embodied in specific laws, procedures, and an appropriate working structure—in other words, "government"—do they acquire practical force.

Figure I.1 The 1297 Magna Carta. "Magna Carta" actually refers to a group of several documents that delineate various principles of law and rule in England.

The Magna Carta is known for championing such ideas as equal justice and that kings and rulers are subject to law as well.

In the Magna Carta, for instance, consider the article that no freeman should be imprisoned "except by a lawful judgment of his peers." The principle set forth here is equal justice, something that was radical in a time when wealth and power usually won the argument. The procedure for attaining that principle is trial by jury. The structure of government required for such a procedure includes an independent judiciary, capable of judging without interference. Only when all of this is in place and operating can we have real justice, with all players competing on a level field. It works exactly like, say, the game of baseball, whose structure mandates two teams of nine players on a diamond-shaped field, whose rules spell out prescribed modes of participation, and whose freedom of action within that structure and those rules can be as fair (and as exciting) on a public park's diamond as in any World Series game.

A well-founded government must take human nature—complex and unpredictable—fully into account. In order to protect its people from each other, the government must know all the ways in which human beings might be victimized. Tougher still, it must be able to protect the people from *itself*, for rulers are also human. To the extent that government is like a machine, the machine must be able to perform a bewildering array of tasks without resorting to oppression. It must heed the majority will without being unjust to minorities. It must spell out rights and duties and give substance to both. It must allow for the equal treatment of some and the unequal treatment of others, based on standards of fairness. It must be able to build roads and dredge harbors and educate children and keep military forces at the ready. Trickiest of all, it must guard human freedom, which by its very nature is at odds with the idea of governance. John Adams once observed that the sound design and effective operation of such a machine were almost beyond the wit of man.

Republic

From the Latin *res publica*, the "public thing," when citizens of the political state govern themselves rather than submit to a despot or an oligarchy.

The founding of the United States of America resulted from one of these peculiar occurrences where great men with good intentions gathered to design a nation of many parts as one truly unique and stable organization. The Founding of our nation represented a rare circumstance in which it was possible to fabricate a government by conscious intention and call a nation-state into being. In the heat of a colonial rebellion, many ideas of the time were brought together and forged into a set of founding documents—the Declaration of Independence, the Constitution, the Bill of Rights—so extraordinary in character as to transform a rebellion against a few major political injustices into a true social, economic, and political revolution. These documents set forth a new theory of statehood, a new concept of rights, a new design of government, a new method of federation, a new form of political interaction, and eventually, a new approach to political economy. More remarkable still was the fact that all this novelty and innovation actually worked!

The American Founding was created for a specific group of people with a common historical experience and an emerging national character. Yet it reached toward the universal, claiming that "all men" were meant to devise their own political societies by such means. Accordingly, the unique and important fact of the American Founding was one of those fundamental principles—as fundamental as the Magna Carta—that would come to underlie virtually everything that transpired in that amazing twenty-five year period closing the eighteenth century. This was the idea of *government by the consent of the governed*.

Government by consent had long been discussed in the western world, but only as a philosophical proposition, never as a working reality. Government that was beholden to its own citizens—as opposed to a set of rulers—was the way things *ought* to be, perhaps, but rarely had been. In classical antiquity there had been a few republics operating on this principle—the term **republic**, from *res publica*, the "public thing," implying popular consent—and in Renaissance Europe there had been one or two brief experiments in republican self-rule. But most republics had proven to be unstable and short-lived, drifting toward anarchy on the one hand or tyranny on the other. Great Britain itself was a sort of republic, with its freely elected Parliament, but there again control by the few was unmistakable. Actually building and preserving a government based on the idea of popular consent had seemed only a dream. Now, suddenly, here it was.

Tied to the idea of consent was the companion idea of freedom for the individual—something else that had merely been talked about by philosophers and theologians. Americans creating their own government and supervising its operations made America's citizens free in ways that most human beings could barely imagine. Government by consent ensured that there could never be systematic oppression, at least for the broad majority, simply because people would never consent to their own tyrannizing. (Minorities would pose a delicate problem, of course, one we still struggle with today. But even for these it was possible to extend and secure freedom.) In the radically different new America there would be no storm troopers on midnight raids, no churches closed for unorthodox beliefs, no dissidents muzzled by the party in power.

As free American citizens, the former subjects of King George now had power—power over their government, power over their lives. Accordingly, they could reshape both as they saw fit and natural circumstances of the times allowed. The reshaping of American government took place at election time, and generally in an atmosphere of excitement. "Throw the bums out!" became an American battle cry around election time, and many an incompetent or law-breaking incumbent "bum" came to know its meaning, humiliated by a landslide of votes for his opponent. The reshaping of lives took the form of searching for opportunity, of seeking out the main chance, of selling stock and buying farms, of studying nights, learning a new trade, getting a better job, moving to a different town, starting over, doing better. Americans could invent themselves, it was observed, and reinvent themselves over and over again.

They could also reshape the Founding itself. This was the ultimate expression of government by consent. Part of the story we will tell here is what Americans made of their own Founding in the course of the nation's history. How did free people of so many varieties, armed with sovereign power, revisit their own institutions? How did they develop founding principles, extend these to new situations, apply them by analogy? How did they work out kinks, compose contradictions, resolve dilemmas? How, for instance, did they reconcile the idea of human freedom with the harsh fact of slavery? In sum, how did the American Founding operate both as a fixed body of agreement and as a growing, changing, increasingly prosperous organism?

Implied within this question is another one, equally important. To the extent that the original Founding succeeded, how have these subsequent modifications and innovations affected that success? If Jefferson or Madison stepped out of a time machine and looked around in the twenty-first century, would they be pleased with what they saw? And if they weren't pleased, should it make any difference to us?

This question can be put another way: What, exactly, is the heritage of the American Founding? **Heritage** is a term historians and others often use, never with much precision but always with emotional impact. A successful founding, one that stands the test of time, seems to validate its own principles. As in the case of Magna Carta, future generations buy into those principles and look back to them as a guide. Precedents, rules, procedures, an entire political culture may grow out of such principles, with the result that we begin to enshrine them. They become the legacy of a particular founding.

Heritage

The traditions, beliefs, principles, events, etc. that we inherit (or choose to inherit) from the past.

And yet legacy is also something with which we interact. Each successive generation revisits the founding principles, reframes them, and may wind up modifying them substantially. If several of the Founders held slaves and protected that practice in the Constitution, does slavery become a part of our heritage? If free speech meant only *political* speech to Thomas Jefferson, does the right to use vulgar four-letter words in public also become part of our heritage? If the Founders made no mention of judicial review (the power of the Supreme Court to declare legislative acts unconstitutional), does the subsequent practice of judicial review become part of our heritage? A true understanding of heritage requires us to be both cautious and thoughtful.

Constitutional growth did not take place in a vacuum. It was the result of an empowered people, and the American people have changed in significant ways in the two centuries since the Constitution was framed. From John Adams' time to our own, U.S. territory has more than tripled in size; the number of states has increased from 13 to 50; America's population has expanded a hundred fold; a tightly structured British Isles culture has morphed into a pluralistic global culture; economic subsistence has developed into economic prosperity; rural living and agrarian ways have been supplanted by urban living and city ways. And during all of these social and economic changes, science and technology have extended life, improved health, reduced the blue-collar work week, increased leisure time and opportunity and brought forth dizzying revolutions in transportation and communication. (Our time-machine Founders would be astounded by all that even if nothing else impressed them.)

And of course all that—and much more—have affected the Founding. Even so, *as* a Founding, the set of agreements and compromises or "compacts" that launched the United States of America as a nation must also be understood as transcending time and historical change. If a founding simply went with the flow, as we say, it wouldn't be a founding at all—as some religious organizations have learned to their cost. The questions addressed by the Founders were essentially the questions of the ages, and the answers they submitted were meant to be abiding:

> What promotes peace and security?
> What gains the respect of other nations?
> What fosters public happiness?
> What creates a sense of justice?
> What invokes divine approval?

Between 1775 and 1800 the American Founders wrestled with these questions in a direct and primary way. The founders were not really "demigods," to use Jefferson's term, but neither were they merely "dead white males," as a sneer of our own time suggests. They were intelligent and accomplished mortals caught in an exceptionally complex and trying situation. They ran risks, pondered alternatives, struggled with baffling difficulties and had no way of knowing how their labors would work in real time, then or for the next two centuries and beyond. They lacked all sense of postmodern political correctness—and all benefit of twenty-first century hindsight. All the same, they created a Founding that worked, that became a marvel to many in the world and something to hate among others, envious of what such freedoms produced. All in all, the Founding is our heritage, one we can justly be proud of. The answers the Founders of our nation came up with still provide order and hope in our lives today.

This book tells that story.

Chapter 1

THE PROBLEM OF GOVERNMENT

Abstract

Most people who have lived on the earth have had to choose between tyranny and anarchy in the great Human Predicament. Very few have been able to enjoy a "good society." Many philosophers and statesmen have tried to form ideas to overcome this problem with government enjoying varied success. This chapter looks at the characteristics of a founding and ancient society's beliefs on good government. Sovereignty, political legitimacy, and alternatives to tyranny and chaos are examined. The experiences of these early societies taught lessons to the American Founders as they tried to create a free society.

In the introduction, we saw that government is a fundamental necessity that often becomes oppressive. In this chapter we will zero in on this problem of government and attempt to see it through the eyes of the founders of the republican form of government practiced in the United States of America.

Figure 1.1 In 1957, the Arkansas governor directly challenged the federal government's power to enforce desegregation by not allowing nine African-American students to attend an all-white high school. President Eisenhower ordered the National Guard to ensure that they attended safely. In this example, the federal government showed that they ultimately possessed the final say, or sovereignty, in the segregation dispute.

Sovereignty

Sovereignty

Ultimate political power—having the final say.

For any government to do its job effectively, it must possess something political scientists call **sovereignty**. Sovereignty is ultimate political power—the final say. A political entity may be found on a map, but if it lacks sovereignty it is not truly a full-fledged nation-state. This is because a geographically delineated area may have a name, a flag, etc., but if its leaders do not have the ability to make final decisions for those living within its established area, the self-sufficiency required of nationhood cannot exist there.

Once in a while we catch a glimpse of what true sovereignty means. After the 1954 *Brown v. Board of Education* Supreme Court decision desegregating schools, Central High School in Little Rock, Arkansas became one of the first white schools in the state to admit black students. Nine black students approached the school on their first day only to find the Arkansas National Guard blocking their entrance by order of Governor Orval Faubus. A reluctant but determined President Dwight Eisenhower, in one of those critical moments of history, sent in the 101st Airborne Division of the U.S. Army with orders to enforce the black students' attendance at Central High by whatever means necessary. He eventually nationalized Faubus's own National Guard to continue to enforce the federal law throughout that year of high school. This event was an object lesson in sovereignty, with federal powers trumping state powers.

Sovereign power, however, carries an implicit and ever-present danger. George Washington captured the essence of the problem. He said, "Government is not reason; it is not eloquent; it is force. Like fire, it is a dangerous servant and a fearful master." Nations exist because of innate human desires to organize themselves to gain power and control over territory and circumstances, and history has shown that human nature can be corrupted by the possession of such power. Human beings have many good qualities, as witnessed through mankind's noble achievements, but they obviously also have a dark side. For every product of beauty and progress created by humankind, there have also been deplorable atrocities—pogroms, gulags, ethnic cleansings, tortures and reigns of terror. It is power and the lust for power that lurks behind most of the horrors of history. "Power corrupts," Lord Acton famously observed, "and absolute power corrupts absolutely."

The Human Predicament

Sovereign power and its potential for ill effects gives rise to the **Human Predicament**. Throughout history, kings, tyrants, **despots**, and dictators have used their power to cause pain and hardship among their subjects. Their subjects had three options: do nothing and remain miserable, leave the country, or revolt. Those who became restive under the despot's yoke began plotting to remove it. If their **revolution** was crushed, even worse pain and hardship were sure to follow. If it succeeded, the result was often no better—and sometimes actually worse—than the tyranny that occasioned it.

The flip side of absolute **tyranny** is absolute **anarchy**. The various **groups** who joined forces to depose the evil king or slay the powerful Caesar could usually never quite agree on a common course of action and often wound up fighting among themselves for control. Sometimes the conflict would go on and on in an endless cycle of violence and terror, as first one group and then another made its desperate bid for power. And sometimes the chaos would be mercifully short—when some *new* tyrant emerged victorious.

Of the two sides of this dreadful coin, tyranny was often preferable. Tyranny, after all, has a constituency in the quarreling groups that backs the tyrant and stands to benefit from the situation. Anarchy, on the other hand, is ongoing chaos that benefits no one. This is why a Saddam Hussein in Iraq and a Pol Pot in Cambodia wound up in absolute power; each was a terrible despot with little regard for the rights of anyone else, but each was supported by *some* of the body politic who stood to gain personally by following their particular tyrant. After all, there were worse things than seeing *someone else* dragged off to the killing fields.

The Human Predicament offers a sad description of certain forces in the ancient world, with its chronicles of war, conquest and revolution. What is even sadder is that even now, in our educated, enlightened world, we still hear constantly about unending wars, conquests and revolution on the six o'clock news. The predicament accounts for the ongoing story of what is happening in Afghanistan, the Congo, Cuba, Haiti, Iran, Iraq, Lebanon, Liberia, Nicaragua, Rwanda and Yugoslavia, to name just a few examples. There will assuredly be another nation or two added to this list by the time this book is in print.

The fact is that people throughout human history had to choose between one of two dismal alternatives—tyranny or anarchy. Indeed, the Human Predicament is a problem in every generation.

THE HUMAN PREDICAMENT CYCLE

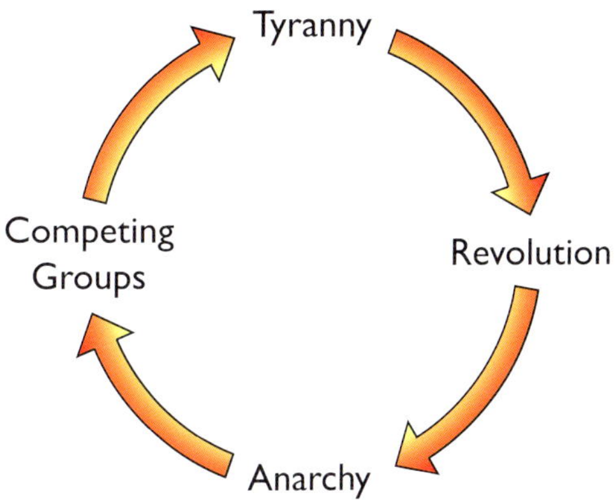

Figure 1.2 Most people who have lived on Earth have had to choose between tyranny or anarchy in the great human predicament cycle.

The Good Society

Yet here and there within history's tapestry of sorrow we see some notable exceptions, societies that escaped the Human Predica-

Human predicament

The cycle from tyranny to anarchy, to which sovereign power and its ill effects give rise.

Despot

A ruler exercising absolute power.

Revolution

A means of removing tyranny from power; part of the human predicament cycle.

Tyranny

Absolute power centralized in one person (or small group); part of the human predicament cycle.

Anarchy

Lack of authority from a failure to agree on a common course of action; part of the human predicament cycle.

Competing groups

Groups that, in a state of anarchy, fight for supreme power and control; part of the human predicament cycle.

Good Society

Reasonably stable and prosperous society without an oppressive tyranny. Usually includes peace, respect, vibrant culture, and personal freedom to live the way one chooses.

Plato

427–347 BC
Greek philosopher and author of the *The Republic*, which extolled civic virtue and the necessity of *areté*.

The painting depicts Plato and Aristotle in Raphael's *The School of Athens*. Athenian philosophers had a particularly large influence on the American Founding.

Political legitimacy

Ruling by a sanction higher than stark necessity; sanction may stem from religion, history, consent, etc.

ment. These mavericks were reasonably stable and orderly, yet under no strongman-style tyranny. They were reasonably prosperous, too, and in general their citizens enjoyed freedom from want. These maverick nations developed strong, vibrant cultures, producing many of the world's memorable achievements. They were peaceful to their neighbors and yet respected. Their citizens were in charge of their own lives, and those lives were enriched with possibility. One may think of Renaissance-era Florence in this regard, or of the early Dutch Republic, or of that improbable collection of cantons that somehow merged to become Switzerland.

But the prime candidate must be classical Athens. True, Athens was not a Good Society in all particulars. The Athenians practiced slavery—boasted of it in fact—and suffered an overmastering itch toward empire. Despite such limitations, what the Athenians achieved was dazzling. They enjoyed a respectable, though not lavish, prosperity. Their lively politics ennobled many citizens and in the resulting flush of freedom creativity abounded. In the high summer of the Athenian democracy, Sophocles wrote *Oedipus Rex*, Herodotus became the father of history, Plato expounded philosophical idealism, Aristotle laid the foundations of science, and Phidias built the Parthenon.

Figure 1.3 The Acropolis at Athens *by Leo von Klenze, 1846. The Greeks' experiments in democracy showed both the successes and dangers of governance by the people.*

The Athenian example is particularly worth recalling because of its influence on the American Founding. Our nation's Founders, virtually to a man, had thoroughly studied classical Greece in the course of their education. They knew Pericles' speech to the Athenians, warning of the dangers of empire. They knew Plato's *Republic*, which extolled virtue, or *areté*, as the backbone of republican morality. They knew what happened at the end, too, when the excesses of democracy led to the Peloponnesian War, when faction contended with faction in the Ecclesia, and finally, when Socrates was tried and convicted on trumped-up charges for preaching high ideals. The Athenian example, alas, became one to avoid.

All of these historical events highlight the questions that must be asked about Good Societies: What made these Good Societies possible in the first place? Why were they so rare? Why were the lifespans of all Good Societies so short? Over the millennia, political thinkers have grappled with such questions and offered their answers. We need to examine some of their theories about Good Societies.

Political Legitimacy

One way used by people with leadership skills and ambitions to get their neighbors to follow them and supposedly escape the Human Predicament was to convince enough people that their leadership was *legitimate*—that the person or family or clique was ruling by some higher sanction than stark necessity.

In the ancient world, **political legitimacy** was usually accomplished by invoking the approval of the gods. Thus, according to myth, it was Zeus himself who founded Minoan civilization, dictating the governing laws of Crete to Minos and Rhadamanthus. And it was the hero Lycurgus who established ancient Sparta, acting under the direction of another god, Apollo. Ancient Rome was famously founded by Romulus and Remus, whom the gods saved from drowning and arranged to be raised by a she-wolf.

In later times, this notion of a god-sanctioned founding developed into a theory called the **divine right of kings**. It went something like this: Just as God organized the world into families and placed each in the charge of a father, He organized society into kingdoms and made each king the "father" of his subjects. Rebellion against the king was thus held to be a violation of the Fifth Commandment—and rebels were shown no mercy.

However, religious precepts could also be invoked against kings. For example, critics of King James I (an ardent divine righter) argued that God spoke to all through Holy Scripture, and it was up to believers to govern themselves according to the written word. These ideas made their way into the American colonies.

There were other answers to the legitimacy question offered by those who wanted to control others on a large scale. Some potential leaders claimed to be divinely inspired, and thus ruled by **theocracy**. (Think of modern Iran, for example.) Other leaders claimed to be in closer touch with the secrets and cycles of history, or with some other metaphysical force, and justified their rule accordingly (as in the former Soviet Union). Some claimed to be smarter, wiser, or wealthier than their fellows, or made much of their supposedly distinguished ancestors, grounding their rule in **aristocracy** (ancient Sparta).

Since the eighteenth century, the most compelling answer to the legitimacy question has simply been "consent." That is to say, a group of people connected by culture, geography or some other reason, legitimize a form of government by the consent of those to be governed, expressed by means of free election. To most modern minds—certainly to most Americans—this seems obvious. But consent was by no means so obvious and generally accepted until quite recently. Modern government by consent comes down to *self*-government. To some, self-government might seem like no government at all—a veritable oxymoron.

Freedom

Ideas about freedom appear to be as old as humankind itself. Yet many of those ideas are elusive, with no settled historical meaning. When Moses begged Pharaoh for the freedom of his people, he didn't have the same thing in mind as, say, a Fourth of July orator would have today. He meant that *the people as a whole* ought to be free—that is, free to do only as God commanded them through Moses and his anointed priestly successors. That another kind of freedom might exist became evident when Moses returned from Mt. Sinai to find the people he had saved from Egyptian bondage had become a host of drunken revelers dancing around a golden calf.

Freedom in ancient Greece had yet a another meaning: the privilege of taking part in the political process. In the Athenian democracy, where every free adult male citizen of the city enjoyed such a privilege, the Greeks began to glimpse still another concept of freedom, one we would appreciate today. In a famous speech to his fellow Athenians, Pericles, the brilliant statesman, general, and acclaimed "first citizen" of that city, pointed out that they not only exercised political power, they also exercised self-sovereignty, for the one quality was necessary for the other. Having a stake in the

Zeus

Greek leaders sought political legitimacy by claiming the blessing of the gods (such as Zeus).

King James I

1566–1625
King James I of England claimed political legitimacy through a "divine right of kings."

Divine right of kings

Political theory that royal lines are established by God and that kings rule by divine decree.

Theocracy

Divinely inspired rule, or rule by religion.

Aristocracy

Rule based on distinguished or wise ancestors and heritage.

Greek freedom

The privilege of taking part in the political process and observing society's rules.

Figure 1.4 The Greeks believed that freedom meant the ability to participate in the political process. Ancient Greeks voted by placing a white or black stone into a pot.

political process required the individual to be in charge of his own life. Only free and autonomous men could participate in the free and autonomous society.

But didn't such autonomy imply anarchy—individuals going their separate ways? The Greeks worried about this too. Athenians were discouraged from amassing private fortunes because it might distract them from public duty, and even acts of heroism counted only if they benefited everyone. The Greeks eventually resolved the contradiction between individual and society by appealing to law. The highest freedom, as they came to understand it, could be realized only by observing society's rules.

Human nature

The fundamental disposition of humans that determines their behavior.

Areté

Greek term for human virtue, the backbone of republican morality. Striving for excellence.

Human Nature

All of these ideas about freedom beg a more fundamental question—are human beings meant to be free at all? Some philosophers—and most rulers—concluded that they were not. The question of **human nature** and how best to define it and understand it is by no means a simple one. While the nature of horses was easy to fathom—all of them behaved in basically the same way—human beings were far more mysterious. So many factors impinged on human behavior, which varied according to time, place, circumstance, and conditioning. If there were rules for explaining the human animal, what were they?

For the Greeks, understanding human nature came down to understanding the virtues of which such a nature was capable. If, for example, the virtue of the horse was to run fast, then good horses ran faster than bad horses, and *all* horses sought to run as fast as possible. The striving of the horse for speed was mirrored in the striving of human beings for their own kinds of excellence. Thus, the athlete aspired to physical prowess. The warrior aspired to battlefield glory. The orator aspired to rhetorical mastery, and so on. Virtue, or ***areté***, explained everything.

The Greeks tied this striving for human excellence both to the question of freedom and to the question of governance. If human beings were made free, so the argument went, they would naturally seek to ennoble their lives by striving for greater and greater virtue, and they would naturally seek to govern themselves by means of the virtues they had gained. Plato identified four cardinal virtues

Figure 1.5 Is human nature one of natural freedom? Or are we meant to have rulers?

in the civic realm, which he called wisdom, courage, temperance (in the sense of moderation), and justice. He believed that free citizens would naturally cultivate all four, especially with the benefit of a proper education. Accordingly, Plato would—at least in the ideal republic—entrust the reigns of power to the wisest, to the bravest, the most moderate, and most just among the population.

Plato's government by virtue was one early and rational way to analyze human nature and solve the problem of government. There were others. The Christian world had a different understanding of humankind and different ideas about governance. Jesus spoke of virtue too, but the qualities he mentioned—meekness, patience, humility, long suffering, compassion, love for one's neighbor—were unlike the heroics of Greek *areté*. Yet Christians demonstrated their own virtues to be equally real.

Christian communities dominated Europe from the time of the later Caesars to the Renaissance, and in many of these the problem of government took care of itself. Ordinary Christians simply did what they were told and left politics to someone else. If that someone else happened to be lord of the manor, or an armed warrior, or a self-appointed magistrate, the very power conferred on him by Christian meekness frequently brought out the kinds of excess we have seen elsewhere in man's historical record. It was not a good solution to the Human Predicament. But then, as it turned out, neither was Plato's.

During the **European Enlightenment**, there emerged yet a third understanding of human nature, and a third set of implications for governing. Enlightenment thinkers, while allowing for Greek excellence and Christian humility, noted that most human beings, most of the time, were motivated primarily by self-interest. (Indeed, the Greek athlete eager to win his laurels and the Christian saint anxious to win salvation had that much in common.) Self-interest meant that, after all was said and done, high ideals were less reliable than ordinary comforts and advantages in explaining the behavior of men and women.

Here was another way of understanding freedom and government. This approach affirmed that people ought to be free to pursue their own self-interest, and that government shouldn't stand in their way. However, just as there was something noble in the idea of Greek *areté* or the Christians' golden rule, there was something ignoble in the idea of "me first," "I want mine." "What's in it for me?" and any of the other ways we now describe self-interest. Many people are obviously short-sighted and irresponsible, desiring creature comforts, sensual pleasures, bright images, passing fads, and a quick buzz. Could such trivial individuals handle freedom? Could they manage self-government?

European Enlightenment

18th century philosophical movement that proposed individual self-interest, rather than Greek virtue or Christian humility, as the motivating factor in human behavior.

Four Alternatives of Government

By the time of the American Founding, thinkers had worked out several different approaches to the problems posed by freedom, government and human nature. How, they asked, given the prevalence of tyranny and anarchy, could a legitimate government and a good society be created given the vicissitudes of human nature? While only four of the possible answers are described below, these four are important. The American Founders considered all of them in the course of their deliberations, and our subsequent experience has often reflected the choices they made.

Autocracy

One of the four alternative forms of government; sees people as children in need of a carefully controlled environment provided by government.

Autocracy

Authoritarian forms of government—including monarchy, dictatorship and other

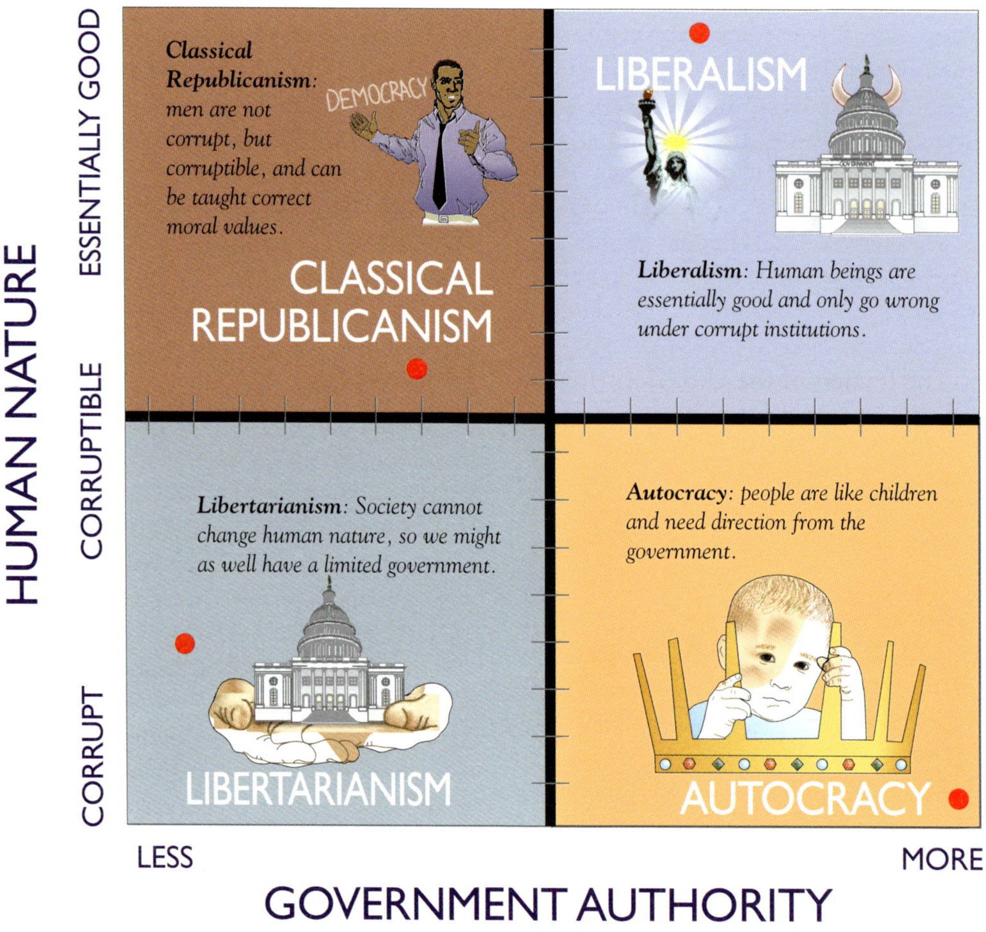

Figure 1.6 Four alternative types of government. The red dots indicate where each type generally falls in terms of views on human nature and government authority.

kinds of despotism—arise from a straightforward analysis of human nature. People are like children, this analysis holds, and hence they must be carefully controlled. Government is not only necessary, it is critically essential, for it alone can bring order to human life. As for individual freedom, that can be forgotten. Indeed, individual freedom is precisely the problem, encouraging chaos and anarchy.

The American Founders believed they were dealing with just such a government in the form of British monarchy. While they reacted strongly against the perceived British tyranny—in fact, mounted a revolution against it—the Founders weren't totally convinced autocratic rule was wrong. They had seen chaos in their own streets before the Revolution—the so-called Boston Massacre being a prime example. The specter of "people in the streets," as historian Gordon Wood put it, continued to haunt them. Brilliant elitist Alexander Hamilton frankly espoused a strong dose of monarchy in the constitutional mix precisely because, as he said, "your people is a beast."

Classical republicanism

One of the four alternative forms of government; sees people (and government) as mostly good but corruptible and so government should have restricted power and try to encourage a good moral climate.

Classical Republicanism

The alternative embraced by the classical republics—and by present-day conservatives—entails a kinder view of human nature. Human beings are not necessarily corrupt, according to this view, but they are corruptible. If they are taught proper moral values, and if this teaching is constantly reinforced, they *might* be able to govern themselves. However, because individuals can be corrupted, so can government, and thus governing constitutions must be carefully designed to constrain governmental power and refine its moral influence. Under the proper constitution, a government can encourage a moral climate conducive to the Good Society.

The framers of the U.S. Constitution generally operated from this perspective. They framed provisions of the **Bill of Rights**—freedom of speech, freedom of religion—specifically to strengthen moral agency. They believed in the virtue of the people as a whole and thought of the Constitution as mobilizing such virtue on behalf of the public good. At the same time, they included many features in the constitutional structure that, frankly, assumed human corruptibility. They recognized the bad intentions of some and did not hesitate to use government power as an antidote.

Libertarianism

On the whole, the prime value embraced by libertarians is that of individual freedom. They believe that all institutions of society, including government, need to be mindful of that value. Libertarians have no illusions about human nature, accepting that some people are as bad as others are good. But they insist that intervention will make little difference in particular cases—people are what they are. Thus, government should be limited to securing the rights of individuals, period. Let people work out good and evil for themselves.

Few of the American Founders could be called libertarians as we now understand the term. But most of them prized the libertarian's strong espousal of freedom, as we will see. When Thomas Jefferson said, "That government is best which governs least," he was embracing a libertarian point of view. Later on, U.S. presidents such as Andrew Jackson would echo Jefferson's sentiments and point out that Americans were at their best when they were free of government restraint.

The economics of the market system is compatible with libertarianism and would be accepted by many as an article of faith. Let individuals alone, said the free traders, and watch them dazzle us.

Liberalism

Of all four alternatives, liberalism—we use the term in its modern sense—has the warmest view of human nature. For liberals, human beings are essentially good, if only their fundamental decency could be freed from poisonous influences. Liberals see the intolerance, greed, poverty and conflict of the world arising not so much from the human heart as from social institutions such as private property or competition.

As with human nature, liberals take a kindly view toward government, which they see both as a weapon against those poisonous influences and as a means of developing human potential. The liberal view of freedom is more complex than that of the other alternatives. For the most part, liberals see freedom as a good thing—indeed, as a battle cry of the oppressed. But too much or the wrong kind of freedom only deepens society's misfortunes. For example, the freedom of, say, a factory owner to pay his workers a starvation wage is certainly not a social benefit. Consequently, liberals generally confine their love of freedom to the political realm, insisting that in economics government needs to take a hand. And of course no one should be free to embrace intolerance, greed, poverty or conflict.

The American Founders knew liberalism by a different name. They saw reflections of it in Plato's *Republic*, which most of them had read, and in Jean Jacques Rousseau's *The Social Contract*, a political best-seller of the

Bill of Rights

First ten amendments to the Constitution regarding basic protections of rights from the government, passed in response to the Anti-Federalist argument against the initial Constitution.

Libertarianism

One of the four alternatives forms of government; sees the most important value as individual freedom and holds that government should only protect that freedom and nothing more.

Liberalism

One of the four alternatives forms of government; sees people in the most favorable light, but institutions or other influences can corrupt them, so government is necessary to protect them from such corruption.

Thomas Jefferson

1743–1826
Jefferson was the third President of the United States, principal author of the Declaration of Independence, and an influential Founding Father of the United States.

A political philosopher who promoted classical liberalism, republicanism, and the separation of church and state, he was the author of the Virginia Statute for Religious Freedom (1779, 1786), which was the basis of the Establishment Clause of the First Amendment of the United States Constitution.

Painting by Rembrandt Peale, 1805.

time. As the result of a dramatic epiphany he had experienced while still a young man, Rousseau had come to see that human beings were born entirely virtuous, acquiring their depravity only through corrupt institutions. Do away with aristocracy, established churches, private ownership and the like and the world might also expunge vice, crime and squalor, Rousseau believed. While few of the Founders would have accepted such idealism wholeheartedly, most of them resonated with Rousseau's ardent sense of justice. The words "with liberty and justice for all," later included in the Pledge of Allegiance, would become a theme of the American Founding, echoing through Lincoln's Gettysburg Address, through Franklin Roosevelt's New Deal, through Martin Luther King's "I Have a Dream."

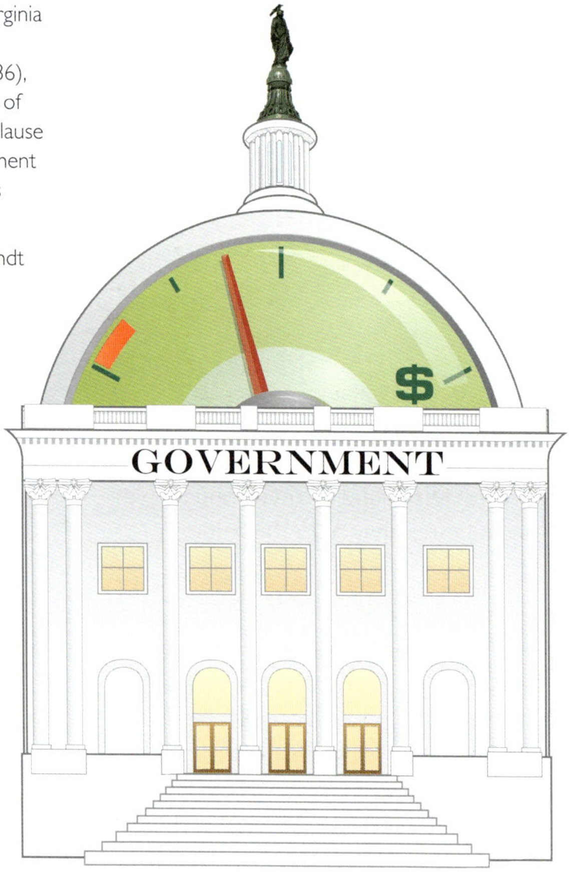

Figure 1.7 Although we separate economics and politics in academic disciplines, in practice the economy and government have always been intertwined.

While the Founders made no conscious choices among these alternatives, all four presented them with options, viewpoints and ways of thinking about nationhood. Questions of human nature and liberty were tied together, and both were tied to the question of government.

Political Economy

While we commonly separate the study of politics from that of economics, there was no such distinction in the world of the American Founding. For example, in pondering the Human Predicament—that choice between tyranny and anarchy—how could the Founders separate the political from the economic? Tyrants imposed their will for the sake of gain as well as power, and the factions that upset the Greek *polis* were often rooted in material interests. To a considerable extent, then, the problem of government becomes a political-economic problem, with both political scientists and economists offering a variety of perspectives and answers.

There is another connection, too. The central problem of government is, after all, a problem in controlling human behavior. If self-interest is the prime mover of such behavior, there are obviously economic as well as political ways of moderating it. Government sets boundaries on human activity by means of laws, courts, and the policeman on the corner. An economic system may accomplish similar ends through incentives and rewards. If the interests of the individual is balanced against interest in the marketplace, for example, the result for society is not much different than if the balancing were prescribed by a government edict.

Some economic systems tend to produce a rough-and-ready equality, others a marked *in*equality. Each condition has political implications. The social justice that liberals seek is usually described in terms of economic equality, the rich and the poor faring more or less the same, while vast inequalities seem not only unjust to some but also politically dangerous. Struggles between haves and have-nots were particularly dicey in Plato's Athens, which is one reason why Plato required the Guardians of his wished-for republic to have no property. In our own time, of course, struggles over economic equality led to the communist revolutions of the twentieth century.

There is one other connection between economics and politics in shaping a founding. Property itself has a double meaning. Economically speaking, property in the form of capital—land, machinery, liquid assets—creates wealth. Politically speaking, property creates power. The rich often get their way in the political world while the poor get ignored and pushed around. Political and economic independence often go together—a fact well understood by the American Founders. For citizens to stand up and be counted politically, the Founders often observed, they must not be under the thumb of bosses, creditors or landlords.

For these and other reasons, the economic dimension must always be considered in addressing the problem of government.

A Founder's Toolbox

The founders of successful political states have a kind of toolbox to go to, full of ideas and institutions that have worked well in the past. We will discuss some of the most important tools in this box:

1. Structure

After carefully studying the constitutions of various Greek city-states, Aristotle concluded that some of them worked a lot better than others, and that political **structure** often accounted for the difference. The philosopher noted that people seemed to behave differently in different structured relationships. A small legislative body, for example, tended to be more careful and deliberative than a large one, which was moved more easily by passion and rhetoric. Same players, different game—and often a different outcome.

Structure

Rules and restrictions designed to better harness virtue.

2. Participation

Encouraging participation of the citizens themselves helps to shape a political society in beneficial ways. The Greek *poleis* (plural of ***polis***) run by tyrants or oligarchs could not match democratic Athens for energy or creativity, though they did tend to be more stable. For when the ordinary citizens can vote, hold office and take part in political deliberations, they acquire a sense of ownership. The public world becomes their own.

Polis

City or city-state, often self-governed by its citizens as the ancient Greek city-states were.

3. Law

The ancient Greeks left the world another important legacy in the domain of law. Law had existed before, of course, notably in Israel and Mesopotamia, but the Greeks learned that laws of a certain kind had a profound effect on the political process. General rules, known to all, made by common consent and applied impartially reduced the scope of arbitrary action in the political arena. When such rules existed, the rulers themselves felt bound by them, and all players had a better chance in the game.

4. Custom and Tradition

Successful foundings can draw upon custom and tradition in order to invoke the authority of the past. For example, there were several English customs that found their way into the American Founding centuries later. Among these was idea of "sanctuary," holding that a fugitive, upon entering some holy

Figure 1.8 A Founder's Toolbox

place or touching some sacred object, could not be taken directly into custody. To be sure, this notion may not have made for good law enforcement, but it did foreshadow the concept of privacy—a realm beyond the power of government—which would shape the American Bill of Rights.

5. Moral Sense

Almost without exception, successful political societies depend on shared values—a common notion of right and wrong, an innate moral sense. Government has a lot less to worry about if there is agreement on such fundamentals. For some, this raises a troubling philosophical question: is there really such a thing as *moral truth*, something to which all polities must adhere? While the question is sometimes answered negatively—think of societies that include cannibals or headhunters for example—history affords few examples of nations enduring with a constitution based on "anything goes."

6. Founding Myths

We have seen that when the gods participated in a founding, it stood a better chance of success. Not all *founding myths* necessarily invoke the divine. Some are simply based on a shared belief, such as that "in America a poor boy can go from rags to riches." While this particular belief may be statistically doubtful, it operates much in the way that myths of old did, fostering a sense of identity, hope and belonging. Founding myths often provide the real sinews of nationhood.

7. Leadership

Could there have been an Israel without Moses or a Roman Empire without Caesar? Foundings require strong leadership. While we shy away from such thoughts in the modern world—where leadership has often spelled tyranny—history recalls few leaderless foundings. It may sound democratic to say "the people came together and decided to act," but on close examination action almost always follows the bold initiative of a few stubborn individuals. The American Republic was no different.

There are other tools in the toolbox and we will have occasion to examine some of them later. Suffice it to say, the creation of a new political society is no simple task.

The Social Compact

Some philosophers of the European Enlightenment envisioned foundings as a **social compact**. That is, people living in a "**state of nature,**" before government, in a sense came together and worked out a common agreement about the sort of political world they wanted to live in. Many of the tools in our political toolbox reflect this key idea.

The social compact remains largely theoretical. With few exceptions, there are no historical instances of free and autonomous individuals convening to forge a new nation-state. Yet the concept is still credible. In the history of, say, France or England, through wars and conquests, revolutions and accommodations, the French and later the English indeed worked out something like a general agreement about their identity, their common purpose, and the manner of their governance. The very fact that they came to call themselves French or English, that they took pride in those labels, and that they were not constantly seeking to overthrow their respective governments suggests a tacit accord among the majority of the French in France and the English in England.

The social compact idea had startling implications for the American Founding, where from the beginning nothing was implied and everything was deliberate. The American colonies were *created*, basically from scratch, by groups with specific purposes in mind, and foundational issues were raised at the very outset. As the American Colonies grew to maturity, their royal charters were revoked in some cases, replaced in others, renegotiated in still others, so that some Americans grew up thinking about constitutional questions. There was evolutionary development, of course, but it was always with an eye to getting matters straight, nailing things down, working out satisfactory and permanent arrangements.

When the governance disputes between the American Colonies and Great Britain became earnest, the air in the New World filled with rhetoric about the social compact. Who had agreed to what? Under what circumstances? By whose authority? With what justification? Over and over, it was "the English constitution" this and "the rights of Britons" that and "the laws of nature" something else. Americans in the coffee houses, in the fields and shops, earnestly asked each other: How were nations established? How *should* they be established? And could such diverse groups of people actually establish a real nation?

The most important implication of the social compact was that the American people, as a people, gradually came to see that they could create whatever political society they chose, according to their hopes and dreams, their ideas and values, their understanding of the past, and their sense of a common destiny. When some fifty-five of the American Colonies' local leaders gathered in Philadelphia in the spring of 1787 and regarded one another across baize-covered tables, there was no thought that they were mere spectators in the drama of life—they were on center stage, before a hushed audience, with the thrill of a rising curtain. They meant to tackle the age-old problem of the Human Predicament once and for all, by constructing a workable, idealistic, beneficent yet practical government acceptable to all.

Social compact

The social concept of a group of autonomous individuals without government making a common agreement about the sort of political world they want to live in.

State of nature

Hypothetical condition assumed to exist in the absence of government where human beings live in perfect freedom and general equality.

Key Terms

sovereignty
revolution
areté
theocracy
autocracy
liberalism
human predicament
anarchy
political legitimacy
aristocracy
classical republicanism
social compact
tyranny
Good Society
divine right of kings
human nature
libertarianism

Key People

Plato

Questions

1. Give a present day example of each stage of the Human Predicament.
2. What is a way in which the Greek view of human nature is seen today? the Christian view? the Enlightenment view?
3. What are some ways in which political legitimacy is accomplished?
4. What elements of the four alternatives of government can you find in our political parties today?
5. What is a real-world example of how structure can affect the political process?
6. What do you think would be important to include in a social compact? Why?

Chapter 2

City upon a Hill

Abstract

After Christopher Columbus discovered North America, two distinct groups emerged to colonize the land: corporate communities seeking profits from the land's abundant resources, and covenant communities seeking religious freedom thousands of miles away from persecution. Each contributed to how later inhabitants of America defined themselves and their objectives. John Winthrop would proclaim this land a "city upon a hill," an example to the world—a view that would influence the Founders and even America today.

The United States of America grew out of a set of colonies "planted" by Great Britain. The colonies were established by a variety of groups for a variety of purposes, and the colonizers learned much and gained significant experience as they adjusted to their new environment. Some of the colonies were business ventures, aiming to promote corporate (and by extension, national) wealth, precisely as a company might set up an outpost in Antarctica today to explore for valuable mineral resources. Others were sanctuaries for groups seeking greater religious freedom. Still others began as feudal fiefdoms for great nobles of the realm. One colony, Georgia, was set up as a refuge for the poor, while another, New York, was captured from the Dutch in a maritime war. In time, the significance of the colonies broadened, deepened, and took on peculiar overtones. To understand these new connotations, we need to reflect for a moment upon the discoverer of the New World, Christopher Columbus, for he cast a long shadow on the American future.

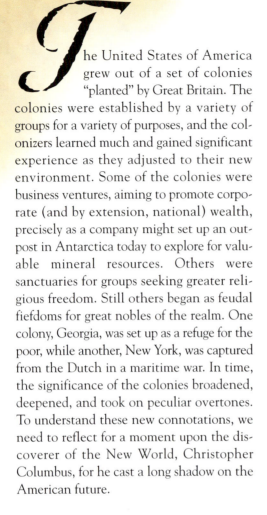

Christopher Columbus

1451–1506
Genoese mariner who discovered the Americas while searching for a new trade route to India.

Portrait by Sebastiano del Piombo (1485-1547).

Columbus

The Genoese mariner we know as Columbus has been accused of power-madness and gold fever, the destruction of Native Americans, the building of an empire and the eventual despoliation of the New World's natural, pristine condition. In some respects, he stands guilty as charged. His character mirrored both the irony and paradox of Renaissance Europe; he was a complex and many-sided man.

Many forget, however, that Columbus was also an idealist and a visionary who believed that God was guiding him. What's more, history has supported his belief. He had amazing luck in catching the right winds, the right tides, and the right ocean currents in crossing the Atlantic and returning to Spain, and on more than one occasion he emerged from fearful scrapes by what seemed like miraculous means. He survived shipwreck, mutiny, blood-soaked rivalries, bad food, exotic illnesses, and political betrayal, and he managed to live out his life and die in his home in his bed.

When it became clear that his plan to reach India had failed, and that he had instead discovered a "new world," Columbus began to emphasize his discovery's beauty, bounty, and desirability:

> This island is fertile to a limitless degree. In it there are many harbors on the coast of the sea, beyond comparison with others which I know in Christendom, and many rivers, good and large, which are marvelous. Its lands are high, and there are in it very many sierras and very lofty mountains. All are most beautiful, of a thousand shapes, and all are accessible and filled with trees of a thousand kinds and tall, and they seem to touch the sky. . . . Some of them were flowering, some bearing fruit, and some in another stage, according to their nature. And the nightingale was singing, and other birds of a thousand kinds in the month of November there where I went. . . . In it are marvelous pine groves, and there are very large tracts of cultivatable lands, and there is honey, and there are birds of many kinds, and fruits in great diversity. In the interior are mines of metals, and the population is without number.

By the time Columbus had reached the mouth of the Orinoco River, he believed he had located the Garden of Eden.

These perceptions also reflected the world from which Columbus had sailed. Europe was tired at the end of the fifteenth century. Corruption, dishonor, and violence

lay everywhere—the turbulent world of Shakespeare's *Romeo and Juliet*. Protracted wars had devastated both England and France. The Catholic Church was divided and discredited, and soon it would be engulfed by the Protestant Reformation. Turkish warlords had pushed their way to the Danube River. Given this troubled state of affairs, it was not surprising that artists and writers of the time dwelled almost obsessively on escapist themes, and especially on the idea of mythic lands beyond the sunset—Avalon and Lyonesse, the Golden Cities of Cibola, the Fortunate Isles, the Isles of the Blest, Utopia. These were all imagined places of beauty and bounty, where life was long and full of promise. Poets supposed that if someone could discover such a paradise, it might enable jaded Europeans to start over again, rediscover innocence, and return to first principles.

Consider, for example, the report of another navigator, describing the islanders of Hispaniola:

> They go naked, they know neither weights nor measures, nor that source of all misfortunes, money; living in a golden age, without laws, without lying judges, without books, satisfied with their life, and in no wise solicitous for the future. With neither ditches, nor hedges, nor walls to enclose their domains, they live in gardens open to all. Their conduct is naturally equitable, and whoever injures his neighbor is considered a criminal and an outlaw.

Inspired by similar dreams, Christopher Columbus came to believe he had been led to America, and that America might have a role to play in the moral life of Europe. We should keep Columbus's legacy in mind as we see how Britain's colonies developed.

Figure 2.1 A depiction of Columbus claiming possession of the New World in a chromolithograph made by the Prang Education Company in 1893.

Corporate Communities

The settling of Virginia was conceived by some British entrepreneurs as a corporate venture. The idea was for a joint-stock company of merchants to send a work party across the sea, set up operations and develop profitable fishing, furring, timbering, or gold mining enterprises. Because there would be no official local government in far-away Virginia (the name referred to the entire coastline), the Virginia Company planned to govern its American operatives, and it acquired the authority to do so through a royal charter. This document, signed and sealed by King James I in 1606, sketched out the organization of the company and described its civil authority. Thus, right from the beginning, British colonists coming to Virginia would live under a constitution.

Corporate communities

Colonial settlements established for economic or financial purposes by various companies. Although usually chartered by the Crown, their remote circumstances helped foster the idea and practice of self-governance.

Life in Jamestown, the Virginia Company's first settlement, proved to be much tougher than anyone had imagined.

The elements were unfamiliar and harsh. Supplies needed to support the colonists were scarce. Incompetence and disorder often prevailed. Many of the colonists who styled themselves as adventurers were not accustomed to hard labor, nor were they eager to work like common laborers, and they weren't sure whether to plant crops or search for gold. After exhausting their immediate supplies, settler after settler perished from illness or malnutrition, in such numbers that there were often few left to dig the graves. The colonists struggled to develop some sort of profitable enterprise, and they failed repeatedly. Costs mounted. The Virginia Company teetered on the edge of bankruptcy. On top of everything, there was an Indian war.

A clash of cultures between Europeans and Native Americans was inevitable. Almost everything prized by the one was scorned by the other. Matters came to a head—as they would in most of the colonies—over land. For the Europeans, land was property, a purchasable and exploitable resource, and the basis of all human wealth. For the Native Americans, land was sacred and free as the air or the water. "Otherness" also became a factor. The Europeans assumed that any people so different from themselves must be inferior, which further justified exploiting the nearby tribes and pushing them off lands the natives had lived on for generations.

When the Indians finally realized the colonists were greedy for their land and resources, they mounted a bold surprise attack and almost drove the invaders into the sea. The English counterstroke was equally vicious, and in the end it pushed the Indians out of the tidewater and west into the wilderness. This victory failed to save the financial situation for the colony's British investors. The Virginia Company declared bankruptcy in 1624 and was taken over by the English crown. That might have spelled the end of the colony. But Jamestown was no longer just a company outpost in a forbidding distant wilderness, nor were the surviving Virginians any longer foppish adventurers.

Figure 2.3 Locations of Jamestown and Roanoke. Roanoke was established in 1585, and after several rocky starts, had a colony of about 115 people by 1587. When supply ships from England returned a few years later, the settlement had been abandoned. While there is no solid evidence of what happened, some theorize that the colonists joined with a local tribe of Native Americans.

John Smith and Pocahontas

Captain John Smith was one of the early leaders of the Jamestown colony. A controversial figure, he was often accused of self-aggrandizement, which makes it hard to separate fact from fiction in his life as the only record we have of many of his exploits is his own. This includes the tale of his rescue from certain death by the young Pocahontas. Whether that story is true or not, Captain Smith played an important role in the survival of Jamestown. He also helped explore Chesapeake Bay.

Figure 2.2 This portrait of Captain John Smith appeared on a 1616 map of New England. The image is colorized by Jamie May from an original engraving by Simon de Passe.

Somewhere along the way, **John Rolfe** and other settlers had finally come up with a business enterprise that worked—the cultivation of a peculiar plant that the Indians grew and used in an equally peculiar way: they burned it and inhaled the smoke. Of course, the plant was tobacco, and Rolfe and a few others saw it as a possible solution—maybe the only solution—to Virginia's financial problems. Growing tobacco (and making it desirable to a European market) was a difficult undertaking, involving a lot of guesswork, and neither Rolfe nor anyone else knew if what they were doing would work or fail. Yet they learned quickly, while their countrymen in the coffeehouses and grand ballrooms back in England took up the smoking habit. The Virginia Company, which had become a burial ground and extraordinarily bad investment, was suddenly in the black, thanks to a crop that couldn't be eaten or made into cloth, and was of value only when it was turned into smoke.

Tobacco was a labor-intensive crop, however, and labor became the basic difficulty of colonial agriculture. One solution to the problem was **indentured servitude**, where wealthy farmers in Virginia would pay the passage of those willing to come over in exchange for an agreed-upon term of service, usually about seven years. After completing the required years of hard and unremitting toil, the indentured servant was released from his obligation, given some land, some seed, perhaps a few tools, and was then free to seek his own fortune.

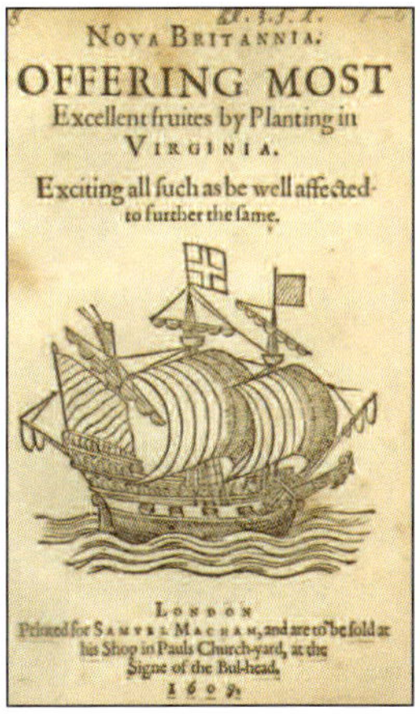
Nova Britannia.
OFFERING MOST
Excellent fruites by Planting in
Virginia.
Exciting all such as be well affected to further the same.

London
Printed for Samvel Macham, and are to be sold at his Shop in Pauls Church-yard, at the Signe of the Bul-head.
1609.

Figure 2.4 Title page of Robert Johnson's Nova Britannia. *London, 1609. Johnson's* Nova Britannia *is an appeal on behalf of the Virginia Company for the purpose of stimulating emigration to Virginia. It was written as a discourse by one of a party of adventurers returned from Virginia.*

Figure 2.5 Tobacco was to become a major export for the American colonies.

Tobacco Output

1615
2,000 lbs.

1629
1.5 million lbs.

1660s
15 million lbs.

1680s
28 million lbs.

Another solution to the labor problem was to import African slaves. These too were at first viewed as indentured servants. But their indentures never quite seemed to run out. A combination of racial prejudice, fears of "otherness," and unabashed greed operated to hold the "blackamoors," as they were called, in perpetual bondage.

Slaves excepted, many willingly came to Virginia in pursuit of their fortunes. No matter how bad things got, the privilege of tilling one's own land, living in one's own house, and taking charge of one's own life was a powerful draw to those who for generations had been landless and essentially homeless and without any hope of improvement. Even after the failure of the Virginia Company, English, Irish and Scottish emigrants continued to come to England's American colonies as simple farmers, humble artisans, or lowly indentured servants. America, with all of its raw land, was seen as a place of opportunity.

It was also a place of self-invention. Many small farmers in Virginia became large farmers, even "planters," for land was cheap and easily acquired. These self-made planters soon began putting on airs of gentility. A big-

John Rolfe

c. 1585–1622
Virginia colonist who pioneered the cultivation of tobacco as a profitable agricultural enterprise. Rolfe also married Pocahontas in 1614.

Indentured servitude

Land owners would pay the passage of those willing to come to the colonies in exchange for an agreed-upon term of service, after which the indentured servant was released from his obligation and was then free to seek his own fortune.

ger house, an imposing barn, furnishings imported from London, a carriage rattling down the dusty country roads—these were symbols of a social station to which only birth could admit one in the Old World. It wasn't a common phenomenon, but it became an important one. Where cheap land was available, virtually anyone, with hard work, determination, and a few good harvests, could pry open the doors of social advancement.

When the Indians had been driven out and men with more practical skills had replaced the adventurers, the Jamestown beachhead grew into the Virginia colony, now presenting its inhabitants with a ticklish problem of governance. The old company governors had been replaced by a royal governor appointed by the king. The governor in turn appointed a council drawn from a few prestigious local families, and together the governor and council ran the day-to-day affairs of the colony; their policies were best described as "what's good for the leading planters is best for everybody."

But there was an important innovation. The Virginians with little or no personal influence with the governor wanted to have a voice in their own governance, and they began electing representatives to meet in an annual assembly they called the **House of Burgesses**. This group passed ordinances, approved taxes, and gave the royal governor—who after all was an outlander—a local perspective on what the general population of the colony thought about colonial developments. Crown officials were never entirely happy with this arrangement, and more than once they tried to throttle it. But the arrangement persisted, mostly because the colonists themselves strongly supported it.

House of Burgesses

An assembly of representatives elected by the common people of the Virginia colony, similar to the House of Commons.

In a rough-and-ready way, the governing institutions of England (now Great Britain) came to be mirrored in New World practice. The crown-appointed governor

Figure 2.6 Patrick Henry before the Virginia House of Burgesses *by Peter F. Rothermel (1817–1895).*

took the place of royal authority back home. The governor's council acted like the British House of Lords, backing up royal authority and exercising a sort of veto power. The Virginia Colony's House of Burgesses behaved quite a bit like the far-off House of Commons in Westminster, discussing, debating, often dragging its feet and jealously guarding the power of the purse.

In time, Virginians began to gain hands-on political experience, just like their counterparts back in London. They learned the standard parliamentary procedures, and with practice, a few parliamentary tricks as well. They learned how to threaten; how to cajole; how to bargain and negotiate; how to hold governors in line by subtle means (such as controlling the amount of their salary); and how to counter the thrusts of royal policy. Above all, Virginians learned how to represent the interests of their various constituencies. Tobacco planters had a clear and cogent interest in policies that were friendly to their particular enterprise, and the representatives they sent to the House of Burgesses came to understand just how to make the weight of that interest fully felt.

So it was that Virginians came to enjoy the blessings of liberty. It had happened without conscious intention, evolving as a somewhat free dividend of the colonial experience. Settlers more or less ran their own political institutions, and in consequence they were able to run their own private lives as well. Royal governance of Virginia was tenuous and too far from home to be harsh or tyrannical. It could only function with the cooperation of those it presumed to govern, and thus it wound up doing their bidding most of the time. Of course there were exceptions, and life in Virginia was still punctuated by the occasional crisis. Gradually, however, Virginians began counting their blessings. They had found something in America that was deeply and primarily important to them.

Covenant Communities

Virginia, the oldest British colony in the New World, became a model for other corporate communities. Indeed, to some extent the Virginia experience was reflected in all of the colonies. Modern Americans still acknowledge the historic importance of Jamestown, but they identify even more deeply with a much smaller colony established on the rocky shores of Cape Cod. Every Thanksgiving that sense of identification is symbolically renewed. It lies close to the heart of something we think of as quintessentially American.

The group we now call the **Pilgrims** was a small congregation of Christian separatists seeking to distance themselves, physically and spiritually, from the Church of England, which often failed to please religious purists. This particular group, the followers of **Robert Brown**, was certainly a purist group.

According to Brown's teachings at Cambridge University, "God's people"—those whom He had specifically chosen for salvation—would always be very small in number, and they would never be mired in the corruptions of an entity like the English Church. Inspired by Brown's teachings, the Pilgrims left England to its fate, crossing the channel to hopefully find a more spiritually congenial refuge in the Netherlands. Their sojourn in the Netherlands exposed them once again to ecclesiastical compromise and corruption so they made a second exodus, this time sailing across the Atlantic to land on a beach near Cape Cod that they christened Plymouth.

Unlike the Virginia Company, which had enjoyed the patronage of the English government, the Brownist Pilgrims were literally on their own in an area where they would face starvation, harsh winters, and problematic relations with the Indians. Like Virginia's early merchant-adventurers, many Pilgrims succumbed to the trials of New World life. "Yet it is not with [us] as it is with other people," wrote William Bradford in his *History of Plymouth Plantation*, meaning that no matter what happened the survivors would never give up their struggle. Their courage in the face of staggering hardships is one reason why we honor them still.

The settlement at Plymouth operated as a "covenant community." One of the Pilgrims' primary beliefs was that God's chosen people covenanted with Him—*and* with

Figure 2.7 The First Thanksgiving 1621 *by Jean Leon Gerome Ferris (1863–1930).*

Covenant communities

Settlements based on religious or moral values, mostly interested in being an example to Europe or to living according to their own moral liberty.

Pilgrims

Small congregation of separatists seeking to distance themselves, physically and spiritually, from the Church of England by emigrating to New England.

Robert Brown

1550?–1630
Writer and proponent of the Separatist movement that demanded separation from the Church of England. His writings inspired groups such as the Pilgrims to emigrate to America for religious freedom.

one another—to live according to the divine plan governed by God's will. As they formed congregations, these covenants tied them together. The salvation of each was bound up with all others in the community.

The covenant was religious in nature but it also had a secular side. It was almost precisely analogous to the social compact discussed in Chapter 1. The Pilgrims' mutual promise to "bear one another's burdens" made it possible for the members of the congregation to form their own government, decide on its organization, and determine how each member would participate in it. In doing so, they made use of many of the tools, such as structure and moral sense, described in our toolbox in Chapter 1. A fundamental equality before God made it possible for them to govern themselves—for there were no permanent rulers set over them.

Puritans

British religious emigrants who wanted to reform the Church of England rather than sever all ties with it; their beliefs of God's Elect, the Christian Calling, and Moral Self-Governance would help shape the Founding and American national character.

Today, one of the deepest meaning of Thanksgiving is its symbolism of divine protection in the face of monumental difficulties. That the Pilgrims possessed such confidence in God was due in no small measure to the strength of their beliefs and to the kind of polity they were able to build on such a foundation. Like Columbus, they found more on the western side of the world than they had expected. They believed they were a chosen people and that God's special protection was a daily manifestation in their lives. That quality of "chosenness" seems to have become part of the broader American experience.

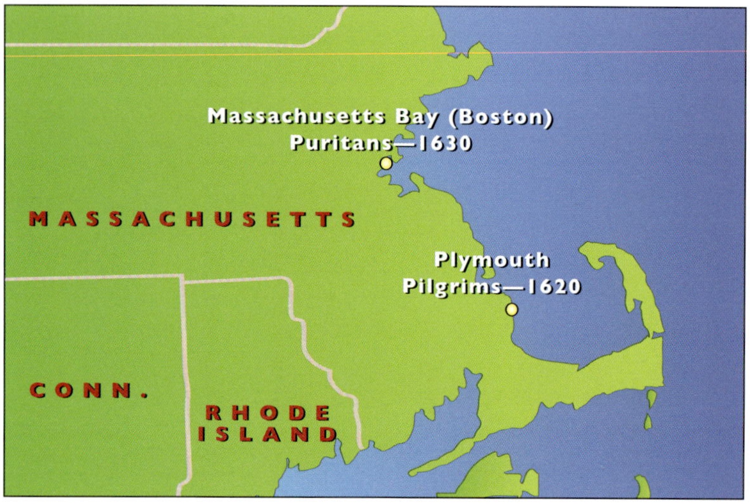

Figure 2.8 Locations of Plymouth and Massachusetts Bay colonies.

The settlers of Plymouth became forerunners of a much larger migration of dissidents from the Church of England, who ten years later settled north of the Pilgrims' little colony. Historians refer to this latter group as the **Puritans**.

The Puritans shared some doctrinal beliefs with their Pilgrim cousins but were not complete separatists. They wanted to reform the Church of England, not sever all ties with it. Even though they felt unwelcome back home, they were not completely on their own in America. Accordingly, the Puritans' model of colonization bore some likeness to that of Jamestown, based in a joint-stock corporation seeking to establish commercial activities across the Atlantic. The Puritans secured a royal charter, just as the Virginia Company had, and with it came a title to a vast stretch of the northern coast of what came to be called New England. It was another case of American opportunity beckoning.

The Puritans made the most of it. Some of them cleared land and started farming the thin and stony soil with a variety of crops. Others embarked on fishing, fur trading, and timbering ventures, which led to shipbuilding and related occupations. Still others took up mercantile enterprises, and Puritan merchants became some of the best in the world, trading local products in an expanding trans-Atlantic network. Like their English compatriots in Virginia, the Puritans eventually discovered ways of making their community into a complete self-sustaining world.

Business aside, however, the Puritans were a people of exceptionally strong beliefs, and many of those beliefs have become asso-

		SETTLERS	LEADER	ARRIVAL DATE
Corporate Community		Jamestown	John Smith	1607
Covenant Community		Pilgrims	William Bradford	1620
Covenant and Corporate Community		Puritans	John Winthrop	1630

Figure 2.9 Table of early colonial communities.

ciated in one way or another with aspects of the American Founding. It is noteworthy, for example, that the Puritans brought their charter along with them rather than leaving it back in London, for once it crossed the ocean their colony essentially became a self-governing republic. Beyond this, Puritan New England—the colonies of Massachusetts Bay, Connecticut, New Haven, and Rhode Island—became in some sense the birthplace of many American ideas and institutions, perhaps even the cradle of the American character.

Puritan towns were all covenant communities, precisely like the Pilgrims' Plymouth, and as such they were politically viable from the start. Each town was formed by a separate congregation and worked out its own power structure, its own machinery of government, and its own body of law. In each there was provision for general participation in the political process—very different from English practice of the time—by way of voting or holding office. In putting their town governments together, the Puritans drew upon biblical precepts, English law, and a strong foundation of shared values. If nothing else, Puritans believed they knew the difference between right and wrong.

American nationhood would later be affected by three basic Puritan religious beliefs: God's Elect, the Christian Calling and Moral Self-Governance.

God's Elect

In the predestination theology of **John Calvin**, who inspired the Puritans more than anyone else, God chooses in advance those who will be saved and those who will be damned. The latter, who accounted for most of the people on Earth, were sensuous and sinful, and nothing could be expected from them but trouble.

The saved, on the other hand, were obliged to illuminate an otherwise dark world. The very fact that Puritans had come together, recognized one another, organized themselves, and were now undertaking this monumental mission for the Lord, attested to their cosmic importance. The elect, as they called themselves, were to build a community that was godly in every way.

This belief meant that for the Puritans almost everything had a moral dimension. Farmers, merchants, and factory owners were enjoined to deal fairly with one another, with employees, with customers, and with the public at large. Sin was everybody's business,

John Calvin

1509–1564
John Calvin was a French theologian during the Protestant Reformation who greatly influenced Puritan beliefs. He taught that the Bible was the final authority for matters of faith and that salvation came through grace only (not works). He also taught the doctrine of predestination.

God's Elect

From John Calvin's predestination theology, the doctrine that God has already chosen those who will be saved. These elect people are to build a holy community as an example.

and vice everybody's problem. If the Puritan community was tainted by drunkenness, fornication, heresy, or witchcraft, no member of the community could turn a blind eye. Accordingly, politics was not so much about power and privilege as it was about good and evil, right and wrong.

The Christian Calling

Calvin scorned those Christians who attempted to reach an exalted state of holiness by retiring into the monasteries and convents. All people were sinners, he taught—at least until they received God's grace—and one of the worst sins was supposing that one could emulate God through a life of self-denial.

Instead, Calvin believed that true Christians should be "workers in the world." They should face up to their flawed humanity and be content to live the life of mere mortals. Instead of emulating God, they were to glorify God by showing forth His great works. Building the godly community was the principal task before them.

Workers in the world pursued a "calling." Some were called to be brick masons, others to be ironmongers, still others to be merchants, and so on. All were obliged to rise early in the morning, work hard, save their money and invest it wisely. And all, of course, were to walk uprightly before the Lord.

If God was pleased with this manner of worship, He would manifest it by enabling the faithful to prosper. If one looked around and saw well-tended orchards and freshly painted barns, if one beheld bustling towns and busy wharves, if the balances on the ledger books remained safely in the black—these were signs of God's pleasure. His kingdom was very much of this world.

Moral Self-Governance

Puritans believed in universal standards of right and wrong. All were to live a righteous life, and they were to do so largely on their own. Each man was responsible for his own actions and those of his family—with an eye on his neighbor as well. There was no penal system in Puritan New England, apart from a few implements of public humiliation, so law and order were more or less up to the individual.

It was but a short step from moral self-governance to political self-government. The very reason that Puritans could trust ordinary citizens to vote wisely and hold political office was due to this similarity. Just as the ancient Greeks had believed all citizens could recognize "the Good" and act accordingly, American Puritans believed that all of God's elect could recognize fundamental truth and shape their lives to it. Political life was held on course by a strong sense of individual accountability.

Puritan ideas and institutions remained influential in America long after Puritanism itself burned out. For one thing, the Puritans had relevant things to say about the problem of government as discussed in Chapter One. For another, the Puritans had mirrored the assumptions of Christopher Columbus about the world he had discovered. It was a blank slate, they believed, a ***tabula rasa***, on which mankind could begin the human story anew—and this time get it right.

Figure 2.10 Examination of a Witch, *by T.H. Matteson 1853. Courtesy of the Peabody Essex Museum.*

The Christian Calling

From the theology of John Calvin—people should pursue a "calling" in some sort of worldly work where they are to rise early in the morning, work hard, save their money, and invest it wisely. Prosperity indicates God's approval.

Moral self-governance

Puritan ideal that all must live a righteous life largely on their own, with each man being responsible for his own actions and those of his family—with an eye on his neighbor as well.

John Winthrop

1587–1689
John Winthrop was elected governor of the Massachusetts Bay Colony before their departure from England, and re-elected many times.

He is known for his sermon "A Model of Christian Charity," in which he stated that the Puritan colony would be "a city upon a hill."

City upon a Hill

Puritan magistrate **John Winthrop**, in a speech to his fellow Puritans while their ship, the *Arbella*, was still making its way across the Atlantic, invoked a remarkable image. "We shall be as a city upon a hill," he said. "The eyes of all people are upon us." This may have been the first iteration of what was to become the idea of America.

What Winthrop meant was that he and his fellow Puritans were going to show the world what God could write upon that *tabula rasa*. His city upon a hill would be nothing less than a vision of the world as God had intended it to be—the world recast according to holy principles. In a later speech, for example, Winthrop delved into the nature of liberty, explaining the difference between natural and civil liberty. The difference would be crucial to that city upon a hill. Given **natural liberty**, men were free to do precisely what they pleased, Winthrop argued, and the sad state of the world reflected the choices that most of them made. In the Puritan commonwealth, by contrast, men would enjoy **civil liberty**, where one was free to do only that which was good, just, and honest.

Winthrop's "city upon a hill" was a little like Plato's ideal republic. It was to be as near to perfection as a flawed and sinful world would allow. It would include many of the attributes that political philosophers had imagined of the Good Society:

- Reasonable order, created by the people themselves.
- Reasonable prosperity for everyone.
- A strong, vibrant culture, prizing science and literature.
- Peaceful toward others, yet strong and well respected.

Figure 2.11 The Puritans believed they were as a "city upon a hill," an example of a faithful Christian community.

- Citizens in charge of their own lives, yet in pursuit of common goals.

Imbued with such qualities, Winthrop's city upon a hill was what we might call a founding myth. It tapped into the power of several other myths, stretching far back into history. One of these was the myth of the Garden—that Eden which Columbus believed he glimpsed at the mouth of the Orinoco—promising a return to some lost golden age. A second was the myth of the Promised Land, the land of milk and honey that God had held before the eyes of ancient Israel. And a third was the myth of the New Jerusalem, that heavenly city of the future in which the Judeo-Christian saga would achieve ultimate fulfillment. Combining all three, Winthrop's vision has for several centuries provided a way for Americans to think about their country, and in the process provided them with a unique sense of identity.

Something like the city upon a hill seems to have been in the back of patriots' minds in the 1770s, when they resisted the Stamp Act, cast the tea into Boston Harbor, and opened fire at Lexington Green. It was reflected again in the determination of the Philadelphia delegates to push ahead with

Tabula rasa

Latin for clean slate or blank slate. Puritans felt that the new world was a tabula rasa on which mankind could begin the human story anew.

City on a hill

Biblical ideal, invoked by John Winthrop, of a society governed by civil liberty (where people did only that which was just and good) that would be an example to the world.

Natural liberty

Where men are free to do what they please, without regard for the moral value of their actions.

Civil liberty

According to John Winthrop, "Where men were free to do only that which is good, just, and honest."

their task of writing a constitution in spite of daunting difficulties. It may have been on Henry David Thoreau's mind a generation later when he wrote about living deliberately, and was remembered by Ralph Waldo Emerson when he wrote about self-reliance. It explains the willingness of Grant's Federal troops to charge the Confederate fortifications at Cold Harbor, knowing in advance that most of them would be killed. It helps us to understand America's participation in the war against Hitler and National Socialist tyranny, and our nation's determination to stop Soviet Communism at all costs. It also helps us make sense of a more recent difficult and costly war against Saddam Hussein and his fellow tyrants.

The city upon a hill concept helps us to better understand culturally English settlements popping up far across the Atlantic. Pilgrim, Puritan, and Jamestown efforts were not simply accounts of colonization. They were stories of the westward movement of something vital that had characterized European civilization. America offered hopes and dreams to ordinary people—those whom the world had almost entirely overlooked. We see this in the eager push into the Virginia tidewater and in the confident towns that sprang up in New England.

Despite the wish to preserve their European heritage, colonists in America were already living a different life. They were working out new kinds of political institutions. They were breaking new ground in economics. And they were applying religious metaphors to both.

Timeline

Contextual Events | **Chapter Specific Events**

1492 Columbus Lands

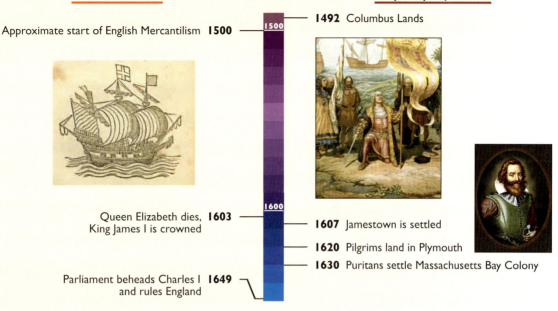

Approximate start of English Mercantilism **1500**

1500

1600

Queen Elizabeth dies, King James I is crowned **1603**

1607 Jamestown is settled

1620 Pilgrims land in Plymouth

1630 Puritans settle Massachusetts Bay Colony

Parliament beheads Charles I and rules England **1649**

Key Terms

corporate communities
Pilgrims
the Christian calling
natural liberty
indentured servitude
Puritans
moral self-governance
civil liberty
covenant communities
God's elect
"city upon a hill"
good society

Key People

Christopher Columbus
John Calvin Puritan
John Rolfe virginia /Tabacco
John Winthrop Puritan
Robert Brown Pilgrams

Questions

1. What is a "founding myth" and what are some examples? How are we influenced today by these myths?
2. How does the Human Predicament apply to the corporate and covenant communities?
3. How did Jamestown self-govern? How did the Puritans self-govern?
4. What are some of the main points of John Calvin's theology?
5. What aspects of corporate and covenant communities are seen in society today?

Chapter 3

The English Legacy

Abstract

Much of what the American colonists sought to establish was derived from their English heritage. The writings of John Locke and formation of British Common Law effectively altered the way ordinary men and women thought about government. Citizens no longer thought to find freedom *in* government; rather they felt they needed freedom *from* government. The Rule of Law gave tests citizens could apply to government law making. The Glorious Revolution and creation of the Whig party changed the structure of society and government. The Founders drew on this English heritage in their design of government.

The irony of the American Revolution was that it claimed to be a revolt against British tyranny, and yet it sought to recover lost or threatened English rights. This irony affirms the importance of English/British ideas and institutions to the American Founding. At the same time, it also affirms a deep-seated fear that the qualities of a free society, be they European or American, may be easily lost.

We saw in Chapter 2 that early in their colonial career Americans began developing rough facsimiles of English political institutions, including such things as representative assemblies. There was much else of importance in the English legacy. In this chapter we will examine three critical elements of that legacy and consider their relevance to the American Founding.

Lockean Liberty

The idea of freedom has long existed in political discourse, going back at least as far as the ancient Greeks. Until quite recently, however, "freedom" had meanings other than the one we use today, with most of them focusing on participation in the political process. During the seventeenth century—the century of America's colonization—there developed in England a wholly new concept of freedom, the freedom of the individual to live his own life and be his own person without interference. This view differed significantly from the older idea of freedom *in* society; this was freedom *from* society.

This development was the result of an intellectual revolution that drastically altered mankind's assumptions about the world. From today's vantage point, we look back and wonder how people with conscience could have practiced slavery in the American South, how indentured servants could have been treated like cattle, and how laborers could have been shot down in the streets for going on strike. The answer in all three cases is that we are looking back across an historic and cultural gulf created by the liberal (meaning freedom) revolution. The practice of this new kind of freedom was slow to catch up with its eloquent theory, but when it finally did so, the lives of Americans were changed forever.

The new idea of freedom—rechristened "liberty"—became bound up with both the American Revolution and the American Founding. Ironically, it flowed from the pen of an Englishman and addressed a situation that could have developed in no other country but England. What happened in seventeenth-century England and why does that long-ago era of turmoil and conflict still shape our lives today?

England's Time of Trouble

When Queen Elizabeth I died without heirs in 1603, the English throne fell to her next of kin, King James VI of Scotland, who became England's James I. The new monarch arrived in Westminster with a pronounced Scottish accent and some disconcerting ideas. James Stuart—and virtually all of his heirs to the throne—had no background in (and still less patience with) troublesome law courts or foot-dragging parliaments. As far as James I was concerned, kings still ruled by the pleasure of God alone and thus never had to answer to anyone.

Stuart claims of ruling by the ancient "divine right of kings," as the doctrine was called, were destined to clash repeatedly both with the English law courts and the English Parliament. Certain judges in the law courts would maintain that established law was primary and fundamental, and that the king himself was bound to obey it. Certain mem-

Figure 3.1 The Royal Line of England. Dates indicate time of rule.

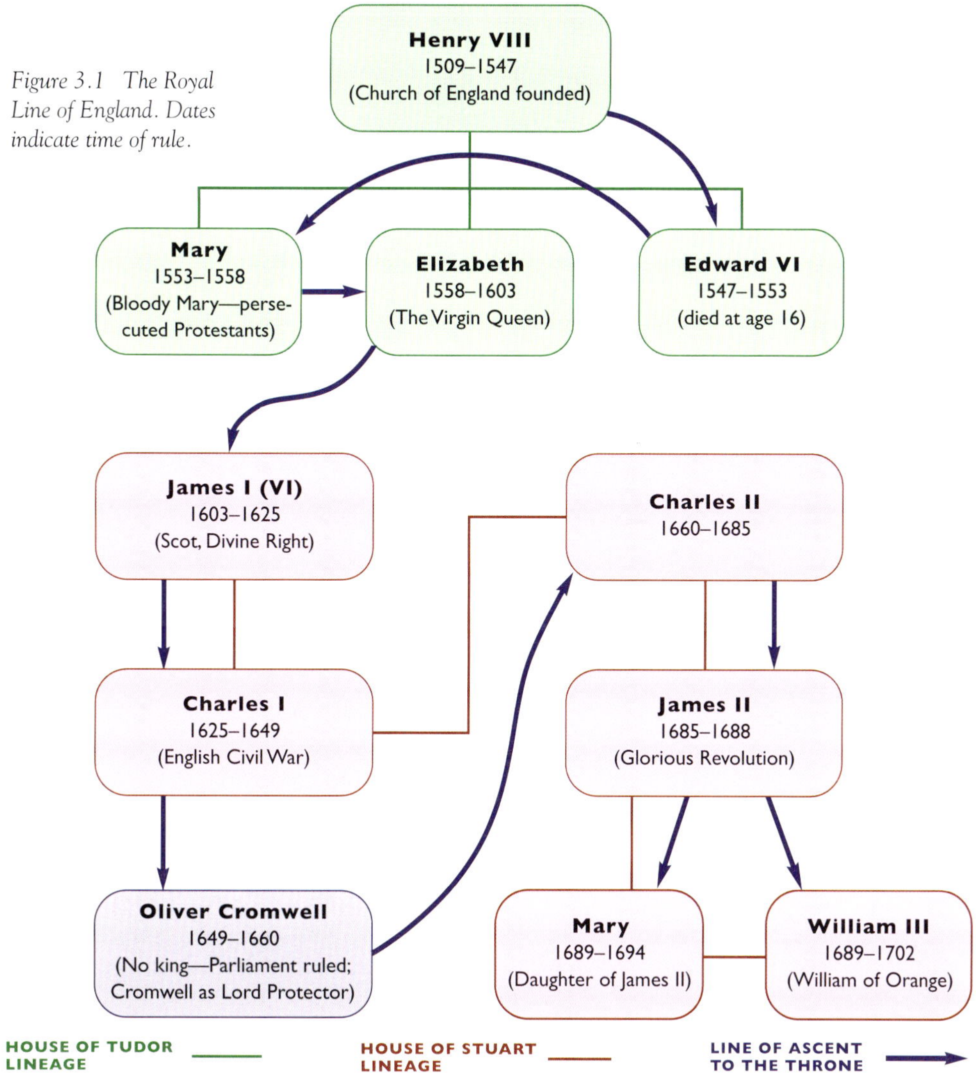

King James I of England

1566–1625
Portrait by Daniel Mytens in 1621.

Charles James Stuart was King of England, King of Scotland, and King of Ireland and was the first to style himself King of Great Britain. He ruled in Scotland as James VI from July 24, 1567, and then from the "Union of the Crowns," in England and Ireland as James I, from March 24, 1603 until his death. He was the first monarch of England from the House of Stuart, succeeding the last Tudor monarch, Elizabeth I, who died without an heir.

bers of Parliament would maintain that their institution alone could make changes in the law and that no taxes could be levied without parliamentary approval. Neither the courts nor the Parliament bought into the concept of the divine right of kings.

Tensions between the two sides mounted slowly, decade after decade, and ultimately led to a civil war. In 1649, Parliament emerged victorious from that conflict and beheaded Charles I—the Stuart monarch. For the next ten years, Parliament ruled England in a kind of legislative dictatorship. However, when Charles II was called back from European exile in 1660 and restored to his late father's throne, the bickering and wrangling resumed. Stuart kings continued pressing for divine right prerogatives and their opponents continued disputing their claims. Thoughtful Englishmen wondered about the positive and negative outcomes of this seemingly irresolvable issue. Some came to realize that England lacked a true constitution, for there was no final authority to embody the realm's sovereignty.

When Charles II died and his son James II assumed the throne in 1685, matters seemed destined for another bloody

John Locke

1632–1704
English philosopher whose *Treatises of Government* espousing natural rights, consent of the governed, and social compacts greatly influenced the Founding Fathers.

Portrait by Sir Gotfrey Kneller (1646–1723).

Whig Party

England's first political party, organized in political opposition to the King; Americans later formed their own Whig party during the Jacksonian democracy era, but the two parties did not hold the same ideology.

Second Treatise of Government

John Locke's work arguing that true political authority comes not from God or precedent but from the people.

showdown. In addition to the divine right bias of the Stuarts, James II had become a Catholic, and his reign threatened a resurgence of the religious wars of an earlier time. At this point a strange thing happened. A nobleman of the realm, Sir Anthony Ashley Cooper, Earl of Shaftsbury, began to organize the opposition to the king *politically*—as we might think of doing today. The **Whigs**, as Shaftsbury's compatriots called themselves, were in essence England's first political party. They were also the spiritual/political ancestors of colonial American patriots of a century later, many of whom would also call themselves Whigs.

But on what grounds did the Whigs of England in the 1680s opposed the king? After all, James II *was* the ruler of England, and the fact that most of his subjects loathed his person and feared his rule might be irrelevant. Lord Shaftsbury's personal secretary, a young man named **John Locke**, went to work on this problem. Locke, highly intelligent and extremely well educated, possessed a philosophical turn of mind. He developed a pair of treatises that addressed the entire question of rulers and their claims to authority. Locke questioned whether the doctrine of divine right was valid. If not, then what was it that legitimized a sovereign's rule?

The Second Treatise

Locke's ***Second Treatise of Government*** became one of the masterworks of Western civilization and was a direct inspiration for the American Founders. It argued compellingly that the authority of all legitimate government is not God or historical precedent or genealogy, but rather the people themselves. It was a careful argument, meticulously crafted, and modern Americans should know its outlines well.

Locke's first point was that in the original "state of nature"—a hypothetical condition assumed to exist in the absence of government—human beings must have lived in perfect freedom and general equality. In such a world, moreover, all had the same rights, for "rights," by their very character, could not be granted by man, but only by nature. All had the right to live their lives, enjoy their liberty and make the most of

Second Treatise of Government

1. In a state of Nature there is no government (no divine right of kings)
2. Men create a social contract
3. Government's only job is to protect people's natural rights
4. Government exists by consent of the governed
5. If Government violates the social contract, the people have the right and duty to revolt

Figure 3.2 A summary of Locke's Second Treatise of Government.

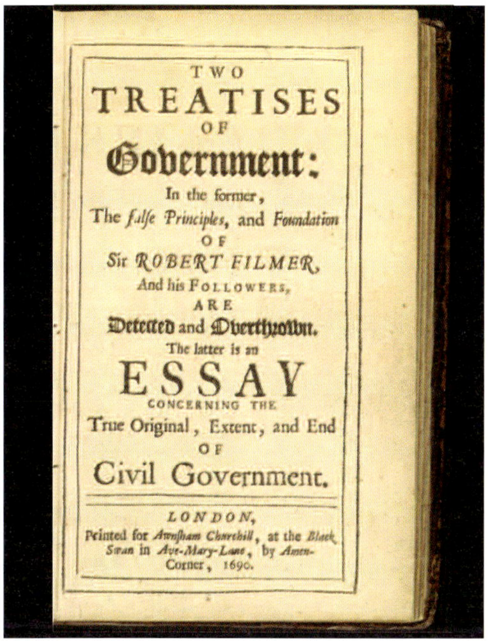
TWO
TREATISES
OF
Government:
In the former,
The *false* Principles, and Foundation
OF
Sir ROBERT FILMER,
And his FOLLOWERS,
ARE
Detected and Overthrown.
The latter is an
ESSAY
CONCERNING THE
True Original, Extent, and End
OF
Civil Government.

LONDON,
Printed for *Awnsham Churchill*, at the *Black Swan* in *Ave-Mary-Lane*, by *Amen-Corner*, 1690.

Figure 3.3 Title page of John Locke's Two Treatises of Government *from the first edition, published anonymously, in 1690.*

their property, as long as they did not disturb the rights of others.

Some few did disturb the rights of others, however, and so there was a manifest need for law and for a common judge to hear and decide disputes. Accordingly, Locke's second point indicated that individuals came together and agreed to invent and establish government. There was no divine mandate for this social compact, only a simple need. Government was a human invention, made to serve a human purpose.

The character of that purpose was defined in Locke's third principle. Government could have but a single end, he argued, and that was to protect the rights of citizens. (After all, why else would they have created it?) Those rights could never be surrendered or abridged, for they had been granted by nature, but they could be disregarded. Government's job—its *only* job—was to make sure that didn't happen.

Government, then, existed only by the consent of the governed—Locke's fourth principle. The government must look to the people for its legitimacy; it could not presume to govern in God's name. There had to be accountability of some kind—elections, representation, parliaments—by which the governed had opportunities to have their say.

Fifth and finally, if government violated the terms of consent, if it lost track of what it was supposed to do and whom it was supposed to serve, the people had the right, indeed the duty, to alter or abolish it, even if that meant revolution. Revolution was a powerful word in Stuart England, but John Locke employed it advisedly. If James II broke his compact with the English people, they had every right to cast him out, which is precisely what they did.

In 1688, shortly after the *Second Treatise* was written (but before its official publication), the English people rose against their king and expelled him from the country in a bloodless revolution. Remembering the endless Parliamentary bickering of the kingless days, they then invited James's daughter Mary and her husband, **William of Orange**, the chief magistrate of Holland at the time, to assume the throne as joint monarchs—*subject to the will of Parliament*. Thereafter, as the invitation had made clear, monarchs would rule England only by the active consent of the people.

Englishmen regarded this second civil revolution, known as the **Glorious Revolution**, as a true founding. It confirmed what judges of the realm had been saying about the supremacy of the law, and what representatives of the people had been saying about the supremacy of Parliament. It cemented the place of rights in English government. Above all, it confirmed Locke's theory of personal liberty, for if government only secured the rights of its citizens, the citizens were bound to enjoy their freedoms in abundant measure. They could do what they wanted with their lives. They could enjoy their lib-

William of Orange

1650–1702
William of Orange acceded the throne with his wife Mary (daughter of James II) in 1689, and became William III of England.

Glorious Revolution

1688 bloodless English revolution against the King, making the King subject to Parliament; considered a true founding of government.

Rule of law

A set of metalegal principles developed by the English legal system as a way of distinguishing whether a particular law supported freedom or not.

Natural law

Law that classical Greeks believed resided in the human heart and reflects our innate sense of right and wrong.

Natural rights

Fundamental rights granted by nature that government can not abrogate and which government is bound to protect.

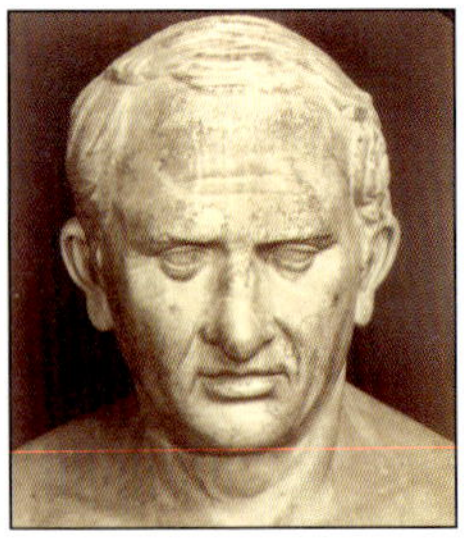

Cicero

106–43 BC
An orator, statesman, political theorist, lawyer, and philosopher of Ancient Rome.

Common Law

Law that is considered to be from natural law principles but that is framed in a form that can be interpreted more concretely.

erty to the fullest. They could buy, own, improve, and sell their property. They could, as Jefferson would put it, pursue happiness.

Later on, in the tumultuous decade that led to the American Revolution, the colonists would look back to the Glorious Revolution and refer in their arguments to John Locke's formulation of political truth. They had come to live in a Lockean world, one in which government literally was created by the people and rights seemed natural and fundamental. When Locke pointed out that the people had the right to revolt if their monarch let them down, many American colonists took that argument seriously.

The Rule of Law

Locke believed that these political truths were derived from nature, reflecting the world as God had created it. This argument went back to the ancient world. Intellectuals of Greece's Classical Age, followed in turn by Roman sages, spoke a good deal about what they saw as **natural law**. They considered it to be the moral law, the law that resided in the human heart, the law that reflected our innate sense of right and wrong. It followed that natural law would protect **natural rights**. If there was, for example, a natural right to life, then there was a natural law against the arbitrary taking of life; if there was a natural right to liberty, there was a natural law against slavery and so on.

Figure 3.4 Justice must be blind.

According to ancient legal theorists like the Roman orator and politician, **Cicero**, the laws handed down by various cultures' great lawgivers, such as Moses or Hammurabi, were accepted by the people because they embodied natural law principles. If that is true, then logically, the laws enacted by legislative bodies should, in principle, do the same. No legislature in the civilized world would dream of passing laws that condoned arbitrary murder, nor would judges smile benignly at theft, robbery, or burglary. In consequence, there came to be something mystical, almost ineffable, in the very idea of law. Obedience to the law, under any and all circumstances, had great political value in ancient republics.

Such ideas persisted in seventeenth-century England. The great law courts that applied and slowly developed the English **common law** often took the view that they were "discovering" natural law principles. In consequence, the law was often regarded as a companion to freedom. But even the "common sense" principles of natural law required some careful thought in their application.

Who, for example, should be able to bring criminal charges against someone? If there is no *prima facie* case of wrongdoing—such as a thief caught in the act—common law practice required that charges should be brought only by a group of disinterested citizens empanelled as a grand jury. Otherwise, there was a clear temptation for political authorities to injure rivals or detractors simply by accusing them.

Or, on what grounds should a person be held in custody? Should a sheriff, for example, be able to arrest someone on suspicion of some crime and then simply lock him away until a trial becomes convenient? Once again, the common law judges said no. They developed the writ of *habeas corpus* as an instrument to secure the immediate release of any jailed individual when there was insufficient evidence to hold the accused person.

Who should decide guilt or innocence? There were obvious dangers in allowing the government to do it. But what about some neutral party, such as a judge, someone who could bring learning and sophistication to the task? While there were arguments in favor of this solution, the common law evolved away from it and toward the use of juries. The "jury of one's peers," as the phrase goes, consisted of ordinary citizens like the accused. If *they* could be persuaded that the accused was guilty, the evidence had to be pretty convincing.

The law of nature applied not only to the legal process but to the political process as well. Should a king, for example, be given a free hand in taxing his subjects? What king wouldn't take full and probably unfair advantage of such a blank check? England's famous Magna Carta laid down the natural law principle that taxpaying subjects should themselves have a say in such matters, and Parliament traced its roots to that day in 1215 when King John reluctantly accepted the demands of the nobles and magnates and signed the fateful document.

Even so, England's political experiences in the tumultuous seventeenth century demonstrated that the law could still be misused. For instance, those who made the law could still use it for political ends, as in the notorious bills of attainder. There was a case in which Parliament, in a desperate move against Charles I, passed a bill pronouncing the Earl of Stratford—one of the king's corrupt counselors—guilty of treason and prescribing his punishment. It was all perfectly legal, wasn't it? asked the authors of the bill. After all, Parliament *was* the voice of the people!

Actually, it wasn't perfectly legal, but it took a long time and much thoughtful reflection before anyone could explain why. The problem, in four words, was "the rule of law"—possibly the single most important concept evolved by the English legal system. The rule of law provided a way of distinguishing between cases in which the law supported freedom and cases in which it didn't. It was not a law itself, but rather a set of metalegal principles that, if respected, ensured that the law would be a beacon of liberty.

The rule of law encompassed a number of principles. The following five were uppermost:

1. Generality

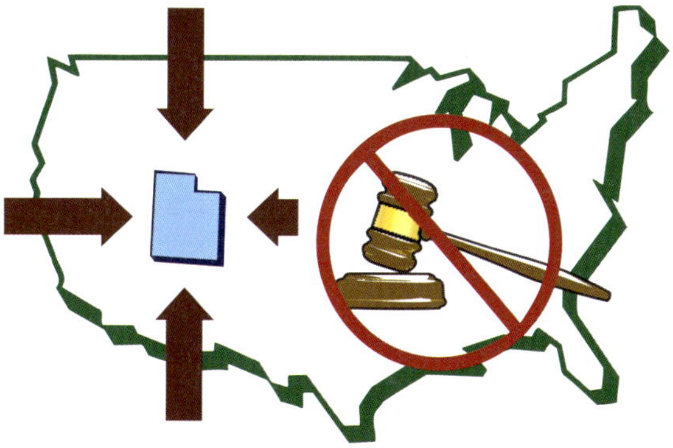

The laws must be general. They must apply to broad categories of people. They must not single out individuals or groups for special treatment. If there is any such designating, it should be done by the people themselves *after* the laws are passed. That is, I place *myself* in the category of drivers when I climb into my car and start the engine.

Generality

Rule of law principle that states laws must apply to broad categories of people and must not single out individuals or groups for special treatment.

We would sense something wrong with a law that singled out the drivers of red Toyotas, but not any other driver. Such ordinances could also target Mormons, Southerners, or liberals. Those who make the laws should never know in advance to whom they apply. Lamentably, the British

Figure 3.5 During World War II, the U.S. government relocated over 100,000 Japanese-Americans, a direct violation of the generality principle of the rule of law.

Parliament of the middle 1700s lacked that sort of blindness when it singled out American colonists and began laying special taxes on them.

Prospectivity

Rule of law principle that states laws must apply to future action and not past action.

In practice we, too, create exceptions. We address some laws specifically to children, for example, forbidding them to purchase alcohol or drive a car; by zoning certain neighborhoods, we deny specific landowners any right of commercial development. Our tax code discriminates between rich and poor. Some of these laws violate the principle of generality and ought to be viewed with mistrust. Others are a practical necessity. There is no hard and fast way to tell which is which.

However, a few categories of designation have become absolutely taboo. These involve ethnicity, gender, religion, political affiliation, and increasingly, sexual preference. Above all, we abominate separate laws for the rich and the poor. "Equality under the law" has become one of freedom's great banners.

2. Prospectivity

The laws must apply to future action, not past. The theory is that the individual must always be able to decide *in advance* whether or not to obey a given law. We would consider a law punishing those who failed to vote in the past election as unjust, because the opportunity for moral decision would be absent.

Tyrants and demagogues don't like this principle. The last thing a Hitler or Stalin would desire is for intended victims to evade sanctions by altering their behavior. *Ex post facto* laws—the term we give to violations of prospectivity—commonly single out an individual or group and in effect say, "We don't like what you did and you must be punished for it." Thus, most violations of generality also violate prospectivity, and vice versa. Together, these two abridgements of the rule of law account for much of the tyranny the world has known.

Figure 3.6 A man commits a brutal murder and the citizens of the town do not feel that the existing laws will punish the murderer severely enough. So legislators pass stricter murder penalties and ensure that this particular murderer gets a heavier sentence, thereby violating prospectivity.

3. Publicity

The laws must be both known and certain. They must be well publicized so that everyone knows of their existence—the laws of tyrants are often kept secret—and their enforcement must be reasonably reliable. Laws that are capriciously enforced, selectively enforced, or not enforced at all, undermine public respect for lawgivers and create contempt for those whose responsibility it is to enforce the law.

But police work and its necessary apparatus are expensive, so governments occasionally protect their operating budgets by funding only token enforcement of the laws, hoping that the possibility of punishment, no matter how remote, will deter potential law-breakers. Some people will accept the risk, betting that legal actions against them for their misdeeds will be too expensive, or burdened police will give up the pursuit for justice. If government isn't serious about its laws, its citizens won't be serious either.

Figure 3.7 A violation of publicity would occur if a speed limit was changed and enforced without any signs being put up to notify drivers of the new limit.

Capricious enforcement was an aspect of British policy before the Revolution. It led to widespread smuggling on the part of colonials, and to customs racketeering on the part of the authorities. Later on, when the British sought to reform the system and more rigorously enforce the Navigation Acts, many Americans were outraged.

4. Consent

The laws must be generally acceptable to those who must live by them. Electing the lawmakers and upholding the constitutional system are examples of the way we give consent.

Yet not all lawmaking is subject to such formal approval. Natural law, customary law and judge-made law are never submitted to the voters. What counts, however, is that the people, if they truly dislike a certain law, have the means at their disposal to wipe it off the books. Theory holds that the people as a whole would never consent to injustice because they themselves would be its victims. That theory only works when the laws are *also* general and prospective.

Consent becomes mandatory in the matter of taxation. Once the people themselves consent to the amount they will be taxed, the entire character of government is transformed. If the lawmakers themselves have to

Publicity

Rule of law principle that states laws must be known and certain, such that everyone knows of their existence and their enforcement is reasonably reliable.

Consent

Rule of law principle that states laws must be generally acceptable to those who must live by them.

Figure 3.8 A colorized 1846 lithograph of the Boston Tea Party. The dumping of British tea into Boston Harbor by the Sons of Liberty was a reaction to taxation without representation. The British Parliament had violated the principle of consent.

pay the taxes they approve, all incentives toward exploitation go out the window.

It was the absence of consent that brought on the American Revolution. Parliament was allowed to make laws for a distant group, the colonists, and was not itself accountable to those laws. There was no one from the colonies in Parliament to represent the American point of view or to challenge Parliament's tax laws specifically focused on the Americans. Why *not* pass a Stamp Act? asked the members of Parliament. It was an excellent way to raise revenue to help pay for the expense of the recently concluded **French and Indian War**, and besides, *they* didn't have to buy the stamps.

French and Indian War

1754–1763 conflict between the French and British/Americans and their respective Indian allies in which the French forces were defeated.

Due process

Rule of law principle that states laws must be administered impartially.

5. Due Process

The laws must be administered impartially. Justice, as the saying goes, must be blind, considering no questions other than guilt or innocence. If the accused is black, or poor, or a communist, or a Christian Scientist, the law must say: So what? And there must be established procedures to insure that everyone is given a fair trial, an adequate defense, and, if found guilty, a reasonable punishment.

Due process accounted for yet another sore spot in British-American relations. The British authorities had few qualms about revoking jury trials and other traditional rights in the interest of tighter law enforcement. If Americans were a threat to others, they said, then by all means let's close up those procedural loopholes. We often say the same thing about threats today.

Society and the Rule of Law

If the laws of a society met all five of these tests, chances were that the rule of law was a working reality, and the result would be freedom. Under the rule of law, the laws of society become very much like the laws of nature—steady, evenhanded, predictable. We use the laws of nature to our advantage because we know in advance what they are

Figure 3.9 Due Process would be violated if a guilty robber is given 5 years in prison while another robber who committed the act under the exact same circumstances is given 10 years.

and how they apply. Just as we don't have to step off a cliff in order to know whether gravity is working, so too we don't have to steal that shiny red Porsche in the parking lot in order to know if the laws against theft are in force. In a society without the rule of law, a person must in effect step off the cliff. He might steal the red Porsche and never get punished because the laws were not working that day. Worse, he might decide not to steal the car and get punished anyway because the laws were working or being interpreted in some unfathomable way. Under the rule of law, we alone determine what happens to us. That's how the law makes us free.

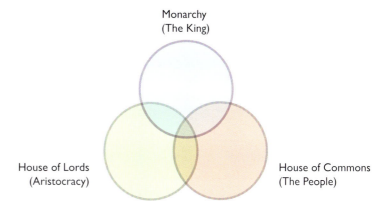

BRITISH GOVERNMENT AFTER THE GLORIOUS REVOLUTION

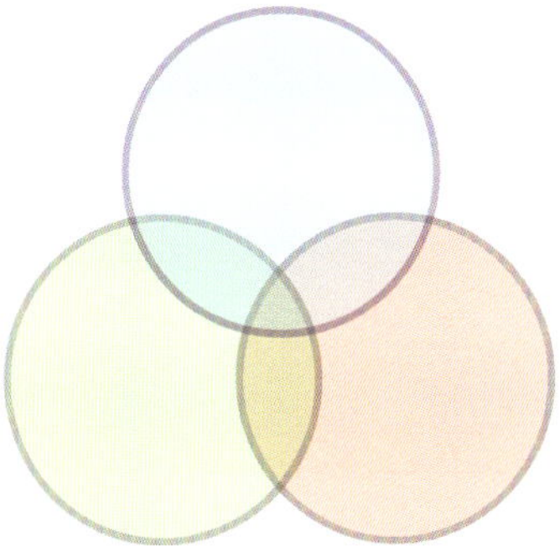

Figure 3.10 The intertwined structure of the British government.

Virtue and Structure

After England's Glorious Revolution, the Whig majority in Parliament became the actual rulers of Great Britain. Kings still sat on the English throne, but Parliament now had the power to approve, turn away, or modify all the royal commands. Whig political thinkers, however, continued to worry. Liberty had shown itself to be a fragile thing. If a monarch's formerly absolute power was no longer a direct threat, other concerns were just as unsettling. Among these was the pattern of corruption that always seemed to appear in high places. There were some spectacular scandals in British politics, and they weren't just about money. Sometimes, the very governing body that checked the king's power mired itself in scandal.

The Whigs had a dark view of human nature, believing that those who gained power, even in their own Parliament, would probably misuse it. For answers to this difficulty, they turned once again to the great thinkers of the ancient world. Their reading of Aristotle convinced them of the value of structure as one way of curtailing the misuse of power.

The idea was fairly simple. Aristotle, a careful student of constitutional forms, had noted that power in a relatively few hands always seemed to spell trouble. By contrast, if one mixed and balanced power among rival authorities in a constitutional system, chances were that no single one of them would accumulate enough power to do the others harm. The Whigs took some comfort in this, noting that in their own government the powers of the king were now shared with those of Parliament, that the House of Lords tugged and pulled against the House of Commons, and that all parties in government had to contend with the will of the judges. "Checks and balances," a phrase that would later become crucial in the construction of the Constitution of the United States, gained currency.

The British government became a model for Europe's premier political scientist, a Frenchman named Charles-Louis de Secondat, Baron de la Brède et de **Montesquieu**. He lived under an often corrupt and occasionally tyrannical monarchy and was filled with admiration for the British system. While it was unwise to disclose his admiration too pointedly, he nonetheless made his preference clear. In his masterwork of comparative government, *The Spirit of the*

Montesquieu

1689–1755
Charles-Louis de Secondat, Baron de La Brède et de Montesquieu, more commonly known as Montesquieu, was a French political thinker who favored the British system of rule and lauded the idea of separation of powers.

Laws (1750), Montesquieu argued for the genius of British institutions virtually on every page. Far from England and even farther from the tangled French monarchial system, the Americans who later became the nation's Founders studied Montesquieu's arguments with care.

All the same, history afforded abundant examples of corruption leading to tyranny in governments that enjoyed mixed and balanced powers. Structure was important, but structure wasn't enough. In time, Whig writers developed an alternative approach to the problem of preserving liberty: the **commonwealth ideology**. The idea of a commonwealth is based on the theory that some individuals will always be drawn to the center of political power, like moths to a flame.

In Great Britain, the center of power, even with an active Parliament, was the royal court, consisting of the king and his cronies, some of whom also sat in Parliament and called themselves **Tories**. If this "**court party,**" as the Whigs called it, was left unchecked, it would draw more and more power to itself. If it couldn't openly challenge Parliament, it could resort to other means, such as the granting of honors, the bestowal of titles, the offering of bribes, to sway, corrupt, or in some other way subvert other authorities. A license here, a franchise there, political favors somewhere else could undermine the integrity of the entire body politic. In the end, the court party would succeed in building up its own power, with or without a legal structure.

The court party, the Whigs concluded, could be held in check only by a political rival of equal weight and determination, "the **country party**." This group consisted of merchants, bankers, manufacturers and especially landed gentlemen out in the shires. These "commonwealthmen" were economically and, consequently, morally independent; moreover, residing out in the country, close to the land and away from the celebrity of court life, would keep their motives pure. Such men would be the members of Parliament who would steadfastly oppose the intrigues and machinations of the court party.

Whig writers vested their new political heroes with a sense of ancient virtue or *areté*. Plato's four cardinal virtues—wisdom, courage, temperance and justice, in addition to England's advocates for a commonwealth factored in some Christian virtues as well, things like patience, humility, personal integrity, and brotherly love.

The virtuous patriot was supposed to behave in carefully prescribed ways. In circumstances where others might be hasty or foolish, he was supposed to be wise. In situations where others might lose heart, he was to be unflinchingly courageous. Above all, when others were being bought off and compromised, the virtuous patriot was the one who couldn't be corrupted, his independence shining like a beacon.

This was a lot to ask of the political world, which had never been known for sterling qualities, but the commonwealthmen were serious. They turned out dozens, even hundreds, of books and pamphlets arguing the commonwealth ideology and its importance for the preservation of liberty. One particularly noteworthy book, *Cato's Letters*, written by two Whig journalists named John Trenchard and Thomas Gordon, became a sort of bible for the country party. It was circulated widely in Great Britain and even more widely in the British colonies.

The Moral Basis of the Founding

Chapter 2 discussed how the American colonies began to gain political experience, and that with political insight came a sense of political maturity. In large measure, these

Commonwealth ideology

The idea that the "Country party" had the best strategy and opportunity to preserve liberty against the "Court party."

Tories

English royal court and the the center of British political power, known also as the "Court party" and characterized by corruption and subversion.

Court party

English royal court and the the center of British political power; known also as the "Tories" and characterized by corruption and subversion.

Country party

English opposition to the "Court party" that consisted of commonwealth men (everyday citizens). The Court party was considered morally independent with pure motives.

inheritances from Great Britain provided the sum and substance of such maturity.

Americans read John Locke and other writers of the European Enlightenment, and they rejoiced in the triumph of the Glorious Revolution. They saw the victory of Parliament over the king and his cronies as the guarantee of their English liberties. They resented any intimation that as "Americans" they were somehow burdensome stepchildren in Great Britain's increasingly powerful empire, and thus not fully qualified for the rights of Englishmen living back in the "home country."

More important, perhaps, Americans had more or less started to live in a Lockean world. No one had planned it that way. Locke himself, who had played a role in colonizing the Carolinas, had not foreseen such a development. Yet here they were, these American colonists, living in a world where freedom seemed to emanate from the rocks and trees. The colonists were coming together, organizing their own governments, taking part in the political process, spelling out their rights in little documents that would soon be called bills of rights, asserting their own kind of moral independence from the Old World. On the colonies' western frontiers, where social controls were particularly weak, adventurous Americans were living their lives in Locke's state of nature.

Brilliant young American lawyers such as **John Adams** had long studied English law and now conducted their own investigations of constitutional theory. Adams became a strong advocate of the rule of law. Indeed, after the infamous Boston Massacre in 1770, the idealistic Adams took on the courtroom defense of the British soldiers who stood accused of murdering Boston colonists. Adams firmly believed that the whole thing had been political, not criminal, and that criminal charges in such a case would undermine the rule of law. Adams demonstrated in court that the soldiers had fired into the crowd only after being goaded beyond human endurance.

Later on, when representatives from the former colonies drafted the Constitution, they carefully inserted provisions for generality, prospectivity, publicity, consent, and due process. When it came to designing the structure of the federal government, they tried to think of structure as a way of promoting rule of law outcomes. The way they separated the judicial branch and insulated it against political interference, for example, spoke of their desire for keeping rule-making and rule enforcement in separate hands, so that the law could not be used to "get" someone.

The American Founders created more checks and balances in the United States Constitution than the British system had, and they invented a whole new structural idea they called **separation of powers**. So complex did the federal system become that skeptics doubted it could work.

John Adams

1735–1826

Adams felt strongly about the importance of the rule of law and so defended the British soldiers involved in the Boston Massacre. He later served as the first Vice President of the United States and the second President of the United States.

Figure 3.11 The Boston Massacre, *Mar. 5, 1770. Chromolithograph by John Bufford. Five colonists were killed when British soldiers fired into a rioting mob.*

Separation of powers

Dividing powers of government between the separate branches.

As for the commonwealth ideology, Americans adopted it as their own. They, too, had read Plato and Aristotle, and works such as *The Republic*, *The Politics*, and *The Nicomachean Ethics* could be found, well-thumbed and annotated, in hundreds of the colonies' private libraries. Dog-eared copies of *Cato's Letters* were read and reread in homes by candlelight, and Cato's political points were hotly debated in village taverns. When General Washington brought entertainment to his troop encampments, it was not dancing girls or stand-up comics but a performance of Joseph Addison's play, *Cato*. This play was the general's personal favorite; it was a story about Roman *virtu*, one Roman's steadfast integrity in the face of overwhelming odds.

Virtually all the American Founders saw themselves as commonwealthmen, and they played that role well. After all, they were men of independent means, beholden to no one, and most of them lived away from the colonies' urban centers, out on their country estates. America itself *was* the country, of course, both in the sense of country living and country as *patria*, far removed from the nefarious dealings of the court. There was one single important exception. Every colony where there resided a royal governor there was to be found a rough facsimile of the British court, right down to the tea-drinking and hand-kissing that made many Americans uncomfortable. When royal governors like Benning Wentworth and Robert Dinwiddie indulged to excess, or took bribes, or placed their henchmen in high offices, Americans knew in advance what that meant. *The court strikes again—the country must stand fast.*

American piety played an important role in this response. Think of the Puritans' "city upon a hill" in terms of "court versus country." Think how a Puritan would react to the news that agents of corruption and tyranny were attempting to corrupt them in order to fasten a yoke upon their necks.

For Americans, the English legacy became all-important. Its three components fit together into a single whole, and the meaning of the whole was this: America had been blessed by God as a land of freedom, those who lived in this land of freedom had also been blessed, and thus it was their duty to God to resist any and all who would imperil their birthright.

Timeline

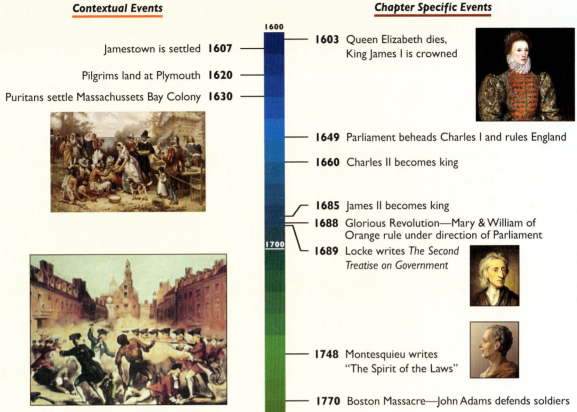

Contextual Events

Jamestown is settled **1607**

Pilgrims land at Plymouth **1620**

Puritans settle Massachussets Bay Colony **1630**

Chapter Specific Events

1600

1603 Queen Elizabeth dies, King James I is crowned

1649 Parliament beheads Charles I and rules England

1660 Charles II becomes king

1685 James II becomes king

1688 Glorious Revolution—Mary & William of Orange rule under direction of Parliament

1700

1689 Locke writes *The Second Treatise on Government*

1748 Montesquieu writes "The Spirit of the Laws"

1770 Boston Massacre—John Adams defends soldiers

Key Terms

- divine right of kings
- Social compact
- common law
- prospectivity
- due process
- country party
- Whigs
- Glorious Revolution
- rule of law
- publicity
- Tories
- commonwealth ideology
- state of nature
- natural law
- generality
- consent
- court party

Key People

John Locke

Questions

1. How does freedom *in* society differ from freedom *from* society?
2. According to Locke, how were rights granted unto men?
3. According to Locke, what is the "job" of government and how does government gain the legitimacy necessary to do its job?
4. Explain the terms under which Mary and William of Orange were permitted to assume the English throne. What was significant about those terms?
5. How does the structure of our government (today) "curtail the misuse of power?" In what ways do you think structure falls short?
6. Is there a modern-day court vs. country political struggle in America? If so, describe it. If not, explain why you think that is the case.
7. What are present day examples/violations of the rule of law?

Chapter 4

A Conflict of Interests

Abstract

The colonists were motivated by economic self interest as well as liberty. The colonization of America brought the promise of wealth and economic mobility. In a world dominated by government control over economics, Adam Smith argued for a free market economy governed by an "invisible hand." The Founders' experience with British economic policy and their knowledge of market principles allowed them to establish a government that would provide a solution to the inefficiency and oppression of mercantilism. This chapter examines the characteristics of a market economy and the limited roles a government should play in such an economy.

The abundant land across the Atlantic presented an unsurpassed economic opportunity for Europeans, especially for the ordinary citizen of limited means and little influence. The contrast could not have been more stark between the British Isles, where every square foot of land was owned (most of it by king and nobles), and England's vast land claims in North America, where millions of acres laid free for the taking. From 1600 to 1770, hundreds of thousands of everyday men and women endured the Atlantic crossing to make their homes in North America. This chapter discusses the colonial economy and the influence of the great British economist and philosopher **Adam Smith** on the Founders.

Adam Smith

1723–1790
Scottish philosopher and economist who wrote *The Wealth of Nations*. He is considered the father of modern economics.

Labor in the Colonies

Migrating from Liverpool or London to Boston or Philadelphia was an economic investment that offered substantial dividends. Labor in the colonies was scarce, and therefore more valuable than labor in England. With higher wages, a migrant could live better and eventually own land and begin climbing the economic ladder. The scarcity of labor also changed social and political relationships in the colonies. A wagon maker in Connecticut had to treat his apprentice with respect and dignity or risk losing the apprentice to a wagon maker in New York. A man who aspired to the colonial legislature had to watch his manners with the lower classes because many of them would soon own property and thus be eligible to vote in the next election. America enjoyed a fluid social environment. There was no nobility by birth. When newly arrived immigrants looked at the wealthy landowner or merchant, they saw someone who had arrived a few years before them, worked hard, and with good fortune, succeeded. They expected to follow that same path themselves.

Indentured servitude offered many impoverished Europeans the opportunity to come to America. Since large numbers of the would-be immigrants could not afford the expensive passage across the Atlantic, poor migrants agreed to be servants for a period of time, say seven years, in exchange for passage and a small amount of money to get started after their service. Unfortunately, indentured servitude and free immigration still did not fill the increasing demand for hard stoop labor in the Americas.

European colonists found the work of growing the plantation crops of rice, sugar, and tobacco to be unfamiliar, difficult, and something to be avoided if at all possible. The general shortage of workers in America's pre-Industrial Age prompted colonies, particularly the Southern ones, to look for new sources of labor. African natives shipped against their will across the Atlantic in European vessels handily filled this demand,

This Indenture witneſſeth, That

doth put himſelf Apprentice to
Citizen and JOINER of London, to learn his Art; and with
(after the manner of an Apprentice) to ſerve from the Day of
Date hereof, unto the full end and term of Years,
thence next following to be fully complete and ended: During
Term, the ſaid Apprentice his ſaid Maſter faithfully ſhall ſerve,
Secrets keep, his lawful Commandments every where gladly
He ſhall do no Damage to his ſaid Maſter, nor ſee to be done
others, but that he to his Power ſhall let, or forthwith give
ing to his ſaid Maſter of the ſame. He ſhall not waſte the
of his ſaid Maſter, nor lend them unlawfully to any. He
not commit Fornication, nor contract Matrimony within the
Term. He ſhall not play at Cards, Dice, Tables, or any
unlawful Games, whereby his ſaid Maſter may have any Loſs. With his own Goods or others, during the
Term, without Licence of his ſaid Maſter, he ſhall neither buy nor ſell. He ſhall not haunt Taverns, or
houſes, nor abſent himſelf from his ſaid Maſter's Service Day nor Night unlawfully: But in all Things as a
ful Apprentice, he ſhall behave himſelf towards his ſaid Maſter, [illegible] during the ſaid Term. And the
Maſter, (in Conſideration of the Sum of

his ſaid Apprentice, in the ſame Art which he uſeth, by the beſt Means that he can, ſhall teach and in
or cauſe to be taught and inſtructed, finding unto his ſaid Apprentice, Meat, Drink, Apparel, Lodging,
all other Neceſſaries, according to the Cuſtom of the City of London, during the ſaid Term. And for the
Performance of all and every the ſaid Covenants and Agreements, either of the ſaid Parties bind themſelves
the other by theſe Preſents. In witneſs whereof, the Parties above-named to theſe Indentures interchang
have put their Hands and Seals, the Day of in the Year of our Lord
and in the Year of the Reign of our Sovereign Lord GEORGE II. King of Great Britain,

Figure 4.1 Papers of indenture for a "James Buckland" from the year 1740.

Figure 4.3 Sale of Estates, Pictures, and Slaves in the Rotunda, New Orleans. *Detail of engraving by W. H. Brooke and J. M. Starling, c. 1860. This sale was conducted in the rotunda of the St. Louis Hotel. Slave traders imported an estimated 420,000 slaves into the American colonies.*

but slavery also caused a deep and divisive challenge to the core values of freedom and equality so fervently espoused by the colonists.

In the first two hundred years of America's tragic experiment with slavery, several hundred thousand black Africans were imported from Africa to the American colonies. The typical slave had been captured and enslaved by other Africans and then marched to the west coast of Africa. There, European or American ship owners purchased the slaves, chained them together, and stowed them in cargo space in appalling conditions. The largest number of slaves were sold to sugar planters in the Caribbean and to the plantations of Brazil.

All of the American colonies used slaves, but during colonial times they were especially important for the production of tobacco in Virginia, naval stores (tar, turpentine, and pitch) in the Carolinas, and rice in South Carolina and Georgia. Slavery, combined with indentured servitude and the labor of free immigrants, partially satisfied the colonies' appetite for work in the fields, in various skills, and in the homes of the newly wealthy. The promise of higher wages and far more economic opportunities in the New World continued to attract free white migrants, and by 1770 the population of Great Britain's thirteen American colonies' had grown to about two and a half million people.

On the eve of the American Revolution, most of the colonists were wealthy, measured by the standards of the day. They ate better and were physically taller than their British cousins. Though the homes of many—especially those who were living on the colonies' western frontiers—were often just huts or simple cabins, the colonists—again unlike their British cousins—enjoyed plenty of space for development. Their incomes were about the same as those in England, but they had much better economic opportunities. The colonists had developed new industries and crops, making them an integral part of the British Empire. Southern colonies exported tobacco, rice, indigo (a deep blue dye), naval stores, and timber (the latter two

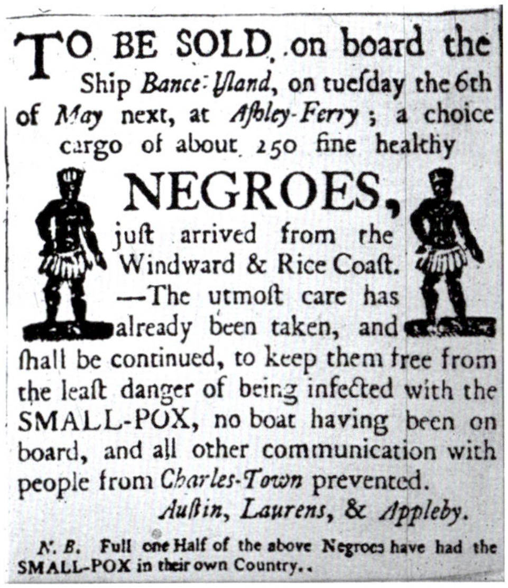

TO BE SOLD on board the Ship *Bance-Yland*, on tueſday the 6th of *May* next, at *Aſhley-Ferry*; a choice cargo of about 250 fine healthy

NEGROES,

juſt arrived from the Windward & Rice Coaſt. —The utmoſt care has already been taken, and ſhall be continued, to keep them free from the leaſt danger of being infected with the SMALL-POX, no boat having been on board, and all other communication with people from *Charles-Town* prevented.

Auſtin, Laurens, & Appleby.

N. B. Full one Half of the above Negroes have had the SMALL-POX in their own Country.

Figure 4.2 Slave ad from the South Carolina Gazette, *c. 1743. Courtesy of Drayton Hall, Charleston, South Carolina. Along with indentured servitude, the slave trade helped meet the high labor demand in Colonial America.*

of enormous importance to the home country's military superiority on the oceans). The middle colonies exported wheat and flour to the West Indies. Because of abundant wood and adequate iron ore, the colonies also exported iron bars to be crafted into tools, machinery, and other high-value products in England and elsewhere. New England, settled on indifferent land with few natural resources, had to look to the sea for their export wealth. They sent good quality cod to Catholic Europe and sold the leftovers to the West Indies as food for slaves. They hunted whales for their oil and bone. With ample raw materials and high wages for skilled craftsmen, shipbuilding became a thriving industry from Boston to Baltimore. Ships and excellent harbors naturally led to a vigorous seafaring trade in the colonies, especially in New England. Shrewd Yankee traders were soon pursuing profits throughout the world.

Mercantilism

An economic theory that emphasized the importance of gold and silver to the economic power of a nation. Mercantilists regulated the economy by encouraging exports and restricting imports.

As the 1770s opened, economic prospects could not have been brighter for the colonists. They had solved their most pressing labor problems through indentured servitude and slavery. They were producing an abundance of high-quality food at a lower cost than their stay-at-home cousins were able to do back in Britain. The colonists had excellent resources to exploit for profit. They would also expand westward if they could convince Britain to ignore the rights of the Native Americans occupying the land. Soon the colonies would become better at manufacturing goods as well.

Britain's King George III and Parliament had different ideas. They saw the colonies as an important cog in Britain's growing economic machine. Parliament and the king did not view their American colonies as an opportunity for ordinary Englishmen to get rich. For them, America, like the other parts of the British Empire, existed to enrich the British treasury.

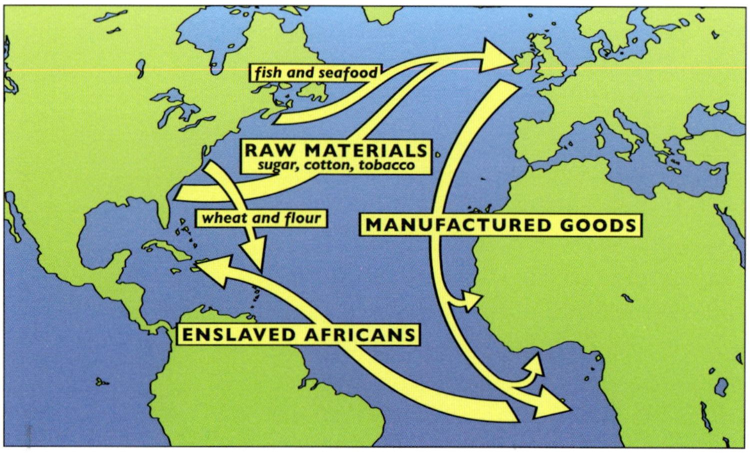

Figure 4.4 Map of Triangular Trade routes. The New World provided raw materials such as sugar and tobacco to England and Western Europe. These countries traded manufactured goods, including rum, to Africa for slaves who were sent to the New World. For the slaves this was referred to as the Middle Passage.

Mercantilism

From 1500 to 1800, **mercantilism**—the idea that the government should regulate the economy to strengthen national power—dominated English economic policy. What gave a nation power? For the mercantilists, the key to national power was large stockpiles of gold and silver to finance the army and navy necessary to build and maintain a far-flung empire. But how could Britain, or any nation, for that matter, build up its treasury of gold and silver? The primary way to bring gold and silver into a country was through a favorable balance of trade, where the value of British-made products exported and sold to other countries was greater than the value of foreign-made products purchased from other countries. Thus, the mercantilists wanted to encourage exports and discourage imports. If exports were greater than imports, other countries had to pay gold or silver to settle their accounts.

Consequently, England's mercantilists tried to manage the economy of the empire, including the economy of the colonies, in a way that would increase the quantity of gold

Figure 4.5 Mercantilists believed the more gold and silver a nation possessed, the more world power it had.

and silver in the king's treasury. They encouraged domestic manufacturing to compete with imported manufactured goods, such as textiles and ironware. Laws were passed to prevent craftsmen and artisans from moving with their skills to other countries. They paid subsidies, or bounties, to encourage exports. In other words, the government managed the economy to further government interests.

As part of this overall effort, the British Parliament passed many economic regulations, referred to collectively as the **Navigation Acts**, and set up admiralty courts to enforce these trade regulations in the colonies. All trade had to go through British or colonial merchants and be shipped in British or colonial ships. Certain goods such as tobacco could only be legally shipped from the colonies to England, rather than directly to other countries. The end goal of all regulation was to generate large exports from England, with few imports, so that gold and silver would flow into the motherland.

The American colonists found the Navigation Acts and other British efforts to manage their economy economically frustrating and demeaning to their concept of freedom of choice. Why couldn't the colonists sell tobacco or naval stores directly to the rich Dutch or to the French? Why were the colonists pushed to purchase everything from England and her other colonies? (Ninety-eight percent of imports to the colonies came from English lands.) George Washington found the regulations so irritating that he stopped growing tobacco and tried to purchase domestic goods and become self-sufficient. The Navigation Acts in themselves were not an overwhelming burden to the colonists, but the acts were a nagging reminder that the Americans were subservient to the distant British Parliament and to British courts in matters of commercial policy. When King George, his ministers, and Parliament decided to raise more revenue from the colonists with a series of taxes and fees, the colonists realized they were viewed as dependents and second-class citizens of the empire. It was one thing to be subjects of King George, just like their cousins back in England. It was quite another to be viewed as subjects of England's merchants and manufacturers.

The Market Economy

As American colonial discontent grew, a Scottish economist and philosopher was writing a book that effectively attacked mercantilism and described the basic operation of what we now call **capitalism**, or a market economy. Adam Smith's *An Inquiry into the Nature and Causes of the Wealth of Nations*, commonly referred to as ***The Wealth of Nations***, was published in 1776. Smith was part of the remarkable Scottish Enlightenment. Smith, a professor of moral philosophy at the University of Glasgow, was already famous for an earlier book, *The Theory of Moral Sentiments*, which gave a description of human nature that included

Navigation Acts

Economic regulations passed by British Parliament to enforce trade regulations in the colonies: all trade had to go through British or colonial merchants and be shipped in British or colonial ships with the end goal to generate large exports from England, with few imports, so that gold and silver would flow into the motherland.

Capitalism

The philosophy of a free market economy in which the government serves only to create an acceptable environment in which to make exchanges.

The Wealth of Nations

Book written by Scottish economist Adam Smith that criticized mercantilism and proposed a free market economy in which the "invisible hand" determined prices.

both self-interest and benevolence. *The Wealth of Nations* pushed Smith to a higher level of intellectual accomplishment and made him the father of modern economics.

In the introduction to *The Wealth of Nations*, Adam Smith immediately broke with the mercantilists by stating that wealth (the goal of a nation's economic activity) is not its treasury of gold and silver, but instead what a nation can produce and consume in a year's time. His definition of a nation's wealth would be what we think of today as per capita income. In some ways, this view of a nation's economic goal was revolutionary, since Smith focused on the happiness or welfare of ordinary people, rather than on the financial condition of the king or the government. Economists would later refer to this emphasis on consumption as consumer sovereignty. Adam Smith put economic sovereignty in the hands of consumers, just as John Locke had earlier put political sovereignty in the hands of the people.

Markets
Divisions of the economy that specialize in certain goods or services.

Market economy
An economic model proposed by Adam Smith in which the government serves only to create an acceptable environment in which to make exchanges.

Exchange
Trade between two parties.

Smith's Description of a Market Economy

Smith began *The Wealth of Nations* by describing an economic model that was free from government regulation and intervention—a **market economy**. Such an economy would start with simple, ordinary **exchange**, or trade, between two individuals. A farmer trades some of his grain to a cobbler for a pair of shoes, or a weaver gives a bolt of cloth to a flour miller for an agreed-on amount of flour. Smith noted very early in *The Wealth of Nations* that exchange was motivated by self-interest:

> It is not from the benevolence of the butcher, the brewer, or the baker, that we expect our dinner, but from their regard to their own interest. We address ourselves, not to their humanity, but to their self-love, and never talk to them of our own necessities but of their advantages. Nobody but a beggar chooses to depend chiefly upon the benevolence of his fellow-citizens.

A simple but very powerful truth about exchange is that both parties involved feel the exchange benefits them. An exchange is a voluntary act between two parties motivated by their interests in improving their circumstances. Both parties can benefit from an exchange because they value the items traded differently. The hungry student exchanges $5 with the pizza maker for a pizza because he values the pizza more than the $5, while the pizza maker values money more than the pizza. By rearranging who has what through exchange, both benefit. This small miracle of exchange happens billions of times each day across the world.

Mercantilism, which was basically the way governments, both good and bad, had operated throughout history, ignored this fundamental characteristic of exchange. Governments had always restricted exchange in ways that would benefit the ruler, in this case King George, by increasing his stock of gold and silver. But his benefit came at the expense of his citizens, who lost substantial benefits because they were legally restricted from trade that would have been beneficial to them. Smith used this simple fact to condemn mercantilism and its restrictions. This

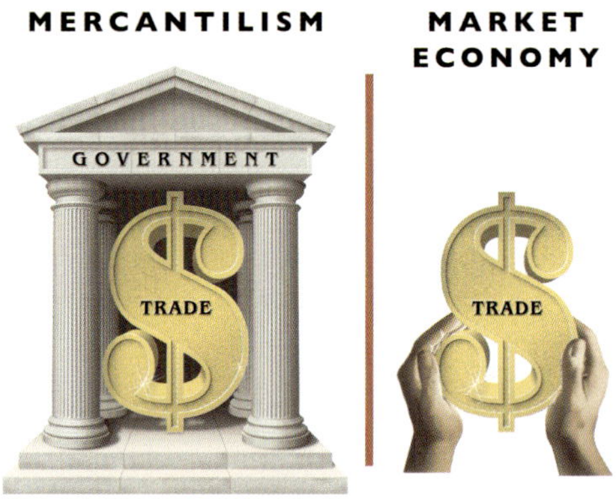

Figure 4.6 In mercantilism, trade is controlled by the government for its own benefit; in a market economy trade benefits producers and consumers.

Figure 4.7 Money allows you to trade with the pizza maker, who in turn can trade for his wants.

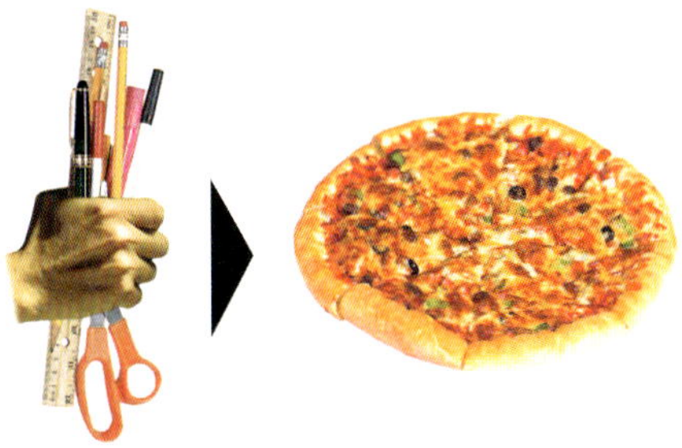

Figure 4.8 Without money, you could only get a pizza if the pizza maker desired the items that you had to trade.

same fact about who benefits in trade remains a powerful criticism of government today. Whenever government restricts exchange, for whatever purpose, some people are going to be worse off because they have been denied beneficial trade opportunities.

Role of Money

One of the oldest innovations of civilized societies still plays a central role in the process of exchange. Money allows individuals to extend the benefits of exchange into a complex pattern of trade, often involving hundreds of individuals. Imagine a world without money. Each of the two parties considering an exchange would have to want the particular good possessed by the other party. The hungry student would have to trade work, or something else of direct value, to the pizza maker for a pizza. If the two parties were fortunate, the pizza maker and the student might be able to work out an exchange that brought in a third party, possibly the flour mill operator. Maybe the student worked for the flour miller who gave flour to the pizza maker, who then gave a pizza to the student. But these **coincidences of wants** are awkward, very localized, and basic. For example, how many auto mechanics are going to want to have a lecture from a humanities professor on Gothic architecture in exchange for a car tune-up?

Money of an established value eliminates the need for a coincidence of wants and separates the exchange process into parts. The college student exchanges money he earned as a groundskeeper for a pizza. The pizza maker combines the student's money with the money of others to buy a new oven. The oven maker uses that money to buy metal to make the oven. The man who refines the raw ore into metal can now send his own pizza-loving son off to college. Money allows us to exchange over a much broader area, which makes the patterns of exchange much more complex. Money rivals the wheel in importance as an innovation of early civilization.

Anything may be used as money as long as everyone agrees on the commodity. Different cultures have used stones, shells, cows, cigarettes, beads, and animal skins as money. The American colonies used tobacco as currency. The most common forms of money have been metals with high value for their weight and durability, like silver and gold. Governments have often standardized money by minting coins or assuming control of paper money.

In modern times, paper currency is used because it is lightweight, convenient and easy to produce. Until relatively recently, paper money was backed by gold or silver. Now, paper money is simply backed by the authority of government. But governments

Role of money

Money facilitates exchange by eliminating the necessity for a "coincidence of wants," functioning as a generally acceptable medium for exchange.

Coincidence of wants

When two parties each possess something desired by the other, promoting an exchange.

are sometimes tempted to use this control over money for their own purposes. Modern governments often create money to finance government programs causing inflation and economic disruption.

However, problems sometimes occur with government control of money. In ancient Rome, emperors occasionally increased the amount of gold in their personal treasury by replacing some of the gold or silver in their minted coins with a cheaper metal. Modern governments do the same with paper, because there is a temptation to cause inflation—seen by some economists as beneficial to the economy—by increasing the amount of paper money in circulation.

Specialization

Specialization

The economic practice of focusing resources on production of one or a few goods.

Once people use money, the patterns of exchange become very complex. The production of any single item could involve hundreds of different exchanges because money allows the separation of the different aspects of exchange. Adam Smith was aware of this complex pattern of trade, and he began *The Wealth of Nations* with a discussion of the value of **specialization**. To illustrate, Smith took his reader on a visit to a pin factory. He noted that one person drew the wire, one straightened it, another cut it, one sharpened the point, another ground the top of the wire to receive the head, and so on. Smith calculated that ten workers could produce twelve pounds of pins in a day with 4,000 pins to a pound, over a million in one year's time. In contrast, ten workers producing pins individually from start to finish may only be able to produce a hundredth of that amount. Smith was describing the great benefits that come from specialization. He referred to this complex pattern of work and trade as the "division of labor."

Specialization, or the division of labor, increases productivity tremendously. The pin makers became adept at their particular tasks. They also developed machines or tools that helped them with each specialized function. As a consequence, the pin makers produced the pins at a very low cost. Indeed, the cost was so low that even the most self-sufficient individuals would not want to make their own pins. They would find it preferable to spend a few pennies for whatever pins they needed. Remember that this specialization is only possible because of the extensive patterns of exchange. The pin makers must be able to trade pins to people across a very wide area to be able to specialize in pin making with a process that produces a million pins.

Why could the pin makers trade with people far away? For one thing, pins were light and easy to carry. For another, transportation costs were declining. Oceangoing shipping was getting cheaper, in part because of all the fine timber and naval stores produced at a reasonable cost in the American colonies. Declining transportation costs would prove to be very important in continually improving America's standard of living from Smith's time to the present. Declining transportation costs made it possible to trade

Figure 4.9 Carmaker Henry Ford in 1927. Henry Ford was a master at specializing the tasks of his workers. Through his assembly line, he made the automobile affordable to the average American.

with people farther and farther away and to trade heavier and heavier objects. More to the point, declining transportation costs meant that there could be more and more specialization, or division of labor.

Smith observed "that the division of labor is limited by the extent of the market." By this, he meant that specialization would be limited if the market were very small, but specialization could be very extensive as the market grew in size. In Smith's day, the market for most British or American-made goods was confined to the British Isles, the more advanced and easily reached parts of continental Europe, the American colonies, and the Caribbean. Today we have a worldwide market in almost all but the heaviest of goods, such as gravel and cement. The invention of new forms of transportation has allowed the U.S. to trade with the whole world and to specialize evermore extensively.

In sum, voluntary exchange is the foundation of free economic activity. We are motivated to trade or exchange because we value goods differently. We value goods differently because we may simply have different tastes or preferences, and because we have different production costs. (The specialized pin makers make their pins much cheaper than pins made by older methods.) Money and declining transportation costs allow individuals to specialize and trade with others across a very wide area.

Economic Competition

All of this exchange and specialization is simply the pursuit of self-interest by individuals as they go about their daily lives. But what will regulate this self-interest? Will not the pin makers want to charge an exorbitant price for the pins that we can hardly make ourselves? Will not the butcher charge us as high a sum as we could possibly pay for a Sunday roast? Isn't it better to be self-sufficient and not at the mercy of others' self-interest? Smith's answer is that these are false

Figure 4.10
Specialization helped make the price of the Model T affordable to the average American. By the 1920s a Model T cost about $300 ($3,300 in 2007 dollars). Photo by Robert Runyon, 1920.

worries. Economic competition will curb the excesses of self-interest, allowing each of us to specialize and trade. If one butcher tries to charge a high price, his customers will simply trade with his competitor down the street.

Smith and economists ever since have used the word "competition" to describe the actions of buyers and sellers in a market for a commonly traded product or service. They say a market is competitive if there are sufficient buyers and sellers so that no single seller or buyer has a significant influence on price. In any market, say the market for bread, buyers want to pay a low price, and sellers want a high price. Clearly, their interests are in direct conflict with one another.

Perfect competition

When buyers and sellers have no influence on price and terms of exchange.

What then prevents exploitation or unfair advantage on one side or the other? The primary safeguards are competition and the interests of others. If there are many sellers and buyers, then an attempt by any single seller or buyer to manipulate or determine the price will fail, because others will undercut the attempt. Suppose a bread maker asserts that he will not sell his bread for less than $5.00 a loaf. If others are willing to sell it for less, say $2.00 a loaf, the first baker sells no bread unless he lowers his price. If there are enough bread sellers, all of them will perceive that the price for bread is largely beyond their control. They compete with one another to sell their bread, but they

Collusion

When sellers are conspiring to maintain a high price and avoid competing with one another.

Monopoly

When one person or group is the sole seller in a market with many buyers; the lack of competition in a market.

Law of Supply

As the price of a particular good or service rises, suppliers will produce more of that good or service.

Law of Demand

As the price of a particular good or service rises, individuals will buy less of that good or service.

accept the price as whatever bread is going for that day. Similarly, customers would like to pay nothing, or pennies, for the bread. But if other customers are willing to pay more, then the price of bread will be higher.

If there are many customers, then all of them believe the price of bread is beyond their control. As long as the bread sellers compete with one another to sell their product, and bread buyers compete with one another to buy bread, no one should have a particular advantage and the interests of all should be well served, yet controlled by market pressures, not by government.

Smith, like economists ever since, worried about **collusion** by sellers or by buyers, or the unusual situation where there might be a single seller or buyer. Collusion implies that sellers are conspiring to maintain a high price and avoid competing with one another. In those circumstances, there would not be economic competition. Bread makers in collusion could set the price of bread high. As long as none of them cheated on their agreement, they could extract more money from buyers. Adam Smith knew that businesses would be attracted to such collusions:

> People of the same trade seldom meet together, even for merriment and diversion, but the conversation ends in a conspiracy against the public, or in some contrivance to raise prices. It is impossible indeed to prevent such meetings, by any law which either could be executed, or would be consistent with liberty and justice. But though the law cannot hinder people of the same trade from sometimes assembling together, it ought to do nothing to facilitate such assemblies, much less render them necessary.

But as this quote suggests, Smith also did not think collusion and **monopoly** would pose large problems for the economy unless government promoted conspiracies to fix prices or monopolize a market. By and large, Adam Smith expected competition to control economic interests and to generate reasonable prices. He saw that prices would generally be determined by the impersonal forces of **supply** and **demand**, with both buy-

Figure 4.11 Computer performance and price comparisons. All prices adjusted to 2006 dollars. Competition creates better quality and lower prices for consumers.

ers and sellers feeling that price was largely beyond their control.

Role of Prices and Profits

In a competitive market in which buyers and sellers see price as a given, what actually determines the market price? Why is bread $2.00 a loaf instead of $2.50 or $1.50? Adam Smith and the economists who followed him observed that markets tended toward an **equilibrium price**, one where everyone who wanted to buy or sell at that price was able to do so.

Suppose the equilibrium price of bread was $2.00 a loaf. If bread were being sold at a price below equilibrium, say $1.50 a loaf, more people would be willing to buy the cheaper bread, but fewer people would sell that bread at the lower price. When particular bakers could not make a profit, they would simply stop making bread. A bread **shortage** would then develop because there would be buyers who wanted to buy bread but were unable to find bread to purchase. Whenever there is a shortage in a free market, the product or service price will rise.

Conversely, a price above equilibrium, such as $2.50 a loaf, creates a **surplus** because there would be fewer buyers who are willing to purchase bread at the higher price, and suppliers would want to supply more bread than they supplied at $2.00 a loaf. This surplus would cause the price of bread to fall. This leaves prices to be determined by natural market forces, causing a market to reach an equilibrium where both buyers and sellers are satisfied with the price. In most cases, equilibrium in a market is easily attainable because demanders (purchasers) and suppliers (sellers) respond in opposite ways to changes in prices.

When the price of a good rises, buyers will demand less of that good; when the price falls, buyers will demand more. On the other hand, a rise in price induces sellers to sell, or supply, more of a good, just as a fall in price

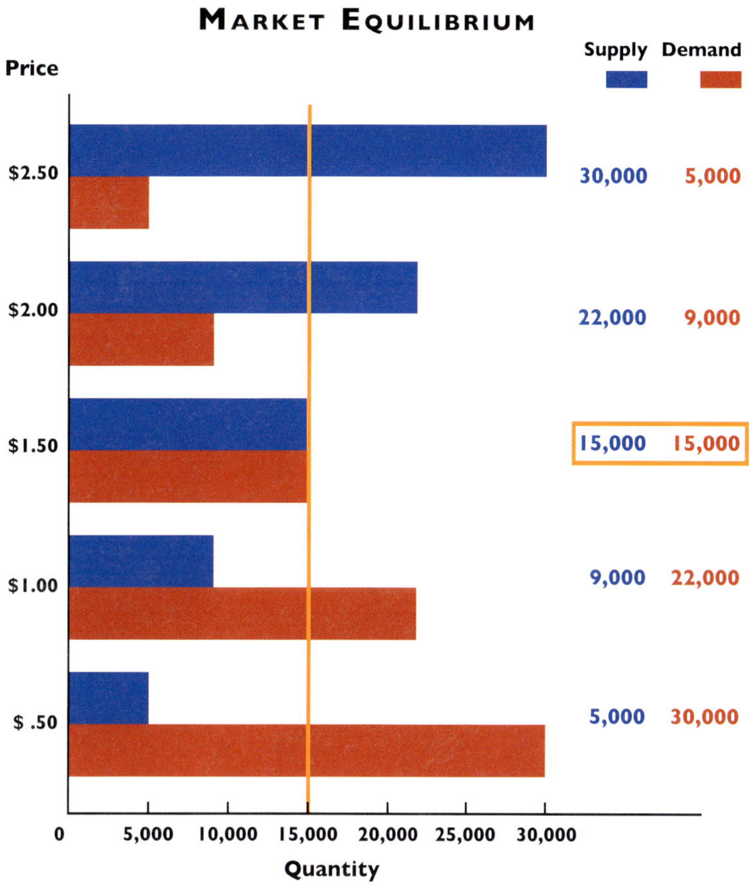

Figure 4.12 In a market the price will move to an equilibrium ($1.50 in this chart) where quantity demanded equals quantity supplied.

induces them to sell less. If the amount demanded has an *inverse* relationship to price while the amount supplied has a *direct* relationship, there must be some price where the amount demanded equals the amount supplied. That price is the equilibrium price. The simple, but profoundly important, feature of free markets is the ability to find an equilibrium price. No single individual knows in advance what the equilibrium price will be. But the actions of all the participants in the market will move the price to its equilibrium level.

There are many markets in any econ-

Role of prices

In a market economy, prices determine the quantity of goods supplied.

Role of profits

In a market economy, as profits increase, the number of suppliers and resources for making that good will increase.

omy. There are markets for all goods and services, markets for labor including markets for each kind of skill, markets for land and natural resources, markets for capital (tools, machinery, and buildings), and even markets for people skilled in bringing all those other markets together. These markets are all linked to one another by prices. These prices act as signals to individuals participating in the market. Suppose people start eating less bread, causing the price of bread to fall. The price change in bread will affect flour millers and the wheat market, causing farmers to grow less wheat and, perhaps, raise more livestock, causing the price of beef to change. The market for baking equipment will be affected. All of these markets are linked together and respond to one another without requiring intervention by government or other institutions.

One element other than price links market economies together. Profits, the excess of revenues over costs, perform an important function in Smith's market economy. Differences between profit and loss send an important signal to a business as it allocates its resources. When profits in a particular industry are high, it is a signal to invest more resources in that industry. New businesses enter production and old businesses expand production. When losses occur, it is a signal to leave a particular industry and pursue some other opportunity in the economy.

Thus it is that prices and profits act as signals for all of the participants in the market economy. Each individual observes the prices and profits generated by market forces and decides on the best course of action. As Adam Smith noted, when working properly, the free movement of market prices and profits, combined with the pursuit of self-interest by individuals, achieves the most efficient outcome for the economy as a whole. To capture this remarkable property of market economies, Smith coined the phrase "**the invisible hand**" and noted:

> . . . by directing [his] industry in such a manner as its produce may be of the greatest value, [an individual] intends only his own gain, and he is in this, as in many other cases, led by an invisible hand to promote an end which was no part of his intention. Nor is it always the worse for the society that it was no part of it. By pursuing his own interest he frequently promotes that of the society more effectually than when he really intends to promote it. I have never known much good done by those who affected to trade for the public good. It is an affectation, indeed, not very common among merchants, and very few words need be employed in dissuading them from it.

The invisible hand is the coupling of self-interest with efficiency-producing movements of prices and profits. Together they manage a market economy by allocating resources to the right ends, by inducing firms to produce the right constellation of goods and services, and by rationing scarce goods and services in the most beneficial way. Instead of an economy managed by the visible hands of the mercantilist bureaucrat signing decrees and approving actions by private business, the invisible hand of prices and profits manages the economy. Moreover, Smith argued this invisible hand does it much more efficiently.

Smith used his description of a market economy and its ability to allocate resources efficiently to criticize the mercantilist wisdom of the day. Gold and silver were of little economic importance compared to the productive capacity of the economy. Mercantilists were concentrating on the wrong goals. Trade with other countries was good and should be fostered, just as trade among farmers and shopkeepers was of value. Government trade restrictions, subsidies or bounties to exports, and interference with the natural development of colonies were of no value. Mercantilism inevitably sacrificed

Equilibrium price

The price at which the amount demanded is equal to the amount supplied.

Shortage

When the amount demanded is greater than the amount supplied.

Surplus

When the amount supplied is greater than the amount demanded.

The invisible hand

Adam Smith's term for the natural self-regulation of a market economy driven by self-interest and efficiency.

Laissez-faire

Policy in which there is little or no interference with exchange, trade, or market prices by the government.

The Role of Government in a Market Economy

1. Prevent Coercion and Fraud
2. Provide Money
3. Provide Basic Transportation and Communication
4. Define Property Rights
5. Enforce the Exchange Agreements

Figure 4.13 According to Adam Smith, government should only have a few functions in regards to the economy.

the interests of consumers to the interests of producers, particularly to the interests of merchants and manufacturers. In general, Smith advocated a **laissez-faire** government policy; in other words, he wanted the government to leave the economy alone. Government should not interfere with exchange, trade, or market prices. The economy worked best when individuals were left free to pursue their own interests restrained by economic competition, but unrestrained by government regulations:

> Every man, as long as he does not violate the laws of justice, is left perfectly free into competition with those of any other man, or order of men. The sovereign is completely discharged from a duty, in the attempting to perform which he must always be exposed to innumerable delusions, and for the proper performance of which no human wisdom or knowledge could ever be sufficient; the duty of superintending the industry of private people, and of directing it towards the employments most suitable to the interest of the society.

Smith would leave individuals free to pursue their own interests within the laws of justice. Additionally, Smith excused the sovereign from the duty of trying to regulate economic activity. Smith bluntly reminded the ruler that he did not have the wisdom or knowledge to undertake that task. Indeed, it would be delusional for the sovereign to think himself capable of managing the economy.

Role of Government in a Market Economy

There are a few functions left for government, even in Adam Smith's market economy. A market economy requires a clear definition of property rights. Unrestrained trade requires that the ownership of property and the rights associated with property be clear. Otherwise exchange is difficult because the characteristics of the items being traded are unclear. Suppose ownership of a piece of land entitles the owner only to farm the property, not to build on it. The value of that property right would be different from a property right that allowed the owner to do anything with his land. Definition of property rights is usually left to government. The government also assumes the responsibility for preventing fraud or coercion. Voluntary exchanges are only beneficial if they are truly voluntary and are based on good information. The government has to develop a context for exchange. Courts are needed to resolve disputes about exchanges or the terms of exchanges. The government usually provides the money used in a market economy. Finally, Adam Smith also suggests government should provide improvements in transportation.

In short, the government should create a political and cultural environment to encourage exchange and the ordinary workings of a market economy. According to Smith, once such an environment is in place, government should not interfere with the operation of the market economy.

Economic Basis of the Founding

Mercantilism and the market economy posed the two basic alternatives for economic organization facing the American colonies as they struggled with their role within the British Empire. Mercantilism represented an economy organized to serve higher social purposes as defined by government. In a market economy there were no higher purposes of economic activity; individuals were simply left alone to pursue their own individual objectives.

After the Revolution, Americans found a market economy consistent with their general values and approach to government. A market economy promoted individual liberty and pursuit of happiness. It left the success or failure of individuals to their own initiative and efforts. Success did not depend on connections within the government or the privileges garnered at court. This view was consistent with the old Puritan moral view that individual self-government was the foundation of a moral society. Adam Smith's description of the workings of a market economy also reassured the Founders that a market economy could control the conflicting interests in economy through economic competition without the heavy, and perhaps tyrannical, hand of government. Smith's *The Wealth of Nations* made a powerful and persuasive case that a market economy could be a key component in the creation of a free society. A functioning market economy would allow a government to be smaller and more narrowly focused, since economic matters would not be part of the government's tasks.

Mercantilism and the later more powerful philosophies requiring the government to control exchange and manage the economy would remain an attractive alternative to the Founders and later government leaders. It is always tempting to use the power of government to placate a constituency, or to pursue some larger goal of government. History, especially modern history, shows this tension between free markets and government control of the economy played out over and over. The Founders created the United States with a base of a free economy, but we have often succumbed to the attractive certainty of a hands-on government managing the economy. Of course, Adam Smith, a towering figure of the western world's spiraling economic growth in the eighteenth century, would remind us that any benefits to a citizenry accruing from their government's management of their complex economy is a delusion.

There may be a tendency to think of the American revolution considered in Chapter Five as a revolution fomented by high principles "Give me liberty or give me death." We should remind ourselves that the revolution was also born in a crucible of self-interest. The Navigation Acts, the Stamp Act, policies on western lands and many other actions of Parliament and the British Crown hit the colonists in the pocketbook. In general, mercantilism and British policies transferred income from the colonies to England. The passion for independence was not simply fired by philosophy. Like most human behavior, the revolution came out of the ideas, world view and self-interests of the colonists.

Timeline

Contextual Events

Boston Tea Party **1773**

Intolerable Acts **1774**

Revolutionary war begins (April) **1775**

Thomas Paine's "Common Sense (Jan) is published **1776**

British evacuate Boston (Mar)

Declaration of Independence adopted (July)

Declaration of Independence signed (Aug)

Washington crosses the Delaware (Dec) River and captures Trenton

1770–1780

Chapter Specific Events

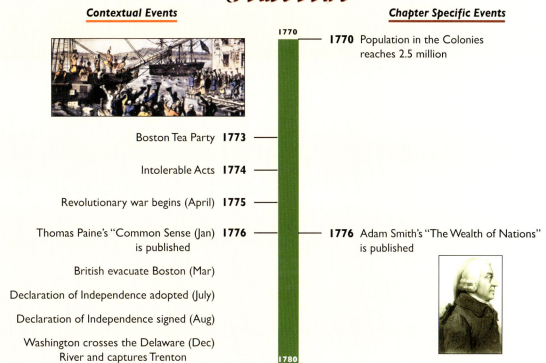

1770 Population in the Colonies reaches 2.5 million

1776 Adam Smith's "The Wealth of Nations" is published

Key Terms

indentured servitude
capitalism
role of money
competition
surplus
equilibrium price
mercantilism
market economy
coincidence of wants
role of prices and profits
supply
the invisible hand
Navigation Acts
exchange
specialization
shortage
demand
Laissez-faire

Key People

Adam Smith

Questions

1. How do the goals and methods of capitalism differ from mercantilism?
2. How does specialization and exchange increase the standard of living?
3. In what specific ways does competition shape our economy today? Has self-interest or personal virtue proven to be the motivation for most of our decisions today?
4. What five things should government do to ensure that a free market remains effective?
5. Compare the current economic structure of the United States with the vision described by Adam Smith's *Wealth of Nations*. In what ways are existing conditions similar and/or different to what Smith envisioned?
6. In what ways was the American Revolution related to economic interests?

Chapter 5

THE AMERICAN REVOLUTION

Abstract

The colonists saw themselves as residing in individual colonies rather than a united country. The Revolutionary War changed that. The majority of colonists identified with the Declaration of Independence, which declared the grievances and injustices laid by King George, justifying the colonists' rebellion. They read pamphlets, such as *Common Sense,* that gave a voice to the common plight of the colonists and unified their cause. They rallied around their leader, George Washington, the American symbol of nationhood. The Revolutionary War forced the colonists to form a government and undertake a war against a world superpower. In the process, a nation was founded.

We saw in Chapter 1 that national foundings typically occur after a political upheaval. The American Revolution certainly was that. The Revolution not only separated the American colonies from the British Empire, it also brought them together for the first time as a single *patria*. It forced Americans in Boston to think about what they shared in common with fellow patriots in Charleston, what distinguished them from their British cousins—indeed, from the rest of the world—and what principles they might stand on if they had to stand together. It gave them firsthand experience in creating their own government and threw them together in that most extreme and demanding of any social group's undertakings: war.

The American colonists all shared the English language and subservience to the British crown, but the colonies had little in common with each other before the Revolution. Their social systems and geographic borders had been established by different groups and for different purposes. Some of the colonies were ethnically and culturally diverse, while others were homogeneous. Their economies operated very differently, some closely tied to the empire, others independent from it. Their societies, even their political systems, were markedly dissimilar. To the extent that they identified with one another, such identity tended to be regional. New Englanders acknowledged, even celebrated their Puritan background, Southerners were proud of their plantation agriculture, the middle colonies made the most of their polyglot diversity, and so on. These regions were no more "American" in their sense of identity than we are "North American" today.

Before 1763 the colonies were bound to the British Empire by economic interdependency, as "hen and chicks," it was said, and by the fear of France and Spain, who were often poised to attack the fledgling chicks from, respectively, north and south. When the French and Indian War ended in 1763, all of that changed. French Canada passed into

Figure 5.1 Map of the original thirteen colonies.

British hands, and the situation in North America was suddenly and dramatically different.

This turn of events sparked a sequence of developments that would soon rend the proud British Empire to tatters. In the space of little more than a decade, chicks and hen would part company for good.

The Coming of the Revolution

The complex events that brought on the American Revolution lie largely beyond our scope of study. A few generalizations are in order, however, because certain aspects of the imperial controversy had a direct impact on the Founding.

England's administration of the Empire had often been lax, and the colonies had grown accustomed to a large measure of independence. Consequently, the colonies had developed political maturity. Additionally, most colonists ignored British economic regulations like the Navigation Acts and smuggling abounded. With the conclusion of the French and Indian War, however, the British government launched a determined effort to tighten colonial administration, close loopholes, throttle smuggling, and weld its global empire into a single strong entity.

This seemingly reasonable, if complicated, task was ultimately doomed because the distant imperial authorities did not view the colonies the way the colonies viewed themselves. Instead of seeing them as mature political societies with local self-rule and a common allegiance to the crown, the authorities saw the colonies as mere possessions that could be administered at will. Administration is very different from the rule of law. To administer, someone orders others around, giving instructions here, changing instructions there, allowing this, revoking that, and revising something else. It is anything but

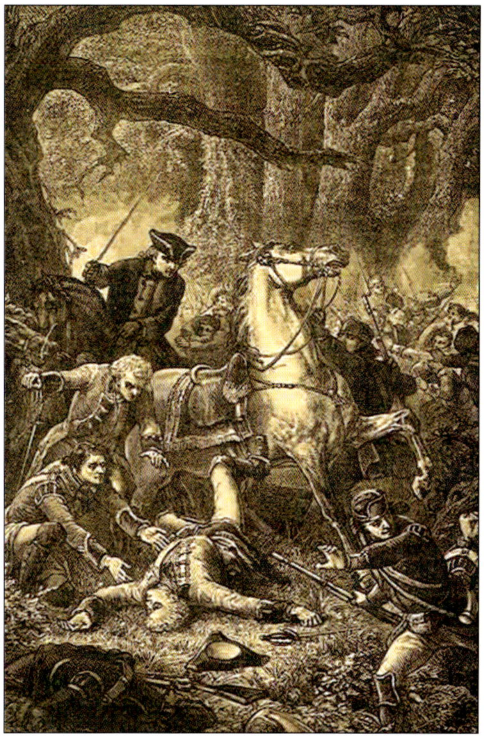

Figure 5.2 A 19th century engraving of the death of Major-General Braddock at the Battle of the Monongahela (1755) during the French and Indian War. The loss of Braddock sealed the British defeat at Monongahela. However, the British won later battles, notably the Battle at Lake George. The eventual defeat of the French decreased the colonies' dependence on England for military protection, opening the door further for ideas of freedom from the crown.

the general, prospective, public rules that are applied predictably. Americans felt their rights as Englishmen were being violated. The British rulers believed in no such rights, at least not for the colonials.

This fundamental disparity gave rise to compounded misunderstandings and perceptions of wrong-doing on both sides. A particular aggravation lay in the matter of **taxation without representation**. Imperial authorities noted that the colonies had largely escaped taxation in the past, and they believed that the time had come to rectify this oversight. Americans at least ought to defray the high cost of their own defense against the French in the recently concluded war. Nor was there any hesitation in turning to Parliament for the necessary authority, for the principle of parliamentary supremacy in the funding of government was now firmly established. But the colonies took a completely different view. The basis of parliamentary supremacy was representation, they argued. Parliament alone could levy taxes

Taxation without representation

Rallying cry of the colonists during the Revolutionary period because of the taxes placed on them by a Parliament in which they had no representation.

Figure 5.3 The repeal—or the funeral procession of Miss Americ-Stamp. *This 1766 cartoon depicts a funeral procession to the tomb of the Stamp Act. Treasury Secretary George Grenville, a principal proponent of the Stamp Act, is shown carrying a child's coffin, marked* "Miss Americ-Stamp born 1765 died 1766."

because Parliament alone represented England's taxpayers. The colonies were not represented in Parliament, so it had no authority whatsoever to lay a tax upon them. That was the responsibility of their own houses of assembly. This was, of course, a thoroughly English response, but it just didn't make sense to the Parliament in England.

Repeated attempts to establish the principle of parliamentary taxation sparked violence and bloodshed in the streets of American cities and gave the colonists a deep-seated suspicion of Parliament. The Americans' intransigence confirmed suspicions on the other side that the colonists just wanted a free ride. The Stamp Act, which required the purchase of government stamps for a variety of legal and commercial documents, became the most infamous of the tax squabbles. The colonists became so heated at the thought of purchasing government stamps that they took to the streets in violent protest. Parliament backed down—but only for the time being.

Other quarrels merged with the tax issue and intensified its effects. The British government closed western settlement to the colonies—dramatically slamming the "free land" door in their faces. Parliament reinvigorated the Navigation Acts and dispatched a new regulatory bureaucracy to poke into the private lives of the colonists. To stem smuggling, which in some seaboard communities was practically a cottage industry, officials broadened the language of search warrants, allowing them to snoop anywhere for contraband. The officials then revoked the right of jury trial for accused smugglers, believing (probably with good cause) that local juries would never convict. They sent British soldiers to troublesome enclaves, notably Boston. Rumor in Boston had it that British officials would soon install an Anglican bishop in Virginia with orders to stamp out religious freedom.

To all of these initiatives, American colonists responded emotionally, dramatically, and with exaggerated paranoia. After all, they had been primed by the English Whigs to expect just such nefarious dealings with the "court party." Instead of seeing a misguided attempt to reform imperial administration, colonists saw an insidious conspiracy to corrupt government further and defraud them of their liberty.

Matters came to a head in 1773 with the passage of the "**Tea Act**," a British government attempt to rescue a failing monopoly, the British East India Company. The Act granted the company new privileges and at the same time forced tax money out of the colonists' tight fists. The Act provided the basis for the British to dump cheap tea on the American market, and would not only undercut other suppliers, but came tightly-wrapped in a Parliament-added tax. It was one last effort by the British government to get the colonies to pay their overdue bill for the costs of the French and Indian War. Many Bostonians viewed the plan as one more outrage on their colonial rights and some took the law into their own hands, boarding the tea ships and dumping its cargo

Tea Act

Legislation passed by the British government in 1773 designed to give the British East India Company a monopoly on tea in the colonies, the Act led to the infamous Boston Tea Party.

THE BOSTON TEA PARTY

After the British passed the Tea Act in 1773, the angered American colonists refused to allow many British tea ships from landing in their ports. In Boston, however, the East India Company had the help of the British-appointed governor. Plans were made to bring the tea in by force, if necessary.

A small group of Bostonians known as the "Sons of Liberty" took the matter into their own hands. On the night of Dec. 16, 1773, they smeared their faces with burnt cork and put on fake Indian feathered headdresses, then marched to the harbor, where they boarded three moored British East India Company ships. To the cheers of others on the wharf, the thinly disguised "Indians" dumped 342 chests of tea overboard into Boston harbor. They were met with almost no resistance. This became celebrated throughout the colonies as "the Boston Tea Party."

The British government responded by closing the port of Boston and putting other laws in place that were known as the "Intolerable Acts." The Boston Tea Party rallied support for colonial revolutionaries and escalated the tension that would lead to the American Revolution.

Figure 5.4 The Boston Tea Party by Nathanial Currier. Although the lithograph shows the tea party taking place during the day, it actually occurred at night. The tide was out and the water was so shallow that tea piled up in mounds higher than the boat decks.

overboard in the infamous "**Boston Tea Party**."

The British concluded that any further appeasement of the unruly Americans would only make a bad situation worse, and in last-ditch desperation, the British opted for a severe response—closing the port of Boston, imposing military rule, and resolving to starve the dissidents into surrender. Divide and conquer, they supposed: single out the chief offender, Boston, and make an example of it.

Figure 5.5 Engraving of the 1770 Boston Massacre by Paul Revere. Five American civilians died after British troops fired into a riotous crowd on March 5, 1770. The incident became a cause célèbre for pro-independence groups.

The American response to the British occupation plan for Boston was remarkably warlike. Many colonists began preparing for armed resistance. In their preparations, we see not only the first signs of open revolt, but also the first signs of the American colonies uniting politically in a common cause. **Committees of correspondence** launched frantic efforts to apprise one another of the situation in their respective colonies. Far-away Virginians, with no direct stake in the Boston controversy, rallied to the side of their besieged compatriots. "Give me liberty or give me death," shouted **Patrick Henry** in the House of Burgesses, his speech treading perilously close to treason. More important, the galvanized local leaders of all the American colonies immediately met together to organize their own, all-colonial government, bypassing and ignoring their British-appointed governors. The Second

Boston Tea Party

On December 16, 1773, American colonists protested the British tax on tea by dumping 342 crates of British tea into Boston harbor.

Committees of correspondence

Groups organized by local colonial governments for the purpose of coordinating written communication with the other colonies. They disseminated the colonial interpretation of British actions among the colonies and to foreign governments. The network of committees would later provide the basis for formal political union among the colonies.

Figure 5.6 "Give me liberty, or give me death!" Patrick Henry at Richmond, Virginia, March 23rd, 1775. *Painting by Peter F. Rothermel, 1851.*

Patrick Henry

1736–1799
Best known for his famous "Give me liberty, or give me death" speech in the Virginia House of Burgesses, Henry was an Anti-Federalist who pushed for a bill of rights to be added to the Constitution after its ratification.

Continental Congress

A body of representatives from the British North American colonies who met to respond to England's Intolerable Acts. They declared independence in July 1776 and later drafted the Articles of Confederation.

Declaration of Independence

1776 document expressing the desire and intention of the American colonies to break ties with Britain due to the injustices perpetrated by King George III.

Continental Congress represented an extralegal—indeed illegal—gathering of delegates from all of the colonies, a Lockean coming-together in the most literal sense.

Armed conflict between Mother England and her difficult American colonials was now virtually inevitable. In April 1775, after several Redcoat thrusts and colonial counterthrusts around the Boston area, the fatal spark landed in the tinderbox at Lexington Green.

The Declaration of Independence

There are several ways that the American rebellion could have ended short of independence. Parliament might have simply accepted representatives from the colonies. The British army, considered by many historians to have been the best in the world at that time, might have overwhelmed the colonial volunteers' resistance on the battlefield and crushed the separatist movement. A dominion status might have been worked out as a compromise, placing the United States on the same political footing as Australia or Canada have today. There were influential groups both in London and the colonies that favored each of these alternatives and forces were set in motion to promote each one.

The colonies' citizens at this time were themselves divided on the issues. It is now estimated that possibly one-half of the colonies' population in the early 1770s didn't care who governed them. The rest were split between "Patriots," those committed to total independence and "Tories," those who were staunch supporters of British rule. The story

Figure 5.7 The Battle of Lexington. *Print shows line of Minute Men being fired upon by British troops in Lexington, Massachusetts. Engraving by John H. Daniels & Son.*

of the Declaration of Independence is thus a story of colonial unification.

Colonists had to decide on their own whether separation from Great Britain was a good idea, whether a war of separation could actually be won, whether the colonies could exist as viable entities, and whether some sort of nationhood among them was possible. These were all uncertain propositions. Accordingly, political battles waged between the outbreak of hostilities in 1775 and the official, final cutting of political connections in 1776 were intense, complex and highly emotional. The battle of ideas was fought out in the Continental Congress, without umpire or referee. This body represented one of the strangest innovations in the modern world, a government created by the governed. It lacked much of that which we would expect from government today, such as careful organization and established rules. With no king to impart a sense of legitimacy, it stood in danger of falling apart. Yet somehow it held together and pushed on, giving voice to a disparate and divided people.

Of the many arguments presented in this early meeting of colonial representatives, most were narrow and particular, reflecting only local interests. As the delegates hammered away, however, an "American" interest began to emerge. It was vague in the beginning. (Remember, the colonies were still very different from one another, and their concerns and sense of identity were almost entirely local.) They came to see that they shared something powerful, and that the word "liberty" had gained great importance to them. It now meant more than just escaping from British tyranny. There was something in it of John Locke and the English Whigs, something of Adam Smith and the nobility of the individual, and more than a little of John Winthrop's city upon a hill. More than anything else, though, there was that compelling connection between national autonomy and personal freedom. Free societies create free people, and vice versa.

Figure 5.8 The Declaration committee. *Lithograph by Currier & Ives, 1876.*

As the battles between trained redcoats and raw colonials grew more intense, delegates became more concerned with the hard facts of separation (not to mention the war that was raging all around them) than with lofty theories of nationhood. They were working politicians, not political philosophers. Yet for American colonials to forge themselves into a single nation, nationhood was a theory that was most needed. A nation, after all, is a group of people united on some acceptable basis. For the thirteen colonies to truly work together as a nation, the people would have to know and believe in the basis on which it was formed. This is where the story of the Declaration of Independence takes on layers of meaning that were never originally intended.

As soon as colonial independence seemed achievable, the Continental Congress chose to make a formal declaration, a gracefully written document that would set forth the American case persuasively and curry the favor of neutrals and bystanders alike, especially the French. Young **Thomas Jefferson** of Virginia, only thirty-three years

Thomas Jefferson

1743–1826

Jefferson was the third President of the United States, principal author of the Declaration of Independence, and an influential Founding Father of the United States.

A political philosopher who promoted classical liberalism, republicanism, and the separation of church and state, he was the author of the Virginia Statute for Religious Freedom (1779, 1786), which was the basis of the Establishment Clause of the First Amendment of the United States Constitution.

old, seemed the logical choice to write an initial draft, for he was nothing if not facile with a pen. There were at least two others on the drafting committee, John Adams and Benjamin Franklin, and looking over their shoulders was the rest of the Congress.

Jefferson later claimed to have no particular source of inspiration, pulling ideas from the very air of wartime Philadelphia. Yet the crucial passages in the first and second paragraphs read like a page out of Locke's *Second Treatise*. There was more to the Declaration than those two memorable paragraphs. It was, after all, a whole bill of indictment against the British crown. But in that brief flight of poetic prose, independence was brilliantly illuminated, and for many who would later read that momentous document, the concept of a United States of America became irresistible.

The final Congressional debates over the Declaration's wording, emphasis, and tone were fierce. The delegates argued passionately about competing sectional interests, the power of local attachments, and the institution of slavery—issues that would live on to haunt American history. By July 4, 1776, Congress was ready with a final, unanimous vote, followed by a formal signing ceremony. The United States of America was a fragile but accomplished fact.

It was only in the aftermath of this event that Americans began to reread the Declaration of Independence and find deeper meanings in its text. Within its flowing cadences was to be found an outline of the agreement that Americans had made with one another as the basis of their social compact:

- All human beings were created equal; there was no ruling class among them.

Locke's Influence on the Declaration

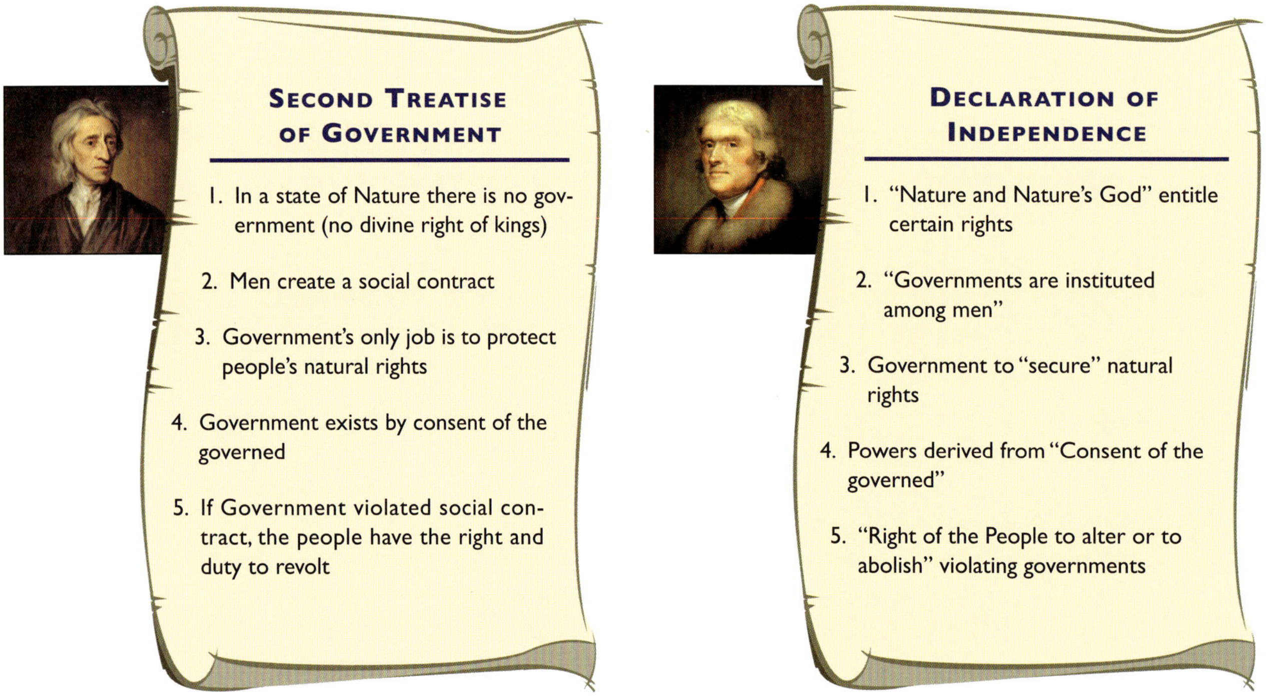
SECOND TREATISE OF GOVERNMENT
1. In a state of Nature there is no government (no divine right of kings)
2. Men create a social contract
3. Government's only job is to protect people's natural rights
4. Government exists by consent of the governed
5. If Government violated social contract, the people have the right and duty to revolt
DECLARATION OF INDEPENDENCE
1. "Nature and Nature's God" entitle certain rights
2. "Governments are instituted among men"
3. Government to "secure" natural rights
4. Powers derived from "Consent of the governed"
5. "Right of the People to alter or to abolish" violating governments

Figure 5.9 A quick comparison reveals that Locke's influence on Jefferson's drafting of the Declaration was more than slight.

Figure 5.10 The Declaration of Independence *by John Trumbull. The painting depicts the "Committee of Five" that drafted the Declaration of Independence—John Adams, Roger Sherman, Robert Livingston, Thomas Jefferson (presenting the document), and Benjamin Franklin—standing before John Hancock, the President of the Continental Congress. The painting includes portraits of 42 of the 56 signers and 5 other patriots.*

- All were endowed with the same rights, granted by nature, not by government, and these rights could never be alienated or abridged.
- The purpose of government was to protect such rights.
- Government was legitimized only by the consent of the governed.

These arguments were fully intelligible to Americans. Each point led logically to the next, and together they provided a complete rationale for the American Founding. There was a way the world had been set up, it asserted, yet few if any of the world's nation-states appeared to grasp that fundamental truth. The United States, among all of them, would be the nation-state that was constructed on the *right* foundation, indeed, a city upon a hill.

The Declaration of Independence foreshadowed both the Constitution and the Bill of Rights. It anticipated the emergence of political parties and the advent of a representative democracy. It implied the eventual triumph of the market system and pointed to the egalitarianism and individualism that would come to define the American soul. It even gave a hint of civil war. More than anything else, the Declaration provided a sense of what American nationhood would mean. The United States would not be a nation of rulers but of people. Human life, not governance, would be its main business. Of all political societies on Earth, it would be the one dedicated, not to war and conquest, not to wealth or power, not to aristocratic brilliance, or cultural excellence or the brave accomplishments of the few—but simply and directly to life, liberty, and the pursuit of happiness.

The Revolutionary War

Thomas Paine

1737–1809

Thomas Paine was an English intellectual, scholar, revolutionary, deist and idealist, who spent much of his time in America and France. A radical pamphleteer, Paine helped foment the American Revolution through his powerful writings, most notably *Common Sense*.

Oil painting by Auguste Millière (1880).

Common Sense

A political tract written by Thomas Paine that helped convince colonists about the necessity to fight against Britain and to become independent.

Just as the Declaration of Independence was the great unifying idea among the thirteen colonies, the Revolutionary War became the great unifying event. For centuries afterward, Americans have come together on Independence Day to celebrate and act out symbolic tableaus of the war era. Folklore would swirl around Paul Revere ("To arms, to arms!"), Nathan Hale ("I only regret that I have but one life to lose for my country!"), John Paul Jones ("I have not yet begun to fight!"), Betsy Ross, Molly Pitcher, Ethan Allen, Francis Marion, the Pennsylvania riflemen, and those sturdy souls who turned out in the dead of night to answer the call of the Minutemen. The new nation's patriotic sentiments were later captured eloquently in a famous, heartwarming painting of color-bearer, fife-player and drummer-boy limping victoriously to the tune of "Yankee Doodle."

The unifying aspects of the Revolutionary War were several. Americans voluntarily came together from the length and breadth of the colonies to fight in the Continental army. The nation's first army was an untrained, ill-equipped rabble, to be sure, and it lost many battles, but it was indeed "continental" in scope, and its officers and men came to know one another as Americans rather than as Virginians or New Yorkers.

The war was fought everywhere, north and south, east and west, in cities and towns, along rivers and inlets, up and down country roads, across farms, in open fields and dense forests. No one was safe from the war's ravages or immune from its sorrows. If the loss of life was modest by later standards, the war's physical destruction was catastrophic. It would take more than a generation for Americans to clean up the mess.

Americans took heart, however, from common sources of encouragement. **Thomas Paine**'s little book, ***Common Sense***, was a masterpiece of political propaganda, and its timely appearance in the spring of 1776, had much to do with swinging popular opinion in favor of independence. Paine also went to work on the sixteen *American Crisis* papers, whose notable beginning rang out: "These are the times that try men's souls." Taken together, Paine's apologetics struck not one but a whole series of responsive chords in the American heart. Paine's writings convinced many fence-sitting Americans that monarchy truly was a thing of the past and that Britain's claims of sovereignty over them

Figure 5.11 Yankee Doodle, 1776. *Lithograph by Clay, Cosack & Co. after A. M. Willard.*

Figure 5.12 The Battle of Bunker's Hill. *Engraving by J.G. von Muller, 1798, after John Trumbull.*

were more than a little absurd. Paine succeeded in placing the American struggle for independence in the context of a human struggle for freedom. There was more at stake here, Americans concluded, than mere separation from the British Empire.

In facing the British Army, Americans knew they were challenging the best military force in the world. Symbolic victories, such as the battle of Bunker Hill near Boston, where a small contingent of determined colonials inflicted withering casualties on disciplined British regulars, fired American self-confidence and turbocharged their sense of patriotism. However, later and more numerous battlefield defeats, such as those suffered at Brooklyn Heights, White Plains, Germantown, and Brandywine, showed Americans that their freedom would not be won easily. The Americans' determination to press on in the face of battlefield catastrophes and constant shortages of military necessities had its own effect on their feeling of nationhood.

The way the British handled the war may have had greater effect in unifying Americans than any other factor. British commanders enjoyed every kind of wartime superiority—economic, military, and technological—but they lacked a clear, coherent policy for conducting operations, and worse yet, they failed to understand the character of their enemy. British commanders were never quite sure whether they wanted to conquer, pacify, or intimidate the American colonists. British commanders tried all three approaches, making themselves seem weak and indecisive.

Almost every dilemma faced by the British commanders went unresolved. If they were lenient with rebel sympathizers, the British embittered the rebels' loyalist neighbors. If the British were harsh and vindictive, they angered the very people they were trying to win over. When the British dug in and fortified an area, the enemy would operate with impunity all around them. If they headed off in pursuit, the loyalists that depended on their protection were abandoned to rebel reprisals. And the British commanders couldn't seem to teach their own soldiers the difference between rebels

Paul Revere

1735–1818

An American silversmith and a patriot in the American Revolution, Revere is well-known for his role as a messenger in the battles of Lexington and Concord. He helped organize an intelligence and alarm system to keep watch on the British military.

Oil painting by John Singleton Copley, c.1768–70

and noncombatants. The colonials, both rebels and neutrals, all looked alike, and so they were all pretty much treated alike—badly. The cumulative effect was to isolate, offend, or betray the very majority of the colonials that the British were trying to win over. Every mistake they made alienated the colonials affected by that mistake. Rebel Americans were quick to write down those mistakes, publish them with crude but effective woodcut illustrations, and circulate the information far and wide, driving the undecided toward the patriot cause.

By contrast, American amateur warriors essentially got the politics of war-making right. In the beginning of the conflict, colonial opinion was divided among three alternatives: support for the cause, opposition to the cause, and indifference to the cause. Most Americans did not understand the issues of the conflict very well, regarding it as a lawyers' quarrel. Moreover, most Americans had little experience with, or interest in, politics. Many saw politics as a pursuit of the privileged classes. Most Americans just wanted to get on with their lives.

Figure 5.14 The Lexington Minuteman *statue in Lexington, Massachusetts. The Statue represents Captain John Parker and was sculpted by Henry Hudson Kitson in 1900.*

Figure 5.13 Artillery retreat from Long Island, 1776. *The Continental Army suffered more than one defeat at the hands of the British. This painting shows the evacuation of Continental troops to Manhattan. They would continue to be pressed by the British all the way across New Jersey and the Delaware river. Lithograph by The Werner Company, 1899.*

Yet events of the war kept pushing the undecided off the fence. There was horrific property damage, most of it sustained by civilians. Noncombatants were often targeted for reprisals, especially by loyalist commanders. Neutrals were harassed by both sides until it became more difficult to walk down the middle than to throw in with one group of partisans or the other. If the real war was a contest for the "hearts and minds" of the undecided, as John Adams maintained, indecision became increasingly difficult.

The single greatest factor in the rebels' efforts against apathy appears to have been the colonial militia. Every town had its militia company, which have been aptly described as a combination fraternal lodge, drinking society, and political cadre. The militias were often driven from the field in hot military engagements, but they played useful roles in many battles, especially as support and reinforcements for the Continental

Army regulars. But their real value was political. Militias were instruments of political education. After a drill on the village green and a visit to the local pub, a company would often be instructed by its leaders on fine points of natural law, the dark doings of the court party, and the importance of human liberty. There may have been scant comprehension of the complex political and economic issues before Lexington and Concord, but by the final battle at Yorktown there was a great deal of comprehension throughout the ranks and among most non-combatants about why the war had been fought.

Indeed, by the war's end, there wasn't much support left for the British cause. Many loyalists had already packed up their belongings, boarded British ships, and left for Canada, England, and elsewhere. Other loyalists had swallowed their pride and accepted the fact that King George was no longer their ruler. Those who stayed behind might not like the idea of no longer being Brits, but keeping quiet about it was the better alternative to getting tarred and feathered by their neighbors. Through harsh redcoat example, through clever rebel propaganda, through thoughtful argument and conclusions, Americans of every persuasion had gained a fairly clear understanding of those principles on which the founding of their new nation was taking place.

George Washington

If the Declaration of Independence was a unifying idea and the war itself a unifying event, George Washington became a unifying symbol of nationhood. For fellow Americans educated in the classics, the distinguished-looking Virginia planter-turned-military commander was favorably compared to the great founding heroes of antiquity—Lycurgus, Solon, Caesar, all rolled into one. Washington had been an aspiring young planter and would-be member of the Virginia gentry. He admired the aristocratic Fairfax family, whom he knew personally, and sought entry into their glittering world. Gentlemanly status meant more to him than land or wealth, but land and wealth were an integral part of the package. As a result, Washington became an aggressive entrepreneur, land speculator, canal promoter, and agricultural experimenter—the complete American.

Washington believed that a gentleman must be first and foremost a person of character. Accordingly, he assiduously cultivated virtue in the classical sense while still very young. He made a list of desirable qualities and worked on developing them one at a time. He wanted to be the best at everything: the best farmer, the best husband, the best father to his two stepchildren, the best rider, the best poker player, the best military officer. Washington was outclassed in that last category by the scions of British families who had the benefit of old-school educations. He was told that he would amount to nothing more than a provincial officer, and he could scarcely build a career on that footing. The

George Washington

1735–1818
Washington led America's Continental Army to victory over Britain in the Revolutionary War and was the first President of the United States, from 1789 to 1797. Because of his central role in the founding of the United States, Washington is often called the "Father of his Country."

Oil painting by Rembrandt Peale.

bitter disappointment did not endear him to things British.

As he grew to maturity, Washington identified with the country party and saw himself as the quintessential commonwealthman. Accordingly, even as he continued to increase his acreage and add to the size of his house, he could not turn his back on public duty. He was elected to the Virginia House of Burgesses in 1759, where he contributed the classic virtue of temperance and stolid wisdom. He was as different as could be possible from his flashy and mercurial House colleague, Patrick Henry.

Washington's military career was short and free of victory. As a headstrong young officer, he actually touched off the French and Indian War in 1754 after being sent to investigate the building of a French fort on the Ohio frontier. Later, as commander of the Virginia Blues accompanying British General Edward Braddock, he was caught in a disastrous French ambush that virtually annihilated the British army. Washington's uniform was ripped and torn by six bullets in that engagement, which tells us something about his courage in the rout, as well as about the role of providence in history.

Politics abounded in the Continental Congress decision to appoint Washington, a Southerner, to take charge of an army in the North. It was a gesture toward American unity, yet it was unpopular with the better trained and more experienced officers, several of them New Englanders, who aspired to the rebels' general command. The disappointed hopefuls did not argue Washington's southernness, of course; they argued his incompetence. What had he done, they asked, except start a war and take part in a debacle?

For all that, George Washington proved himself to be without peer in either army. He was high-minded and occasionally aloof, but never erratic, arbitrary, or unjust, which were the failings of his many rivals on both sides. He learned from his mistakes, acquired a sound sense of tactics, and knew how to deploy his inferior forces to good advantage. He had a better sense of strategy than his acclaimed British opponents, which was critical to the survival and eventual success of his small volunteer army.

He concluded early on that by holding his rag-tag army together, he might force the British into a no-win situation. But how could he hold such an army together? After the British attack on New York in August 1776, the Continental army's numbers dropped from 20,000 to less than 4,000 by Christmastime, and the remnant was perilously close to dissolving away. Challenging times, however, magnified Washington's deeper qualities.

He was absolutely beloved by those who served under him, his famous temper notwithstanding. It was one thing to be respected, but George Washington was

Henry Knox | Joseph Reed | Nathaniel Greene | John Sullivan | Israel Putnam

Figure 5.15 Some of General Washington's trusted generals and officers.

Figure 5.16 Washington Crossing the Delaware, *by Emanuel Leutze, 1851.*

adored. He was a man of stoic patience in the face of many battlefield reversals. Having been called by the people to serve a cause that was just, he was evidently willing to fight to the very last ditch and he communicated that resolve to everyone around him. Finally, he was a canny politician, one who knew all the wiles and stratagems of infighting among both soldiers and politicians.

In addition to the superior forces that faced him on the battlefield, Washington had to deal with a host of other difficulties. Rival officers Charles Lee, Horatio Gates, James Wilkinson, and Thomas Mifflin conspired to oust him as commander-in-chief. Congress played shameful politics with the war effort, often shorting Washington's army in the process. American suppliers frequently sold to the highest bidder, which was always the British. There were logistical tangles, sagging spirits, intelligence leaks, and traitors in high places. Poorly cast field guns blew apart. Gunpowder stored poorly became damp and useless in wet weather. Britain's European enemies, automatically friends of America, had minds—and of course agendas—of their own. Militias melted away at the first sound of enemy bullets. Regulars deserted by the hundreds, often leaving to harvest their crops or to plow their fields.

Washington bore all of this and more, never flagging in his devotion to the cause. Where others blew hot and cold, switched sides, and jockeyed well behind the lines for personal advantage, he soldiered on, heeding only the call of duty, his patriotism flowing from some bottomless well. His vision of America as a nation was far in advance of his time.

A major British assault in the summer of 1776 drove Washington's army from its fortifications on the Hudson and pushed it deep into New Jersey. There he rallied his depleted forces and mounted bold counterstrokes at Trenton—the famous Christmas night attack on the Hessian garrison—and Princeton. These victories boosted a sagging American morale.

After this first year, the war settled into a dynamic equilibrium. The British continued to win important victories, but so on occasion did the colonists. One such colonial victory was the battle of Saratoga in October 1777, which brought British General John Burgoyne's promising Hudson River Campaign to a cataclysmic end. French observers' reports, on the strength of this achievement, convinced France to enter the war. Age-old enmity between France and

England now made an American victory actually possible.

With France on his side, Washington was now able to obtain more munitions, call upon the superior experience of French officers, and more important, gain the support of a French naval squadron. At the same time, the British generals fatefully shifted their strategy once again, sending large numbers of their soldiers south to pacify the lightly-defended Carolinas. While this move proved effective in the south, it also led to the entrapment of a large British force on the Yorktown Peninsula in Virginia during October 1781. With French and American armies before him and French ships blocking his escape, the British commander, Charles Cornwallis, had no choice but to surrender, effectively ending the war.

Military historians have pointed out that with a few exceptions (and these largely symbolic) Washington did not really participate in the war's key turning points. The victory at Saratoga, for example, was credited to Horatio Gates. Washington's role had been to engage the main thrust of British attention and keep an army on the field that proved worthy of such attention. While they faced Washington, the British were unable to sustain successful operations elsewhere.

Ordinary Americans knew that Washington had contributed much more. They lovingly recalled a series of events that embodied the revolutionary cause: Washington's strategic crossing of the Delaware River on a snowy Christmas night to surprise the Hessians in Trenton; his mustering of beleaguered forces at Monmouth Courthouse, pulling victory from the jaws of defeat; his enduring the hardship of Valley Forge with his starved and freezing soldiers; and finally, facing down a mutiny of his officers by appealing to their dignity and pride. He had grown old in the service of his country, he told the mutineers, and so he had.

Demigod

Being half human and half godlike, a trait sometimes wrongly attributed to the Founders.

This was the man Americans came to revere as a **demigod**. They named cities, towns, and a state in his honor, wrote his name on bridges and highways, erected monuments to him, hung his portrait above their mantels, and even commissioned an heroic marble statue of him clad in a Roman toga. It was, however, not for his military qualities but rather for his embodiment of virtue in all of its forms, for his willingness to forsake home and hearthside to serve the public, for his patience and endurance in the face of daunting obstacles, for his wisdom and ability to inspire others, and for his humility in the face of arrogance and pride.

Indeed, it was partly because they believed so devoutly in George Washington that the former colonials-turned-Americans came to believe in themselves.

Figure 5.17 Surrender of Cornwallis at Yorktown, Virginia, October, 1781. *Lithograph by N. Currier, 1846.*

Timeline

Contextual Events

Adam Smith's *The Wealth of Nations* is published **1776**

Chapter Specific Events

1754 French and Indian War begins

1760

1763 French and Indian War ends

1765 Stamp Act passed
(Oct) Stamp Act Congress meets

1770 Boston Massacre—John Adams defends soldiers

1773 Boston Tea Party

1774 Intolerable Acts
1st Continental Congress
(Mar) Patrick Henry "give me liberty . . ."

1775 (Apr) Paul Revere's Ride
Revolutionary War begins
Battle of Lexington and Concord
(May) 2nd Continental Congress
(June) Washington becomes commander of the Continental Army
Battle of Bunker Hill

1776 (Jan) Thomas Paine's "Common Sense" is published
(Mar) British evacuate Boston
(July) Declaration of Independence adopted
(Dec) Washington crosses the Delaware River and captures Trenton

1777 Articles of Confederation written
Continental Army wins Battle of Saratoga
Washington's army camps at Valley Forge

1780 Southern Campaign begins
Benedict Arnold is caught

1781 General Cornwallis surrenders at Yorktown
Articles of Confederation adopted

1783 Treaty of Paris signed
Washington resigns as Military Commander

1787 Constitutional Convention

1790

Key Terms

taxation without representation
continental congress

Key People

Patrick Henry
Thomas Paine
Thomas Jefferson
George Washington

Questions

1. How did the self-image of the colonists differ from the British leadership's opinion of them?
2. How do John Locke's ideas in his *Second Treatise of Government* compare with the Declaration of Independence?
3. What did the Declaration of Independence provide to the average American colonist? What does it provide to citizens today?
4. Name the unifying idea, event, and symbol of nationhood in the Revolutionary War era.
5. What effect did Thomas Paine's *Common Sense* have on the colonists?

Chapter 6

Designing Government

Abstract

The Revolution gave rise to the need for a new government. The new nation's first attempt was the Articles of Confederation. However, the Articles were not sufficient for the successful governance of the states. James Madison and other Nationalists wanted a stronger national government to solve these inefficiencies. They called for a Constitutional Convention in Philadelphia, in which delegates from the various states debated, negotiated, and outlined the functions of a new, stronger national government. This chapter examines the politics of the convention, the compromises in the final document, and why the Founders structured the federal government the way they did.

As the rebellion of the colonies morphed into the American Revolution, America's former British colonists turned their attention to the problem of governance. They had difficulty imagining a single American government, just as they had difficulty imagining a single American nation. As colonies, they had always been separate, and they assumed that would continue; they would vest sovereignty in the states as states, rather than in the nation.

Their model for this assumption was classical antiquity. Americans at first saw themselves in a situation somewhat like that of ancient Greece. The various republics would have amicable relations with one another and would bind themselves into a strong alliance against foreign attack. They would be a "nation" in the sense that ancient Greece had been a nation, a league with friendly rivalries, cultural exchange, and similar institutions. But the main purpose of a "republic" was to preserve a body politic that was small and cohesive.

Oligarchy

A form of government where most or all political power effectively rests with a small segment of society, typically the most powerful, whether by wealth, family, military strength, ruthlessness, or political influence.

Republican problem

The question of how the benefits of self-government can be enjoyed without incurring its inherent problems.

State Governments

Republican theory became important to Americans as they considered the reality of their independence. The principal idea of a "republic" (from the Latin *res publica*, the "public thing") was for citizens of the political state to govern themselves rather than submit to a despot or an **oligarchy**. In extreme forms of republicanism, such as in the democracy of ancient Athens, the citizens literally handled the daily business of government. Most republics, however, were governed by chosen representatives, which in practice was far more workable.

It was assumed that a republic needed to be small in size, no larger than the classic Greek city-state, because citizens must remain close to the governing process and keep a watchful eye on it. Republics, however desirable in many ways, were known for their instability. Their histories in the ancient world were fraught with wars, revolutions, palace coups, and a pandemic factional turmoil the Greeks called *stasis*. Republics had their share of tyrants, but tyranny was often preferable to anarchy, where no one won and everyone lost.

Philosophers of the ancient world had spent a good deal of time and energy attempting to resolve the **Republican Problem:** how could the benefits of self-government be enjoyed without incurring the problems inherent in that self-government? Plato, in his famous work of political philosophy, *The Republic*, argued for *areté* (virtue) as the answer. He laid out a system for recruiting and training the best and brightest in Greek society to hold the reins of government, and for educating them in the highest performance of virtuous conduct. Aristotle, by contrast, still believed in virtue, but he tended to emphasize structural solutions to the Republican Problem. By mixing and balancing elements of monarchy, aristocracy and democracy in a government, Aristotle supposed that power could be fragmented and shared among various groups and interests, the result being balance and stability.

All of this became relevant to American constitution makers, for they soon came to realize that the Republican Problem was their problem. In addition to reading about the ancients, the Americans had their own experience to draw upon. They thought of the colonies as laboratories for the development of republican practice, with each state sharing its experience with the others. John Adams became so excited about this prospect that he began writing a comparative analysis of American state constitutions. What worked well among them? he asked. What didn't work at all? What led toward disaster?

Because most of the states were used to operating from charters, they favored written constitutions, with all provisions clearly spelled out. Some states, in fact, merely revised their old charters and struck out all references to the king. Most of them, however, worked up their state constitutions from scratch.

Many of the U.S. Constitutional framers were men of learning, well read in the European Enlightenment, and virtually all of them were schooled in the art of politics. Even so, Americans began to learn that there was a big difference between theory and practice when it came to designing government.

Sometimes the smallest details in a constitution could produce large future consequences. A given mechanism might have unanticipated side effects, defeating the very purpose for which it was included. (For example, the Constitution stipulated that the Electoral College could only vote for President, with the runner up being made Vice President. When Thomas Jefferson and Aaron Burr ran together for President and Vice President, respectively, during the 1800 election, they tied with 73 electoral votes apiece, meaning the House of Representatives could vote for Jefferson *or* Burr as President. The Twelfth Amendment fixed this by stipulating that electors distinguish between the two offices.) A bias might suddenly pop up in the **constitutional structure**, or a new opportunity for malpractice, or an unfair advantage to one group or another. Human nature could also be a surprise. In political situations, people often behaved at variance to their professed ideals, even at variance to their normal daily conduct; though angelic in their ideals, men could become monsters in practice. Constitutional governance was a new experience.

Some state governments could be counted successful, but others displayed conspicuous weakness and outright failure. Some became stained with corruption. Others proved to be unworkable. Still others became known for high-handedness. Particularly troublesome was the phenomenon of **constitutional drift**, when power in the government did not remain where it was originally placed. Legislatures, for example, proved to be adept at stealing power from governors and state courts, so that in time only the legislature's power remained. In the worst cases, such as Pennsylvania and Rhode Island, state governments seemed to behave like Old World tyrannies as a particular group or interest would gain control and then use its power to thwart all rivals. As for the rule of law, it was nothing for a state government to violate generality or prospectivity, enacting legislation requiring creditors to accept worthless paper currency in payment of debt.

Some of the Founders, reflecting on the difficulties of the ancient republics and the then present-day state constitutions, were moved to observe how little had really been learned. Was this what they had fought and died for in the name of liberty?

Figure 6.1 The Founders had learned by experience. Being thousands of miles away from England's direct control, the colonies were able to form their own virtually-independent state governments like the Virginia House of Burgesses shown above.

Constitutional structure

The nature and arrangement of mechanisms in a constitution that organize the government.

Constitutional drift

When power in the government does not remain where it was originally placed.

The Confederation

Confederation

Defensive alliance among sovereign equals.

Articles of Confederation

Document outlining an alliance of sovereign, equal states in which there was a weak central governing Continental Congress.

Just as the American states found ancient precedents for republican government, they also found precedents for cooperation among sovereign entities. The term for it was ***confederation***: a defensive alliance among sovereign equals. Confederations were never intended to be true governments, for they lacked the sovereignty that government requires.

Even so, those who drafted the **Articles of Confederation** (the "constitution" of the Confederation) in 1781 wanted the American Union to be more than just a circle of friends. For one thing, they believed that the former colonists, whether city dwellers or farmers, north or south, already had a great deal in common and felt some sense of nationhood. The framers also feared that rivalry among the American states might lead to an endless cycle of conflicts, alliances, and diplomatic intrigues. Ancient Greece had known just such sorrows.

The structure of the Confederation was based on its legislative body, an outgrowth of the old Continental Congress. After all, the earlier colonial Congress hadn't done so badly. It had united the states against a common foe, conducted a war, and forged a peace. Why not just carry on that tradition?

As an alliance of sovereign equals, the Confederation wasn't a failure, but it had all the weaknesses and shortcomings of its type. There was no executive, and thus no voice of American leadership. There was no national court system either, and accordingly no way to resolve the growing number of disputes among the states. Conflicting land claims alone had already sent state militiamen reaching for their muskets.

THE CONFEDERATION

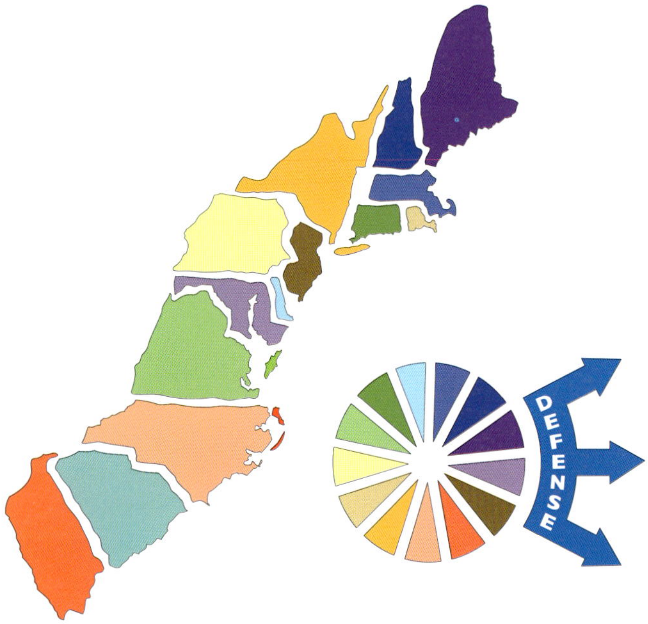

Figure 6.2 The Confederation was an alliance among sovereign equals for defensive purposes. However, it lacked any central authority.

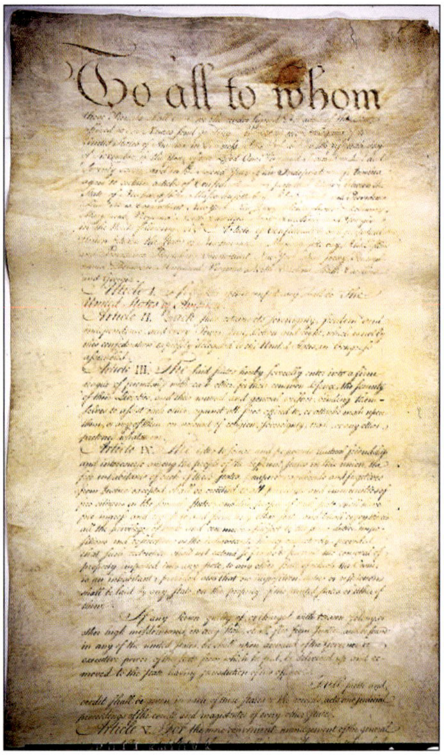
To all to whom

Figure 6.3 The Articles of Confederation were ratified in 1781. This was the format for the United States government until the Constitution.

THE ARTICLES OF CONFEDERATION

Weaknesses	Strengths
Individual states were sovereign	Experience gained helped in the creation of the Constitution
No executive	States acted as experimental labs
No power to enforce conflicts between states	Provided some solutions in convention
No power to tax	Got through the Revolutionary War
No common currency, exchange disputes	Ended western land claiming by states
Not binding if passed	Was a product of the people
No judiciary to resolve disputes	
State wars (Penn-Con)	
Conflicts between state laws	
State trade problems	
States made own international treaties, states were played off each other	

Figure 6.4 An examination of the Articles of Confederation reveals that a change was indeed necessary. However, the Articles served as a vital step in the development of a constitutional government.

There was no authority for trade regulation among the states. As a result, the states conducted economic warfare among themselves. They slapped tariffs on imports and duties on exports. States with port facilities gouged neighbors without them, and those neighbors found ways to retaliate. America was one of the largest potential import markets in the world, but under the Confederation its benefits went almost wholly unrealized.

There was no centralized authority to conduct diplomacy. Individual states began sending out their own envoys to foreign capitals, often with conflicting agendas. They also commissioned their own military forces, which proved dangerous. The outcome was that the diplomacy of the American "nation" was often confused, contradictory, and self-defeating. Few foreign creditors thought the United States was worth much of a risk. Great Britain became so annoyed with the disarray that it began breaching terms of the peace accord.

The Confederation was particularly hobbled by the way power was apportioned in Congress. The fact that all states had equal representation regardless of size or population underscored **state sovereignty**. Many confederations had foundered on this very point. When enough small, weak members can outvote large and powerful ones, the small ones are tempted to gang together to get their way, forcing the big members to depart.

Finally, the Articles of Confederation were virtually amendment-proof. A unanimous vote was required for any amendment, leaving scant hope of resolving its difficulties. A sense of impotence and futility pervaded the American psyche. The fact that the Confederation functioned almost without funds (having no power to lay taxes) simply underscored the pall of defeat that hung over the fragile Union.

The lessons of the Confederation experience resembled those of the state government experience. Constitutional details could cast very long shadows on real-world politics, with the rule of law often lost in the process. Whether the nation was an oppressive government in the hands of a faction or an ineffective league of disorderly state sovereignties, the problem was the same. By 1787, Americans came to realize that their city upon a hill would require better urban planning.

State sovereignty

When ultimate political power resides in the state rather than the federal government.

The New Constitutionalism

Thoughtful Americans analyzed this situation and discussed possible reforms in both state governments and the Confederation. If Americans were sobered by the failures of constitutional structure, they were appalled by the failures of virtue. Americans considered themselves a uniquely virtuous people before the Revolution, and the challenges of the war effort greatly enhanced that feeling. But recent developments had undermined their confidence. Why, they asked, would groups want to take over state governments? Why would individuals seek tyrannical power or unfair advantage? Why would corruption and jobbery pop up here as in some Old World capital? What had become of American innocence?

John Adams

1735–1826
Founding Father and proponent of a bicameral legislature, Adams served as the second President of the United States.

Bicameral legislature

A legislature in which there are two separate divisions or houses.

Figure 6.5 John Adams' Thoughts on Government *would strongly influence the framers of the Constitution.*

Strengthening State Government

John Adams, who was exceptionally well-versed in political theory, hit upon the beginnings of an answer. As a grandchild of Puritans, he shared something of their dark view of human nature. He believed that people responded to situations, not exhortations, even though he exhorted his own children to the highest standards of virtuous conduct. Thus the answer to America's version of the Republican Problem would be found in structure. Adams was an Aristotelian.

In a pamphlet entitled *Thoughts on Government*, Adams set forth a number of ideas on how state governments could be strengthened, stabilized, and made more responsive to public duty. For example, legislatures ought to be made **bicameral**, with two separate houses elected on different principles, one of them more democratic, the other more aristocratic, making it difficult for a single group to exercise tyranny of the majority. The judiciary ought to be isolated from political interference—legislatures often trimmed judges' salaries or shortened their tenure—so that court decisions would reflect true justice.

Adams's most controversial suggestion applied to the executive branch of government, the office of governor. Governors were not in high favor at the time, after the abuses inflicted by the colonies' crown appointees. But that, said Adams, was precisely the problem. Weak governors, like weak judges, were becoming lackeys of the legislature. A strong governor would add a dash of monarchy to the structural mix and operate as a check on the legislature. Specifically, Adams proposed to hand the veto power back to the governors so they could hold the legislatures in bounds. Republican governors would not misuse the veto, Adams predicted, especially if they had to stand for *annual* elections.

Adams's proposals could be understood in terms of the rule of law. His structural modifications were aimed at eliminating confusion and willfulness from the legislative process and securing laws that would be more general, more prospective, and more blind. Such laws could still reflect the will of the people, Adams believed, but they would do so as *law*, not as arbitrary whims.

In 1780, Massachusetts overhauled its original constitution, and when the new delegates sat down to their task, they had *Thoughts on Government* directly in mind. The new constitution implemented virtually all of Adams's suggestions. It worked so well that it became a model for other states and an inspiration for those who would draft the U.S. Constitution seven years later.

Figure 6.6 Shays' Rebellion was a motivating factor in creating a stronger federal government.

Strengthening the Confederation

Where Adams's attention had been focused on the failures of state governments, others focused on failures of the Confederation. Both Alexander Hamilton of New York and **James Madison** of Virginia were prominent in this movement. So was George Washington, who had never lost his continental perspective or his sense of American patriotism. In a series of informal gatherings, these "nationalists," as we might call them, argued the case for a stronger American union, urged on by a vision of the United States as a sovereign nation. They believed such a polity would enhance freedom, expand opportunity, and strengthen the rule of law. They also believed it would be more likely than the state governments to reflect the influence of virtue. A government of real sovereignty, so the argument went, would enlist the participation of America's most virtuous citizens—the Jeffersons, Hamiltons, Adamses, and Washingtons. It would speak for the dignity of all Americans and the achievements of their Revolution. It would be a nation among nations, to be admired, respected, and feared.

With the blessing of Washington, the nationalists engineered an interstate conference at Mount Vernon in 1785. Its official purpose was to resolve difficulties in navigating Chesapeake Bay. Nationalist feeling was in abundance, and the participants wound up calling for a wider conference to be held the following year at Annapolis. While this second meeting was not a success, the nationalists used it as a platform from which to call for yet another assembly, a grand convention this time, to consider ways of improving and strengthening the Articles of Confederation. The host city would be Philadelphia.

The cause of the nationalists was immeasurably strengthened by events in Massachusetts during the winter of 1787. Debt-ridden farmers in the western part of the state rose in open rebellion and shut down the local courts in order to escape foreclosure. Shays' Rebellion, as it was called, raised the specter of American anarchy, a dreadful jolt to those who had recently fought for freedom. "I feel infinitely more than I can express for the disorders which have arisen," wrote a dispirited George Washington. "Who besides a Tory could have foreseen, or a Briton predicted them?" The colonists had not yet resolved the Republican Problem.

James Madison

1751–1836

Fourth President of the United States and Founding Father, Madison is often called the "Father of the Constitution." He co-authored *The Federalist* with Hamilton and Jay, and helped Jefferson create the Democratic-Republican Party.

James Madison. Lithograph after an original painting by Gilbert Stuart, circa 1828.

Creating a Federal Government

The Grand Convention in Philadelphia might well have failed. For a number of reasons, however, Shays' Rebellion large among them, the states took the call seriously and sent some of their ablest statesmen as delegates. The fifty-five delegates who arrived in April 1787 could be generally described in terms of Plato's cardinal virtues. The delegates were courageous; most had fought in the Revolution. They were wise, among the most learned in the Western Hemisphere. They were temperate, always searching out the moderate, the possible, and the do-able. They were just, and the injustice of reckless or impotent governments bothered them a great deal. They were also practical men of affairs, with long political and administrative experience. While they have been called "aristocrats" and "a master class," they are better described simply as America's best and brightest.

A handful of truly exceptional individuals sifted themselves out of the rank and file and assembled into an informal corps of "primary framers." These included James Wilson of Pennsylvania, Roger Sherman of Connecticut, **Gouverneur Morris** of New York, **Charles Pinckney** of South Carolina, William Paterson of New Jersey, and **George Mason** of Virginia. The most important member of this group was another Virginian, James Madison, who had thought long and hard about the weaknesses of the Confederation and the kind of government that ought to replace it. Some of the primary framers were visionaries, inspiring their colleagues with scope and possibility; others were innovators, facilitators, manipulators, and more important, negotiators. Virtually everything in the new Constitution would be hammered out by compromise.

Gouverneur Morris

1752–1816
Pennsylvania representative at the Constitutional Convention, Morris is credited with authoring large sections of the Constitution, including the Preamble.

Charles Pinckney

1757–1824
A South Carolina representative to the Constitutional Convention, Pinckney was a strong promoter of Federalism and helped persuade ratification of the Constitution in South Carolina.

George Mason

1725–1792
Virginia delegate to the Constitutional Convention, Mason refused to sign the Constitution because it did not contain a declaration of rights.

George Washington's leadership was imperative to the success of the Convention. The American people watched him carefully. They were likely to echo his feelings about the new government. Washington himself took little part in the convention's tedious deliberations, but he presided with great dignity. His very presence reminded fellow delegates that this was serious business.

Troubled Politics

James Madison, the most ardent of the nationalists, arrived in Philadelphia with a proposal for a national government that featured **proportional representation** (representation by population) in the congress. Such a plan, he pointed out, would represent people more than states, and hence would reflect a truly American sovereignty. (State governments in his plan would fade into subordi-

Figure 6.7 The Pennsylvania State House, now known as Independence Hall, Philadelphia.

nate administrative units, much like counties today.) The new government would be powerful, with authority to tax and spend, conduct foreign affairs, raise an army, and settle all internal disputes. Madison read widely in the areas of political theory and historical practice, and was thus able to persuade other members of the Virginia delegation to support what was called the **Virginia Plan**.

From the beginning, however, it became clear that there would be no easy victory for Madison and the nationalists. Many delegates had been sent expressly to amend the Articles of Confederation, not abolish them. What Madison was proposing was far beyond their mandate. Also, a few delegates liked the Confederation the way it was. All of them had been sent by sovereign states, none of which wanted to surrender its power to a national entity.

These difficulties had been more or less foreseen by all the delegates. The one problem that threatened to wreck the Convention, on the other hand, popped up by surprise. Madison's proportional representation would give large amounts of power in the national government to big states like Virginia, Pennsylvania, and Massachusetts and little to no power to small states like Delaware or New Jersey. The smaller states, which were in the majority, already saw themselves as threatened by their outsized neighbors. Madison's plan, as they saw it, would render that threat into a working tyranny. Accordingly, the small states asked for time to regroup and come up with a plan of their own. In the so-called **New Jersey Plan**, presented by **William Paterson**, the small states proposed only minor changes in the existing Confederation.

Neither side would budge. For months in that exceptionally hot and sticky summer of 1787, the delegates slugged it out intellectually in the stifling confines of the Pennsylvania State House (now known as

Conflict over Representation

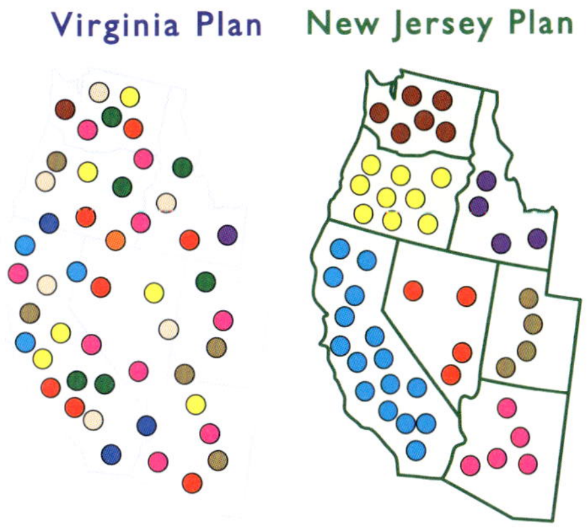

Figure 6.8 James Madison felt that the Virginia Plan would represent American interests on a nearly individual level, each state containing a fair mix of varying interests. The small states, however, viewed each state as an independent and unified political body. Under the Virginia Plan less populous states would not have a significant voice in a national government to defend their own interests.

Independence Hall). There was no air conditioning, of course, and the windows had to be locked in the interest of tight security. Matters were made even worse by a plague of black flies that swarmed the conference rooms, biting the delegates and fraying nerves to the breaking point.

The Great Compromise

Ultimately, **Roger Sherman**, Connecticut's master negotiator, proposed a compromise. Proportional representation, he suggested, would apply only to the *lower* house of the bicameral legislature. This would insure Madison's basic principle of **popular sovereignty**. Equal representation of states would apply to the *upper* house of the legislature, as it had in the Confederation. This would address the small states' concern for state sovereignty. Since every bill would have to pass *both* houses, both principles could exist side by side.

While Sherman's compromise may strike

Proportional representation

When representation is determined by population.

Virginia Plan

Plan presented during the Constitutional Convention in which each state would have proportional representation in the Congress.

New Jersey Plan

Plan presented during the Constitutional Convention in which each state would have equal representation in the Congress.

William Paterson

1745–1806
New Jersey representative at the Constitutional Convention who presented the New Jersey Plan, which gave equal representation to states regardless of size or population.

The Great Compromise

Proposed by Roger Sherman, it brought together the New Jersey and Virginia Plans by having the upper congressional house representation equal by state and the lower house representation proportional by population.

Roger Sherman

1721–1793
Connecticut delegate to the Constitutional Convention, Sherman proposed the great compromise of one legislative house having proportional representation while the other had equal representation.

Popular sovereignty

The idea that power is created by and subject to the will of the people. It was the basis for Madison's proportional representation in Congress and a justification by the South for the continuance of slavery.

Benjamin Franklin

1706–1790
One of the most well-known Founders, Franklin was also a leading printer, scientist, inventor, and diplomat. He helped secure France as an ally during the Revolutionary War.

us as eminently reasonable today, it sounded bizarre in the extreme to most of his listeners. Nothing remotely like this had ever been tried before. It seemed impractical, unworkable and a shortcut to disaster. However, because neither side would give an inch, it remained the only ground of accommodation.

While rhetoric heated and tempers flared, the wearying debates dragged on. On more than one occasion some delegates had to be restrained from physically coming to blows. A pall of gloom settled over the proceedings, and a few delegates packed up and headed for home. The breakup of the Grand Convention seemed imminent. It was in this context that **Benjamin Franklin** pleaded for prayer. Though the delegates did not heed Franklin's advice (they lacked funds to pay a chaplain) Franklin's speech rang in their ears. He placed them at the judgment bar of history, with future generations praising or scorning the outcome. "And what is worse," he added, "mankind may hereafter from this unfortunate instance despair of establishing government by human wisdom and leave it to chance, war, and conquest."

Whether by divine intervention or otherwise, events took a sudden turn. In the last week of June there was a puzzling absence during one crucial vote, switched votes on two other occasions, and the invocation of an obscure rule that nullified the ballot of New York. The majority that Madison and the nationalists had nursed through thick and thin was suddenly reduced to a tie, leaving no other alternative but to embrace Sherman's compromise.

The delegates had no way of knowing it at the time, but **federalism**, as it was ingeniously called, would turn out to be the Convention's single most brilliant achieve-

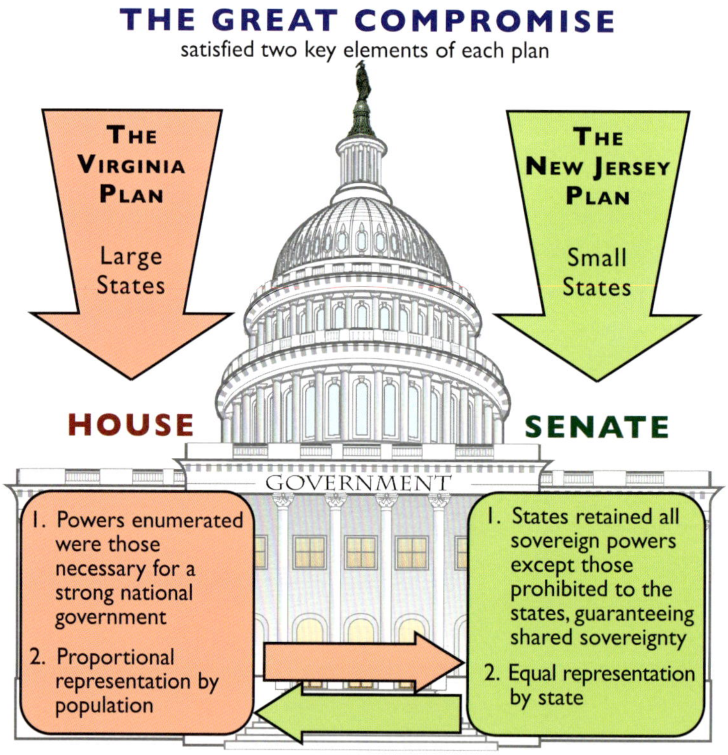

Figure 6.9 Sherman's idea was to have each half of the bicameral legislature embrace representation from a different plan, thus satisfying both plans.

ment and one of history's great structural innovations. Federalism would actually divide sovereignty between the national government and the state governments. It would serve up all the advantages of Madison's unified national polity, while at the same time preserving the smaller units cherished by conservatives and sanctioned by republican theory. The genius of the Constitutional Convention lay precisely in such an outcome.

The Compromise on Slavery

The Great Compromise was not the only anxious moment for the Philadelphia Convention. The disagreement over slavery was almost as traumatic. Although the "peculiar institution," as it was called, had died out in most northern states by 1787, slavery survived in the plantation-heavy South. All parties at the Convention recognized slave-holding as morally dubious, and most of them acknowledged its dissonance with founding principles. (All are born *free* and *equal*. . . .) Yet, the practice could be defended by similar, although somewhat twisted, principles. The people of the South, for complex social and economic reasons, chose to retain the peculiar institution as a matter of their own "popular sovereignty."

No one supposed that the Convention had a mandate to abolish slavery. Yet southern delegates knew that a federal government armed with commercial power might conceivably harm the peculiar institution, deciding, for example, to abolish the slave trade. The issue was further complicated by southern demands that slaves be counted in state populations for the purpose of representation in the lower house—and the northern belief that slaves shouldn't be counted at all. Conversely, should slaves be counted or not counted for the purpose of levying a given state's taxes?

Once again the voice of compromise prevailed. After much debate, the delegates agreed on a "**three-fifths** rule," allowing three-fifths of a state's slave population to be counted both for representation and taxation. The slave trade would remain unmolested for a period of twenty years, which gave the South a bit of breathing space. And authority would be included for fugitive slave legislation, without which slaves might flee from the South to freedom.

Figure 6.10 Effects of the Fugitive Slave Law. *The image shows a man obviously dressed as a gentleman being gunned down along with several presumably escaping slaves. Several historians claim that the slavery compromise made by the Founders inevitably caused the Civil War. However, they often overlook the fact that without the slavery compromise, there might not have been a Union to preserve. One could also argue that the Founders simply delayed what was already an inevitable conflict over a questionable practice. Lithograph by Off & Bloedel, 1850.*

COMPROMISE ON SLAVERY

1. **Slave trade could not be abolished for 20 years.**
2. **Slaves counted as 3/5 of a person for taxation and representation.**
3. **Fugitive slaves were to be returned to their owners.**

Federalism

Dividing powers between the national and state governments.

Three-fifths compromise

Part of the compromise on slavery, where 3 out of every 5 slaves were counted as part of state population for taxation and representation.

Major trouble would ensue from the slavery compromise. The protection of a morally doubtful enterprise was written into the nation's charter. The fugitive slave law would someday license kidnapping. After all, who could tell that an African man or woman walking the streets of Boston or New York was *not* an escaped slave? At the same time, the three-fifths rule would invite Southerners to expand their political power by importing more slaves. The twenty-year reprieve for the slave trade rolled out the welcome mat to slavers around the world, foreclosing any possibility of the peculiar institution just dying out. Americans, on the record for human freedom, were now on the record for human bondage. Yet, as the delegates well knew, compromise on the slavery issue was *the* price of an American union, period.

Auxiliary precautions

Structure in the government to make it more difficult for power to become concentrated in any one group's hands, seen by the Founders as a backup system to virtue. Madison talks about this in *Federalist 51* (see Appendix C).

Constitutional mechanism

Parts of the Constitution that help organize and control power.

Behind the Constitution

Whatever else it did, the Constitution had to confront the Republican Problem. The government it created needed sufficient power to govern effectively, lest anarchy ensue. But the government also had to be constrained from drifting into tyranny, following the example of the government of Great Britain.

James Madison, who became the Constitution's chief architect, agreed with Plato that the virtue of the people was the greatest single check against the abuse of power. He also believed, however, that virtue would often fail. His view of human nature, shared by many other delegates and borne out by historical experience, was that power could and would corrupt. There would always be a "court party."

The framers turned their attention to what Madison called **auxiliary precautions**, a backup system to virtue. (See *Federalist 51* in Appendix C) The idea was to structure the government to make it more difficult for power to become concentrated in anyone's hands, especially those of some future tyrannical majority. Here, of course, they were following Aristotle.

As a guide for their structural architecture, the framers read Adams's *Thoughts on Government*. They also read prominent writers of the European Enlightenment—Hobbes, Locke, Rousseau, and of the more recent Scottish Enlightenment, especially David Hume, whose jaded view of human nature argued for virtue's fragility. More than anything else, they read Montesquieu's *Spirit of the Laws*. What elements of structure, what **constitutional mechanisms**, they asked themselves repeatedly, were conducive to free government?

Three Structural Devices

One of the answers was the bicameral legislature. The lower house would represent the people as a whole and be responsive to their desires. Its members would serve short terms of a mere two years, and they would have to return to the people repeatedly to renew their mandate. The upper house, representing the states, would be far different. It would be distanced from the people, and its members would serve terms of six years, with staggered elections. Where the lower house would be "hot" in its responsiveness to public opinion, the upper house would be "cool" with wisdom and reflection, and it was assumed by the framers that many a measure passed in democratic enthusiasm by the House of Representatives would fail in the more dispassionate Senate.

A second device was indirect election. Their reading of David Hume convinced them that the consent of the people could be filtered to good purpose in ascending tiers of representation. Voters, Hume argued, would always choose representatives from the wisest and most virtuous of their fellows, and

THREE STRUCTURAL DEVICES

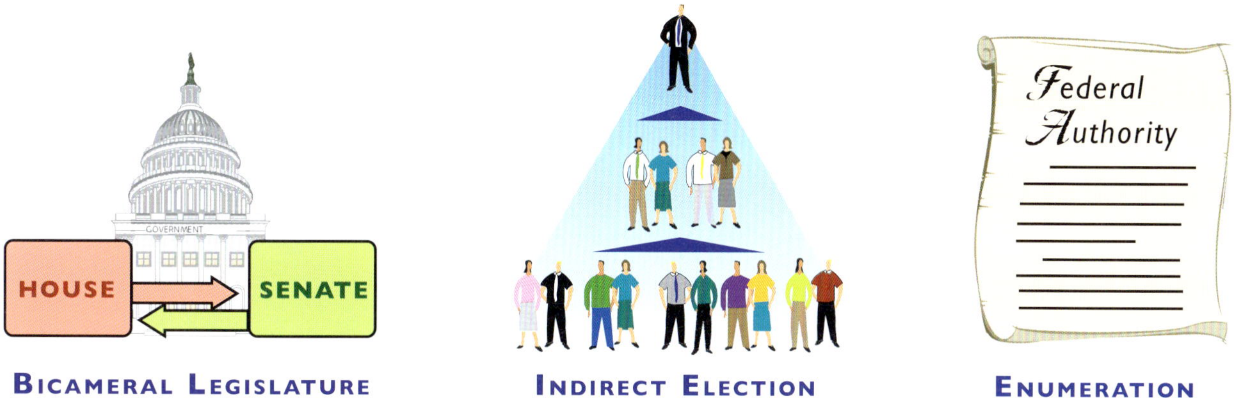

Figure 6.11 Three structural devices were employed to ensure a balanced, free government and prevent corruption and abuse of power.

if these representatives chose representatives of their own, the latter would be wiser and more virtuous still. The process would reflect the consent of the people, but it would be filtered through wisdom and virtue. Senators would be chosen not by the people directly but by their representatives in the various state legislatures. The President would then be chosen by special electors in each of the several states. Federal judges would be chosen by a President who himself had been elected indirectly. **Hume's filter** was a complicated system, and it seemed to some that in a real world setting it would never work. But it did work.

A third structural device was called **enumeration**. The powers of the federal government were enumerated, listed in black and white. Accordingly, unlike any government before it, the federal government's sovereignty lay only in certain areas. Congress was given the authority to lay and collect taxes, regulate commerce, coin money, set up a postal service and a patent office, declare war, raise and support a military establishment, and attend to certain other national concerns. All other powers of government remained with the equally sovereign states.

Separation of Powers

Beyond these elementary devices, the framers set about to fragment power and place its components in separate hands. The term for this unusual approach to governmental structure was **separation of powers**. Montesquieu had discussed it, and provisions for it had been included in several state constitutions, but no one really knew how to make it work. How could the executive power actually be separated from the legislative power, while still allowing for their cooperation? How could the judicial power be separated from the other two? Toughest of all, how to make the separation actually work? The experience of state governments had shown how easily an aggressive legislature could cross that parchment barrier and invade the other branches.

Yet there was a compelling theory behind the concept, having to do with the rule of law. If the rule-making power and the rule-enforcement power were placed in separate hands, then the rule makers would perforce have to operate blindly; the rule makers could never use their authority arbitrarily. The same went for the rule enforcer. If there really was a way to bring it off, the separa-

Hume's filter/indirect election
When the people select the most virtuous representatives, who in turn select even more virtuous government officials.

Enumeration
The written listing of the powers of government.

Separation of powers
Dividing powers of government between the three branches.

tion of powers promised laws that were more likely to be general, prospective, public, and so on.

The delegates in Philadelphia stumbled onto their method by accident while designing the Presidency. They supposed at first that the executive should be chosen by Congress, that he should be a kind of servant to "execute" the congressional will. Yet they also wanted the executive to be a leader, like the British prime minister, capable of rallying public opinion and focusing common effort. The two requirements didn't fit together. Then **James Wilson** had a strange idea. What if it were possible for the executive to be elected by the people rather than by the Congress? This would take some doing, to be sure, in a world without rapid transit or mass communication. An indirect election would have to be utilized, but if it *could* be accomplished, think of the result. The executive could be made as strong and independent as Congress itself, yet be responsible to the people.

James Wilson

1742–1798

A primary framer of the Constitution, Wilson proposed the three-fifths compromise for slave representation and election of the President by the people. He was also key in Pennsylvania's ratification of the Constitution.

It required tediously complex negotiations for the Committee on the Executive to put all the pieces together. We now take the American presidency for granted, forgetting how improbable the office really is, and how deftly it combines enormous authority with humble submission to the popular will. That was precisely the point of the Founding Fathers. With separation of powers, the President would be given powers greater than those of most monarchs. As the nation's chief diplomat, he would conduct foreign affairs. As the nation's top military officer, he would be commander-in-chief. He would appoint high officials. He would execute the will of Congress and implement the laws. He would run the federal establishment and wield awesome powers of patronage. He would be the country's foremost political figure. He would represent all of the people. As we scroll back through our history, the names of great Presidents identify eras and ages: Reagan, Kennedy, Franklin Roosevelt, Wilson, Teddy Roosevelt, Lincoln, Jackson, Jefferson, Washington. We build monuments to them, set up Presidential libraries, and chisel their features on Mount Rushmore. Yet when their terms of office expire, they quietly step out of the limelight.

The framers managed this partly by giving the nation's top executive his own enumerated powers (and of course his own separate election), but also by giving him the power to exercise conditional veto. Together, these two devices made it possible for the President to be both strong *and* independent. John Adams had suggested as much in his *Thoughts on Government*. Congress could never push the executive around or usurp his authority as long as he could protect himself with the veto. Conversely, the executive could never terrorize or blackmail Congress as long as Congress could override the veto by a two-thirds vote.

Such was the genius of separated powers. Next to federalism, it was the Founding's most important contribution to political theory and constitutional practice: a strong

SEPARATION OF POWERS

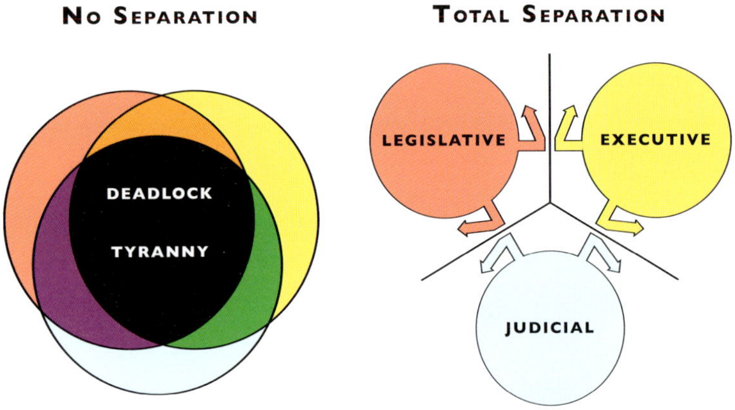

Figure 6.12 Without the separation of powers tyranny and deadlock could result. The question of how to separate powers and still allow balance and cooperation remained.

executive and a strong legislature yoked, yet with independent wills and *both* responsible to the people.

The Founding Fathers also applied the separation of powers logic to the federal court system. Unlike those state courts where the legislature was constantly meddling, the federal courts were shielded. Once appointed and confirmed, justices of the Supreme Court (and later on, other federal courts as well) had tenure for life, barring bad behavior that could get them impeached. Their salaries could not be reduced, nor could their bureaucratic establishment be tinkered with, for they controlled it themselves. Those who enforced the rules were their own bosses. Accordingly, those who *made* the rules had to make them blindly.

Checks and Balances

Separation of powers created a tendency for each branch of government to go its own way. The framers did not want structural anarchy, of course, nor did they want gridlock. So they bridged their walls of separation with a system of **checks and balances**.

With both separation of powers and checks and balances, each branch of government has a check upon the other, and the two must balance one another to consummate the action. Even if each branch eyed the other skeptically, they were motivated to cooperate.

In the U.S. Constitution, the bicameral legislature was a check and balance. Both halves of the Congress had to agree for a bill to become law. There were other examples of this mechanism. The appointment of many high-ranking officials, Supreme Court justices, ambassadors and ministers, Cabinet officers, and the like, was to be made by the chief executive but subject to the advice and consent of the Senate. Advice and consent considerably altered the psychology of appointment. The President would have a hard time installing mere cronies in high places if he was to go before the Senate and explain each nomination.

The diplomatic and war-making powers both included checks and balances. Congress alone could declare war, but once that was done, it was up to the president (as commander-in-chief) to fight it. If Congress didn't like the way things were going, it could scotch any war effort simply by refusing to fund it. When it came to foreign affairs, the President was more or less given a free hand, but any treaty he negotiated had to be ratified by two-thirds of the Senate.

Congress was given unspecified powers of investigation and impeachment, which was yet another element in the check-and-balance machinery. Congress could investi-

Checks and balances

Bridging the separation of powers between branches of government by placing part of each power within two separate branches.

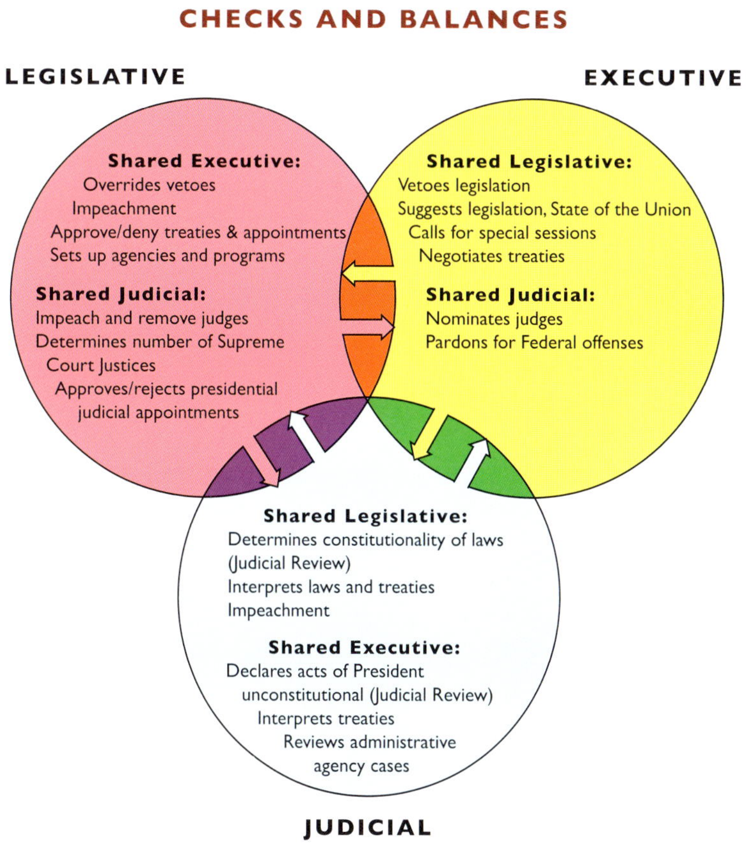

Figure 6.13 To effectively separate powers, a system of checks and balances was implemented to create balance and foster cooperation. To a certain extent, each branch is dependent on the other two.

gate virtually anything it chose to and bring corruption to light. For treason, bribery, or other high crimes and misdemeanors, Congress could impeach (by a majority of the House) and convict (by a two-thirds majority of the Senate) high officials in the federal establishment, including the President himself.

The framers, James Madison later explained, had tried to design their machinery in such a way as to give various authorities both the constitutional means and personal motives to seek justice, serve the public interest, and resist incursions into their respective domains.

Faction

A group of individuals who share the same specific political agenda.

The Extended Republic

Some of the Constitution's auxiliary precautions were extremely subtle. One of these deserves special mention. Like federalism itself, which created a whole new set of checks and balances by counter-posing the sovereignty of the states against that of the federal government, the device in question was a happy accident. It derived from the size of the American Republic, which had been a major stumbling block to the delegates going into the Philadelphia Convention.

Classical republican theory had held that republics must be compact in size and manageable in population, like the *polis* of ancient Greece. The founders supposed that too large a nation-state would quickly succumb to factional infighting. In the ancient world, **factions**—groups organized around influential politicians or competitive interests—had been troublesome enough, creating endless turmoil and confusion. The Founders feared that in an extended republic such as the United States, factions might become so large and powerful that they could never be brought under control.

James Madison began to rethink this idea in the course of the constitutional deliberations, and by the end of the Grand Convention he had reached a surprising conclusion. Republican theorists had gotten it backward. In the extended republic, as Madison was soon to reason in *The Federalist*, there would be many factions, but for that very reason they would render the body politic not *less*, but *more*, stable. (See

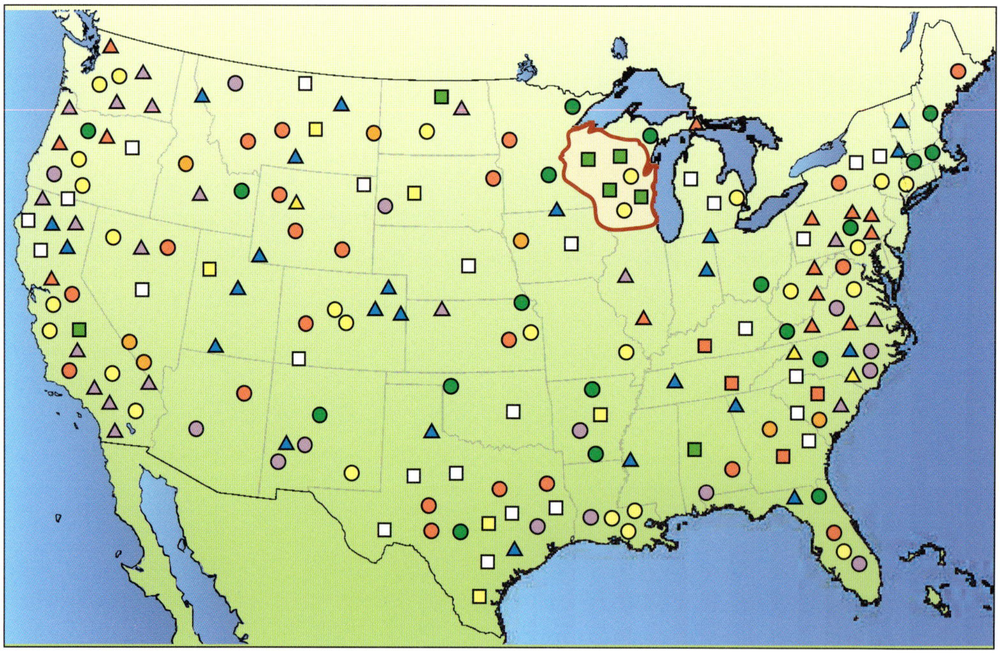

Figure 6.14 Factions might exercise majority tyranny in a small republic like Wisconsin (outlined in red), where one faction clearly dominates the state. However, when a republic is enlarged, the chance of one faction taking control is less likely.

Federalist 10 in Appendix C) In a small *polis*, any given faction might be large enough or powerful enough to take over, as had been the case in tiny Rhode Island. In a sprawling, continental-sized republic, no single faction could come close to possessing such clout. The factions in an extended republic would contend with one another in an endless game of "king of the mountain," pulling each other from the pinnacle of power the moment any single contender threatened to succeed.

The United States Constitution was the product of nearly two centuries of historical development in which a variety of constitutions all played a role. It was inspired by the English constitutional experience and by the experiences and theories of an ancient and far different world. The Constitution of young America was designed for a free people and a virtuous people, as John Adams stoutly asserted, and many of its mechanisms were based on the idea that virtue, properly arranged, should play a decisive role in political outcomes. Yet it was also designed for self-interested human beings and even potentially corrupt ones, for its mechanisms depended on counterpoise, pitting interest against interest, ambition against ambition.

Perhaps the greatest achievement of the Constitution was its embodiment of the social compact. Americans had come together freely, analyzed their constitutional difficulties, and through wisdom and reflection worked out a plan of government that addressed the Republican Problem. At the signing ceremony in September 1787, emotions, which had been so ragged in the course of the long deliberations, ran high. Benjamin Franklin, who often cast amorphous feelings into memorable words, quipped about the image of the sun that had been carved into the back of the convention president's chair. He had long wondered, he said, whether it had been a rising or a setting sun, and was pleased now to conclude that it was a rising one. There was a chuckle or two and a light patter of applause. All the same, there were tears in the old man's eyes when he signed the historic document.

Figure 6.15 The sun chair in Independence Hall. Benjamin Franklin felt that the sun was rising, symbolic of the beginning of the new government.

Timeline

Contextual Events

- Revolutionary War begins **1775**
- Articles of Confederation adopted **1781**

To all to whom

Chapter Specific Events

1700

- **1776** Declaration of Independence
- **1777** Articles of Confederation written

1780

- **1783** Treaty of Paris signed (official end of Revolutionary War)
- **1785** Mount Vernon Conference
- **1786** Annapolis Convention
  Shays' Rebellion
- **1787** Constitutional Convention

1790

Key Terms

sovereignty
European enlightenment
auxiliary precautions
bicameral legislature
Hume's filter/indirect election
republic
constitutional mechanisms
constitutional drift
proportional representation
separation of powers
republican problem
constitutional structure
confederation
federalism
checks and balances

Key People

John Adams
Roger Sherman
William Paterson
Benjamin Franklin
James Madison
Gouverneur Morris
George Mason
James Wilson
Charles Pinckney
George Washington

Questions

1. In what ways did the Articles of Confederation fail?
2. Why did the small states not like Madison's original plan of proportional representation?
3. What were the two biggest political conflicts at the Constitutional Convention and how were they resolved?
4. What examples do we see today to support or disprove the framers' idea of a "cool" senate and "hot" House of Representatives?
5. What examples of factions do we see in today's society? Has Madison's Extended Republic Theory held true throughout US history, or have majority factions ever gained power despite the extended size of the country?
6. Does the theory of Hume's Filter hold up in modern politics? Is it better to have direct elections or to have indirect elections through filters? What do you think are the advantages and disadvantages of each?

Chapter 7

Starting the Engine of Government

Abstract

The signing of the Constitution by the delegates was only the beginning. Ratification by the states had to follow, and a fierce debate accompanied each vote. A Bill of Rights was proposed and created to protect the natural and civil rights of individual Americans. More challenges awaited the new nation as they tried to implement the Constitution. George Washington set many important and sound precedents as the first president, while Chief Justice John Marshall and the Supreme Court established judicial review.

The signing of the Constitution was an auspicious and poignant occasion. Still, the drafted document was merely a proposal, not an accomplished fact. Just as few nation-states had ever worked up their own plan of government, few had ever faced the task of putting such a plan into action. It was a little like taking a rough sketch of some enormously complex machine and then figuring out how to build it.

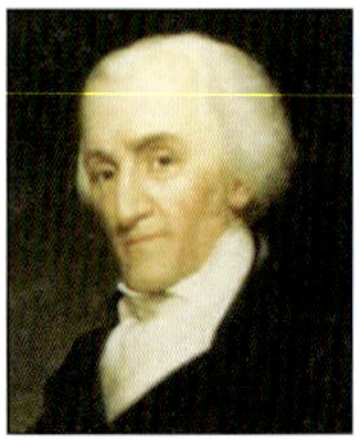

Elbridge Gerry

1744–1814
A Massachusetts delegate to the Continental Congress and a signer of the Declaration of Independence, Gerry was one of three men who refused to sign the Constitution because it did not contain a bill of rights.

Poleis

City or city-state, often self-governed by its citizens as were the ancient Greek city-states. (Plural of *polis*.)

Ratification

The approval of the new Constitution was anything but a foregone conclusion. For one thing, the Grand Convention had vastly exceeded its mandated authority, which was only to modify the Articles of Confederation. Then, too, several of the delegates attending the Philadelphia gathering had left the proceedings in disgust and promised to fight any proposal that came forth. Even more embarrassing, three framers of the document—Edmund Randolph, George Mason, and **Elbridge Gerry**—had refused to sign it, believing that too much had been compromised.

The American people posed a still greater difficulty. While most of them generally recognized the shortcomings of the Confederation, they accepted the basic premise of classical republicanism, which was local (i.e., state) sovereignty. Their vision was of a loose federation of independent republics, like the **poleis** of ancient Greece. The idea of an American nation as such would strike them as radical and perhaps even dangerous.

As Locke imagined the social compact, it was the people themselves, not simply a few leaders, who must decide on their form of government. Americans took this injunction seriously. They understood, for example, that it would not be up to the Confederation government to accept or reject the proposal made in Philadelphia. Additionally, Americans came to see that the state governments were irrelevant as well, so far as approval or disapproval went. The states as such could not create a constitution for the American people. Only the people themselves could do that. They might ratify the document on a state-by-state basis for the sake of convenience, but the actual work of ratification, debating the Constitution's pros and cons, needed to be accomplished in separate conventions, gatherings of the people *beyond* government.

At no time before or since have the American people been called upon to perform a more extraordinary feat. The vast majority of the former colonists were neither political scientists nor constitutional scholars, and yet they had to perform the work of both. They had to decide for themselves if the mechanisms of the plan before them would really work, would truly deflect the danger of tyranny, and would actually hold a sweeping continental expanse together as a single *patria*. For this purpose, they needed Plato's cardinal virtues: wisdom, courage, temperance, and justice.

Opponents of the Constitution quickly emerged. Among them were some of the most accomplished and well-regarded statesmen in America: **Samuel Adams** in Boston, Patrick Henry in Virginia, and Melancton Smith in New York. Nor were they lacking in substantive arguments, which boiled down to the following bill of indictment:

First, the Constitution proposed an "aristocratic" government, far removed from the people. The President would be as powerful as a king. The Senate would be like the British House of Lords. The Supreme Court justices, with their guaranteed salaries and lifetime tenure, would be politically untouchable. Didn't this all reek of that hated tyranny across the Atlantic from which the American people had recently escaped?

Second, there could be no such thing as an "extended republic." The term was an oxymoron. Republics were by definition "the public thing," as the term derived from Latin. Republics were small, local, particular, and run by the people themselves as friends and neighbors. What the Constitution proposed was no republic at all but an empire, like that of ancient Rome.

Third, the Constitution's carefully contrived mechanisms were all smoke and mirrors. Take the enumeration of powers in Article I, Section 8, for example. It looked impressive—laying out the boundaries of congressional power—but, on second glance, appears vague. Congress could also make "all Laws which shall be necessary and proper for carrying into Execution the foregoing Powers." This was a weasel clause if there ever was one. Who got to decide what was "necessary and proper" for building a navy or regulating commerce?

The most telling charge against the Constitution, and the one that packed the most weight, was that it lacked a bill of citizen rights. Ratification would ultimately be decided on this issue. If the framers in Philadelphia had been so innocent behind their closed doors, why hadn't they included any guarantees of all Americans' rights and privileges? The Founders trusted that constitutional procedures would protect rights better than an enumerated list, necessarily incomplete, of citizens' rights.

The **Anti-Federalists** were more than just "anti." They had their own vision of America,

Figure 7.1 The Looking Glass for 1787: A House Divided Against Itself Cannot Stand. *The ratification of the Constitution was not as inevitable as many Americans think. In reality, it was a vicious battle between the Federalists and Anti-Federalists to convince Americans to ratify the document. In this 18th century watercolor, Connecticut is symbolized by a wagon sinking into the mud. Its driver warns, "Gentlemen this Machines is deep in the mire and you are divided as to its releaf."*

Samuel Adams

1722–1803
Second cousin to John Adams, he was a Massachusetts statesman and organizer of the Boston Tea Party. Adams served in the Continental Congress and signed the Declaration of Independence, but was opposed to a strong federal government.

Portrait by John Singleton Copley.

ANTI-FEDERALISTS OF NOTE

George Clinton

George Mason

◂ Patrick Henry

Samuel Adams ▸

Figure 7.2 The Anti-Federalists certainly had strength in their ranks. Many colonial statesmen that we today associate with liberty and patriotism were opposed to the ratification of the Constitution.

Anti-federalists

Political group that was against the ratification of the Constitution.

Alexander Hamilton

1755–1804
Hamilton served as the first Secretary of the Treasury under Washington and founded the Federalist Party. He also co-wrote *The Federalist* and championed a strong central government.

Federalists

A political group that was for the ratification of the Constitution.

and it was quite as compelling as the other side's vision. To begin with, it emphasized a healthy diversity. Anti-Federalist America would be a patchwork of local cultures, each of them vibrant and distinctive. The Anti-Federalists emphasized virtue. Americans would live a pastoral life close to the soil and close to the primary verities, a life of republican plainness and simplicity. Finally, the Anti-Federalists emphasized personal sovereignty. Americans would exercise power themselves, not pass it along to some distant capital. Theirs would be an energetic, town meeting style of governance, under the guidance of decent, God-fearing citizens. If the Federalists' vision was of an impressive city upon a hill, that of the Anti-Federalists was more like a neighborly *Our Town*.

The Constitution's opponents were numerous and well entrenched, but its advocates were on the whole younger, more energetic, and better organized. Moreover, they enjoyed stronger leadership, a factor that was to become crucial, sparked by many of the framers themselves. There were the Pinckneys down in Charleston, drawling softly as they outlined the Constitution's advantages to fellow planters. There was the brilliant **Alexander Hamilton** in New York, intense, obsessive, bursting with energy, anxious to face down his political nemesis, Melancton Smith.

That the Federalists could perform deft strategic maneuvers became evident by their choice of the word *federalism*. By calling themselves "**Federalists**," they made it seem as though they were advocating the very dispersal of power that their adversaries favored, which was certainly not the case. With this choice, they forced their adversaries into adopting the negatively connotative term of Anti-Federalists.

The Federalists soon hit upon a winning strategy. They would avoid rhetorical displays and empty bombast (the Anti-Federalists would become famous for both) and simply argue the merits of their case. On street corners, in alehouses, at public rostrums, the Federalists patiently answered their critics point by point, explaining how the Constitution not only created a strong, workable republican government but also adequately addressed the long-standing Republican Problem.

When it came to the question of a bill of rights, the Federalists ran circles around

FEDERALISTS OF NOTE

James Madison

Alexander Hamilton

Edmund Randolph

James Wilson

Figure 7.3 While most of the Founders defended their work in their own states, these men had an especially notable role in the ratification of the Constitution.

their opponents. Anti-Federalists had hoped to use the bill of rights issue as a distraction. Their strategy was to call for another convention to insert the necessary clauses, knowing that any follow-up would surely fail. So instead of arguing that a bill of rights was unnecessary—which the Federalists truly believed—they promised voters that if the Constitution were ratified, a bill of rights would be the new government's first item of business, to be added by way of amendment.

Smaller states ratified the Constitution with less fuss, for once the Great Compromise had been forged, the small states saw themselves as benefiting from unification. The battlegrounds would be New York, a colonial center of banking and commerce, and the three most populated states, Virginia, Massachusetts, and Pennsylvania. Without these four, no federal union could succeed.

Pennsylvania provided the first real test for the Constitution. Under the best of circumstances, the state's political atmosphere was a power-grabbing free-for-all, with partisan bushwhacking, mass demonstrations, and even the occasional riot. Friends of the Constitution had a majority in the State Assembly, but their Anti-Federalist foes thought they could block the call for a ratifying convention by keeping enough of their own members away to prevent mustering up a quorum. This plan did not stop the vote in Pennsylvania, however. Federalist "bully boys" actually broke through a door where several anti-Federalist state legislators were hiding out, roughed up two of the truants and physically carried them to the State House, where they were held in their seats for the roll call.

The convention itself was much calmer. But James Wilson, one of the Constitution's primary framers, delivered a masterful performance. Holding his head high to keep his glasses balanced on his nose, Wilson patiently answered objection after objection. The convention sat for a respectable five

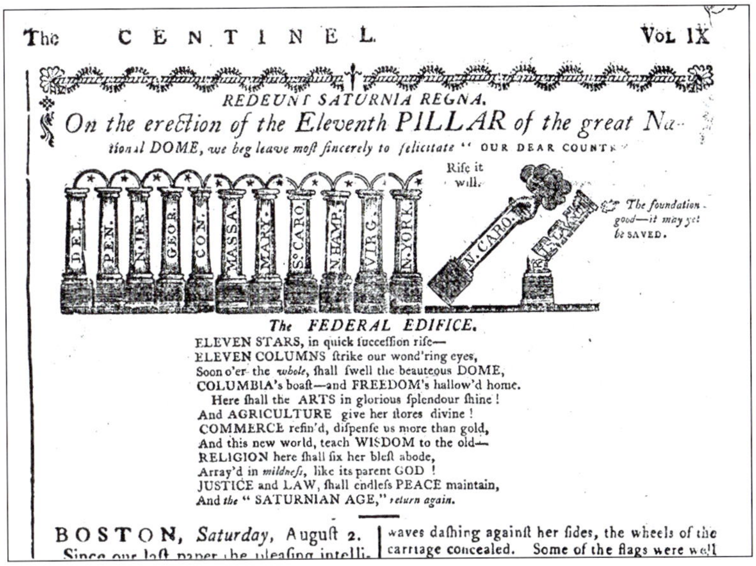
The CENTINEL. VOL IX

REDEUNT SATURNIA REGNA.

On the erection of the Eleventh PILLAR *of the great National* DOME, *we beg leave most sincerely to felicitate* " OUR DEAR COUNTRY"

The FEDERAL EDIFICE.

ELEVEN STARS, in quick ſucceſſion riſe—
ELEVEN COLUMNS ſtrike our wond'ring eyes,
Soon o'er the *whole*, ſhall ſwell the beauteous DOME,
COLUMBIA's boaſt—and FREEDOM's hallow'd home.
Here ſhall the ARTS in glorious ſplendour ſhine!
And AGRICULTURE give her ſtores divine!
COMMERCE refin'd, diſpenſe us more than gold,
And this new world, teach WISDOM to the old—
RELIGION here ſhall fix her bleſt abode,
Array'd in *mildneſs*, like its parent GOD!
JUSTICE and LAW, ſhall endleſs PEACE maintain,
And *the* " SATURNIAN AGE," *return again*.

BOSTON, *Saturday*, Auguſt 2.

waves daſhing againſt her ſides, the wheels of the carriage concealed. Some of the flags were well

Figure 7.4 This image depicts the ratification of the Constitution by New York, making a total of 11 states to do so by July 7, 1788. Each state is represented by a pillar supporting a building, symbolic of the strong foundation upon which the "federal edifice" was established. While a unanimous ratification, 13/13, seemed nearly impossible because of states like North Carolina and Rhode Island, a mere majority, 7/13, probably would not have been solid enough to survive national infancy. The Founders decided that ratification required 9 of 13 states.

weeks and ratified the Constitution by a vote of 46 to 23. Wilson's reward was to be mugged and nearly killed by a band of ruffians while he was out celebrating the document's acceptance in Pennsylvania.

The next test came with Massachusetts. The largest of the state conventions, with some 355 delegates, gathered in Boston's Brattle Street Church, while controversy stormed in the press. As in Pennsylvania, local politics played a role in the proceeding, for backcountry farmers rattled into town to take on the seaboard merchants and their newfangled government. One of these farmers, a rough-hewn ploughman named Amos Singletry, discoursed at length:

> These lawyers and men of learning, and moneyed men that talk so finely, and gloss over matters so smoothly, to make us poor illiterate people swallow down the pill, expect to get into Congress themselves. They expect to be the managers of this Constitution, and get

all the power and all the money into their own hands. And then they will swallow up us little fellows.

Once again, though, the Federalists replied in kind. One of them also happened to be a farmer, similarly unlearned and similarly eloquent. Addressing the convention president, Jonathan Smith replied,

> My honorable old daddy there [referring to Singletry] won't think that I expect to be a Congress-man, and swallow up the liberties of the people. I never had any post, nor do I want one. But I don't think worse of the Constitution because lawyers, and men of learning, and moneyed men, are fond of it. . . . These lawyers, these moneyed men, these men of learning, are all embarked in the same cause with us, and we must all swim or sink together.

As happened in Pennsylvania, local politicos Rufus King, Nathaniel Gorham, and Josiah Strong, who had been among the framers in Philadelphia, took to the floor in defense of their work, fielding question after question. On February 5, the opposition broke ranks and Massachusetts ratified the controversial document 187 to 168. Had ten votes gone the other way, the American Union would likely have died.

Massachusetts had been the largest of the state conventions, but Virginia's was the ablest. The Assembly Hall of the House of Burgesses was packed with members of the tidewater elite. Madison was there of course, along with George Washington, but so were Patrick Henry and George Mason. Henry had boycotted the Philadelphia Convention when he "smelt a rat." Mason had sat through every hour of it and denounced the outcome.

Henry showcased the Anti-Federalist approach to the constitutional debate. "Whither is the spirit of America gone?" he cried in rhetorical lamentation:

> Whither is the genius of America fled? . . . We drew the spirit of liberty from our British ancestors. But now, Sir, the American spirit, assisted by the ropes and chains of consolidation, is about to convert this country into a powerful and mighty empire. . . . There will be no checks, no real balances, in this government. What can avail your specious, imaginary balances, your rope dancing, chain-rattling, ridiculous ideal of checks and contrivances?

Against such oratorical eloquence, the physically diminutive James Madison must have felt even smaller. He stood before the delegates, his hat in his hand (with his notes in his hat), his voice so frail that those in the rear had to strain to hear him. In the way of most Federalists, he simply plodded along with the dull, prosaic facts of the matter, answering questions, allaying concerns, and deflecting Henry's oratorical thrusts.

The real star of the show was **Edmund Randolph**. He too had been at the Phila-

Edmund Randolph

1753–1813

Governor of Virginia and delegate to the Constitutional Convention, Randolph proposed Madison's Virginia Plan but ultimately refused to sign the Constitution in Philadelphia because it had no bill of rights. Later, however, he was instrumental in persuading Virginia leadership to ratify it.

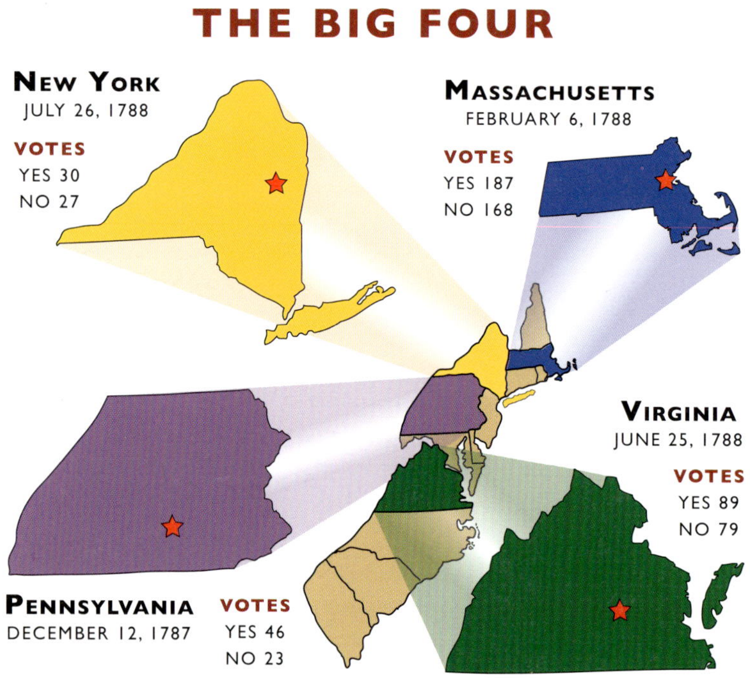

Figure 7.5 Although the Federalists needed 9 of 13 states to ratify, the Constitution would have been a failure without the support of these four states.

delphia Convention and had refused to sign the Constitution. However, unlike his friend George Mason, who now hurled thunderbolts against the document, the stage-handsome governor had somehow undergone a startling conversion and was now in favor of ratification heart and soul. When Madison's strength finally gave out, Randolph rallied the flagging Federalists and took command—giving the political performance of his life.

The Virginia convention made its decision on June 25. The Constitution battered but intact, won by a slender ten-vote margin. News of the victory reached Poughkeepsie on July 2, where the New York delegates had been convened for two weeks. As matters stood, Virginia was in the Union and the Constitution had sprung into life. The New York vote, taken the following month on July 25, saw ratification squeak through, 30 to 27.

It had been feared early on that New York might prove to be decisive—its delegation was split down the middle. Three staunch Federalists, Alexander Hamilton, James Madison, and **John Jay**, decided to collaborate on a series of newspaper essays to enlighten the New York electorate. ***The Federalist***, as they called their work, stands today as a monument both to insightful political philosophy and understanding of political science. It provides us with one of the deepest and most penetrating inquiries ever made into the nature of republican governance. It gives us some idea of the towering level on which the debate over ratification was carried. Scholars and students today still read it with awe.

The Federalist answered all the charges leveled at the Constitution. It explained both the republican and democratic nature of the document, arguing that the framers had not set out to defeat popular government but to create an example of it that actually worked. The empire decried by the Anti-Federalists would be more stable, more just, and provide more freedom than any republic in the ancient world. As for the Constitution's intricate machinery, according to *The Federalist*, this would purify consent so that public policy would truly reflect the public interest.

The essays were written in haste and were difficult to understand. But readers followed the authors' arguments, absorbed the documents' fine points, and debated the conclusions, as students continue to do today. For these defenders of Federalism it was not Political Science 101, but a question that would shape the very lives and destinies of the citizenry. To the lasting credit of the former British colonials, farmers, bankers, and merchants turned new Americans, they considered the merits of the case humbly and prayerfully. And in the end, they ratified the Constitution.

First Captain of the Good Ship USA

No one was surprised by the election of George Washington as President in 1788. It was done by acclamation. Indeed, the ratification of the Constitution may well have turned on the assumption that Washington would be the first head of state. The framers had laid elaborate mechanisms in place to secure the President's election by the people as a whole, believing that the people—not Congress or some other body—would choose the wisest and most virtuous of all citizens as head of state. The system had worked, at least on this first go-around.

George Washington's classical virtues would be taxed to the limit by his calling as the nation's first President. For the *first* President, unlike his successors, would superintend the actual building of a new nation. Imagine for a moment what the job would entail.

John Jay

1745–1829
A Founding Father, Jay served as a President of the Continental Congress, co-wrote *The Federalist* with Hamilton and Madison, and served as the first Chief Justice of the United States Supreme Court.

The Federalist

Series of essays published in New York newspapers under the pseudonym Publius for the express purpose of gaining support for ratification of the Constitution. Written by James Madison, Alexander Hamilton, and John Jay. (See Appendix C.)

Much of the constitutional text had been intentionally left vague. The framers believed that no plan of government should, or could, spell out details of institutional organization, much less describe how these would work in daily practice. In many cases, they had been unable to agree on important points and had left the text vague by way of compromise. In still other cases, the framers hadn't the slightest idea how a given concept would actually work. It was all left for Washington and the Cabinet he would select to make those critical decisions.

One thing the President could call upon was established tradition. Thus, some constitutional phrases could be taken as coded references to familiar practice. Moreover, because precedent was very powerful in the Anglo-American tradition, the President realized that whatever precedents he set might well be honored indefinitely. Small matters also could assume large symbolic importance. Consider the question of how to address the President: Your Excellency? Your Highness? Your Lordship? Your Majesty? Any of these phrases might have set the Presidency drifting toward monarchy, a real danger much too be feared. Washington settled on the very republican, but still respectful, "Mr. President."

If Washington happened not to like some feature of the Constitution, he needed only to ignore that feature or give it his own spin to banish it. He was negotiating his first treaty, for example, when he faithfully honored the wording of Article II, Section 2 that he should do so "with the Advice and Consent of the Senate." The first time he sought advice, however, he found members of the Senate to be so officious and meddlesome that he never consulted them again. Nor have any of his successors.

Our impression of the federal government today is generally one of high organization and cool professionalism. It was not so in those early days. There was a hesitation and tentativeness in the national establishment that we would scarcely recognize, like actors on a stage in their first rehearsal, reading their lines mechanically and wondering where to stand. It was anything but certain to these political experimenters that such a government could be made to work at all.

Figure 7.6 Much of our government's style and workings were a result of Washington's presidency.

Had anyone but George Washington been the first President, we might still be arguing today about the shape, tone, and style of our governance. As it was, the United States was exceptionally fortunate to have a precedent maker in whom it could wholly trust.

The Bill of Rights

As the President and his Cabinet were reshaping the Constitution in one way, Congress was reshaping it in another. Against all probability, the first Congress of the United States decided to press ahead on a bill of rights. The Federalists had promised to add this to the Constitution by way of amendment. However, there was nothing at all holding them to the promise, and because the whole issue had been trumped up in the first place, they might simply have forgotten about it. But this would be reckoning without Congressman James Madison of Virginia, a man who did not take campaign promises lightly. Madison volunteered to chair a committee to draft the amendments in question, and soon he was soliciting proposals from the states. These ten amendments were added to the U.S. Constitution in 1789.

The framers in Philadelphia had carefully considered the possible inclusion of a bill of rights and had unanimously rejected it. There were important reasons for this. Bills of rights, which were very popular in the eighteenth century, had been tacked on to several state constitutions and promulgated elsewhere as well. In rhetorical defiance, bills of rights challenged kings to remember the Lockean truth that the people had fundamental rights granted by nature, rights that government could not abrogate—indeed, rights which government was bound to protect. The framers accepted all this—it had justified their own Revolution—they simply didn't believe that it pertained to *republican* government. Why, they asked, would the people need to be protected against *themselves?* After all, the government's power lay with the people.

A second difficulty lay in enumerating the rights to be included in any bill. What were they, anyway, those rights of man? Some listings were short and concise, others lengthy and elaborate. Some included only a few basic items, such as Locke's "life, liberty, and property," while others delved into a luxuriant array, including the right to be taxed in proportion to one's means, and the right "to require of every public agent an account of his administration." The framers knew that any listing of rights would necessarily privilege the specific items named at the expense of all others. Who knew what the future might bring? Could Madison have imagined, for example, how twenty-first century electronic technology might undermine rights of privacy that in his time were taken for granted?

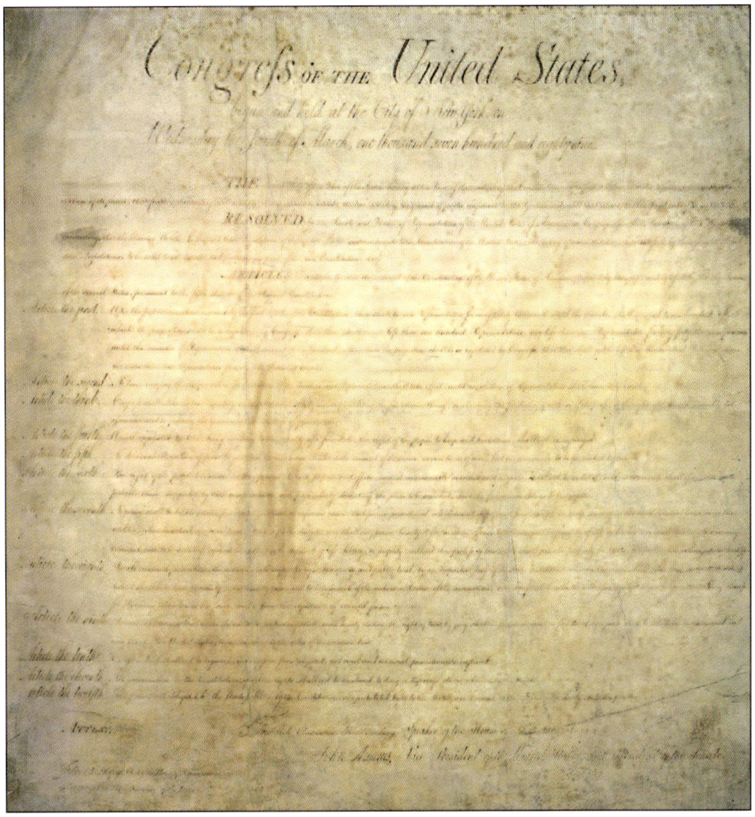

Figure 7.7 A photo of the original Bill of Rights containing 12 proposed amendments. The first two (not ratified at the time) detailed the number of constituents for each Representative and the compensation of Congressmen.

A Slight Delay

Originally, James Madison proposed 12 amendments for the Bill of Rights, but two were never ratified. While researching a paper on amendments in 1982, a University of Texas at Austin student named Gregory Watson discovered that these amendments were technically still pending. His paper on the subject received a C because the professor felt that Watson had not sufficiently argued that an amendment could still be passed after the elapsed 193 years. Watson immediately went to work on a letter-writing campaign to get one of the amendments ratified. This amendment, proposing that no congressional salary adjustment could be made until an election had taken place, was ratified as the 27th Amendment of the U.S. Constitution on May 18, 1992, 203 years after it was proposed.

A third problem was enforceability. How was the vague language of natural rights ever to be applied to real-world situations? Affirming rights was one thing, actually protecting them quite another. Rights as such had never been respected, especially by Old World governments; rights had merely been *claimed*.

Finally, the American framers realized that the very notion of natural rights, however useful in a stand against tyranny, might also undermine legitimate governance. Should *all* people, for example, have an absolute and unqualified right to liberty? Or to privacy? Or to property? Or even to life itself? And if not, how do we deal with the various exceptions and qualifications? Put another way, the rights of the individual must always be seen in the context of society.

Madison the astute lawyer came to believe that these difficulties could be overcome. At the same time, Madison the canny political theorist came to believe that a bill of rights might indeed be necessary, even in a republican government. In spite of all the carefully contrived constitutional mechanisms, unpopular minorities still might be exposed and vulnerable. For example, what would stop a fear-driven Congress from eliminating jury trial for accused spies, or denying poor immigrants right of counsel?

Madison's strategy was to draft his document in careful legal language, trusting that judges of the land would see to its sane, moderate, and wise enforcement. He would create a bill of rights that could stand up in court.

It was a tricky assignment. The drafting of a legal document requires the writer to carefully choose the kind of language to use. Broad, general, or abstract language might have rhetorical value, but it is difficult to apply to cases. Imagine, for instance, a court trying to interpret a general declaration such as: "All have a right to justice!" All who? Justice for whom? Justice from whom? Justice according to whom? What constitutes justice anyway? It might be a great battle cry, but it is weak, even dangerous, legal language.

Fortunately there is another choice. Rights described in narrow, concrete language, full of specific terms and qualifiers, may have little rhetorical value and usually sounds legalistic as well, but it is much easier to apply to cases. For example, instead of "all have a right to justice," think of drafting that general idea into a document for the United Steelworks that could stand up in to court:

> All members of the United Steelworkers, in good standing, with dues current, shall, in the case of a job dispute, have the right to an arbitration hearing in the presence of one company representative, one union representative, and one arbitrator agreeable to both.

It wouldn't make a great battle cry, to be sure, but if a steelworker were to lose his or her job because of someone else's mistake, our long paragraph filled with specifics and

ARGUMENTS AGAINST A BILL OF RIGHTS

- ⃠ States already had bills of rights—it was unneeded
- ⃠ Enumerating rights was very difficult
- ⃠ Enforcing rights was very difficult

ARGUMENTS FOR A BILL OF RIGHTS

- ✔ The Federalists promised during Ratification debate
- ✔ A rather large, powerful central government was just created
- ✔ Could not hurt to enumerate things the government cannot do (especially in the shadow of King George)

qualifiers might at least get the unfairly unemployed individual a hearing.

Madison understood that natural rights had always been conceived in broad, abstract language, which was the essence of the problem. Accordingly, he proposed to render them as **civil rights** wherever possible, toning them down and tightening them up, and thereby making them enforceable. Note Madison's wording of the Third Amendment, for example:

> No soldier shall, in time of peace be quartered in any house, without the consent of the Owner, nor in time of war, but in a manner to be prescribed by law.

The sentence is full of concrete terms such as *soldier, house*, and *quartered*, and uses qualifiers like "in time of peace," "in time of war" and "without the consent of the Owner." Coded references are also used. While these may sound vague, as "in a manner to be prescribed by law," they actually refer to procedures that were already well established.

Madison could have drafted the Third Amendment in the language of natural rights rather than civil rights, in which case it would go something like: "Government shall not interfere with private property," but think of the implications. Such broad and abstract wording would prohibit quartering, but what else would it prohibit?

Madison's civil rights strategy explains why many of the rights listed in the proposed amendments were drafted narrowly and concretely, and why the situations they addressed were not general woes of mankind but specific difficulties that Americans had encountered before: the right to keep and bear arms, security from unreasonable searches and seizures, compensation for property taken by the state. In addition to these, Madison included a list of procedural guarantees, for those accused of crimes, in the Fifth to

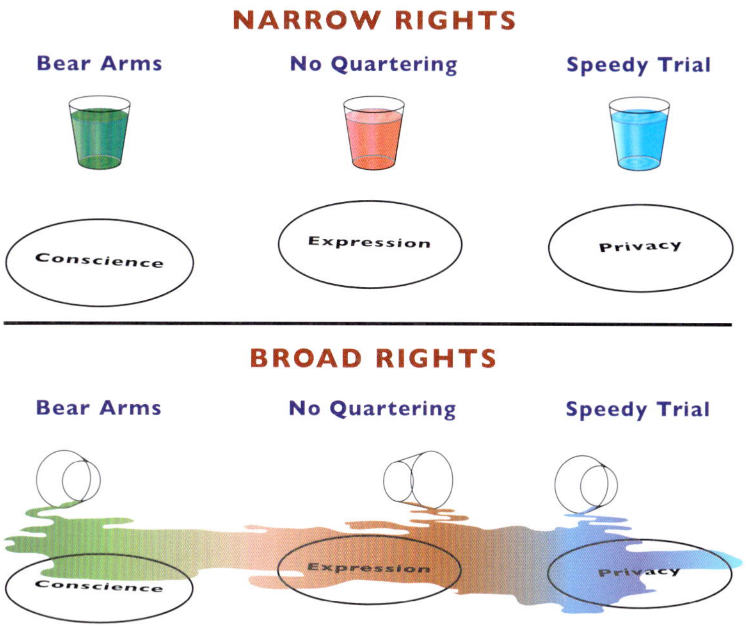

Figure 7.8 Without the constraining effect of narrow and concrete language, enumerated rights tend to expand and may include areas of social behavior never dreamed of by the Founders.

Eighth Amendments. Since most of these protections were found in the English common law, they, too, could be described by coded references. These amendments included everything from grand jury indictment, to trial by jury, to the right of counsel, to reasonable standards of punishment.

Three items in the Bill of Rights were not drafted narrowly and concretely but given a broad treatment in the language of natural rights. These are often regarded as the most important rights of all, the ones that quickly come to mind when the Bill of Rights is mentioned. Clearly, Madison did not approach them in the way he approached the others. He approached these in terms of natural, inalienable rights. They are the following:

- *Freedom of conscience*. Found in two separate clauses of the First Amendment.

Civil rights

Rights defined using narrow, concrete language, full of specific terms and qualifiers.

- *Freedom of expression*. Found in four clauses of the First Amendment.
- *Right to privacy*. Implied by language of the First, Third, and Fourth Amendments.

The obvious question is why, after Madison drafted most of the rights so narrowly, did he leave these three vague and general? Part of the answer is that there is simply no way to discuss religion, expression, or privacy in narrow and concrete terms. Words fail. Beyond this, the topics themselves are of exceptional importance. Freedom of conscience, freedom of expression, and freedom to live one's own life may be considered the fundamental ends of republican government—the "happiness" that free people have always sought. Natural rights have been called **"the Great Oughts,"** meaning that they don't proclaim an "is" so much as an "ought" about mankind's real and imperfect world. People *ought* to be free. They *ought* to seek their own pleasure, and so on. Perhaps a good term for Madison's three exceptions to the will-it-stand-up-in-court rule is the Three Great Oughts.

The Great Oughts

Natural rights that don't proclaim an "is" so much as an "ought" about the world—the way things "should" be.

Judicial review

Political power of the Supreme Court to rule on the constitutionality of laws.

Alien and Sedition Acts

Laws passed by Congress in 1798 to try and stifle the "seditious" writings of French propagandists against the neutrality of the United States with regards to the French and British War.

Did Madison suspect that the inclusion of broad and abstract language in the Three Great Oughts might cause difficulties down the road? He must surely have known that courts would interpret and reinterpret terms like *religion* and *speech*, and courts might ultimately vest them with meanings that to him would seem bizarre. Satanism as religion? Nude dancers as free speech? This is not necessarily to say that Madison would have regretted his work. He felt strongly about republican government and the ends it served. Quite possibly he would allow his thoughtfully crafted Amendments to stand as they were written, offering equal protection to flag burners and those saluting that same flag.

Judicial Review

The single greatest problem in putting the Constitution into practice lay in the question of how its text ought to be interpreted. Some of the blank spaces could be filled by common practice and others by executive decree. But in the early years of the constitutional experience, questions arose that could not be settled by either means.

During the administration of John Adams, for instance, France put the United States in a precarious situation. The French, who after all had bankrupted themselves helping out in the American Revolution, now sought aid against their old foe, Great Britain. When such aid was not forthcoming, they took extraordinary measures to sway American opinion, including issuing scalding denunciations of U.S. policy. Soon French privateers and American merchantmen were shooting it out on the high seas and the new republic found itself drifting toward war.

In order to stifle the "seditious" writings of French propagandists, Congress passed a set of laws known as the **Alien and Sedition Acts**, prescribing heavy fines and jail terms for those who spoke out against the government. But wasn't this a clear violation of free speech? No one knew for sure, although opinion abounded on every side. Nor did anyone know whose opinion ought to prevail. Should the President have the final say? How about Congress? Maybe the courts should decide. Perhaps even the individual states each had the right to make the decision, according to angry manifestos penned separately by Jefferson and Madison. With a written constitution, it was becoming clear, *someone* had to have the final say.

The framers in Philadelphia had discussed this problem, but they hadn't solved it. If a question of interpretation came up in colonial times, it was submitted to the Privy Council in London, which acted in the capacity of a law court. The framers may have assumed that the U.S. Supreme Court would play a similar role.

That was the thinking of John Marshall, who had been named the fourth chief justice of the Supreme Court in 1801. Marshall was about to pull off a stupendous piece of legal precedence: filling in of some constitutional blanks by unprescribed means. He "wrote" judicial review into the Constitution all by himself. Marshall believed that courts were the logical agency for resolving constitutional disputes. The newly formed federal judicial branch, having neither will nor energy of its own, was unlikely to exercise tyrannical authority, he reasoned. Therefore, when a question of interpretation arose, it was the judges who ought to answer it.

Marshall plotted his strategy with care. Sooner or later, he knew, a case would come along to provide the perfect pretext for judicial review, and when it did so he would be ready. Suddenly there it was: ***Marbury v. Madison***. It emerged from the charged atmosphere in which the Alien and Sedition Acts had been passed, and in which John Adams the Federalist had been defeated for a second term by Thomas Jefferson.

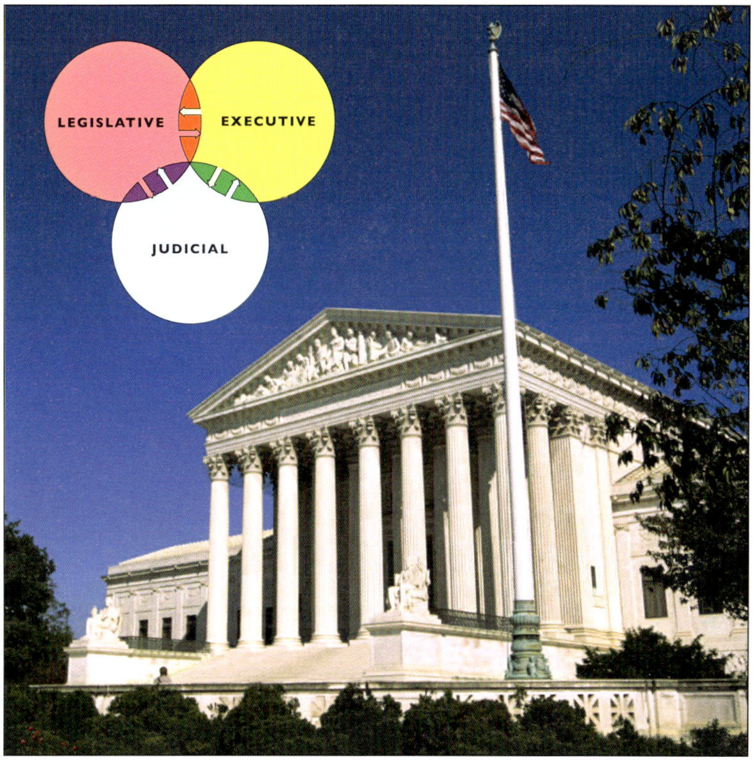

Figure 7.9 The United States Supreme Court can use the power of judicial review to declare a law unconstitutional. Yet, the Founders did not give this power directly to the judicial branch. It was not until Marbury v. Madison *in 1803 that this came to be accepted practice.*

When it became clear that Adams and his party, who still called themselves Federalists, had lost the election of 1800, the out-going President paused to exercise his patronage rights. He appointed a host of fellow Federalists to judgeships throughout the country. The triumphant Jeffersonians felt no great compulsion to deliver these

Marbury v. Madison
Supreme court case in which judicial review was established.

John Adams

Thomas Jefferson

John Marshall

William Marbury

James Madison

Figure 7.10 The major figures involved in Marbury v. Madison.

Midnight appointments

Judiciary appointments of Federalist judges made by Federalist president John Adams shortly before he left office, in response to the Democratic-Republican victory in the Congress and Presidency.

William Marbury

A "Midnight Appointment" by John Adams, Marbury sued Secretary of State James Madison for delivery of his commission, which was being withheld by order of President Jefferson.

Writ of mandamus

A court document forcing an action by a certain party.

Judiciary Act of 1789

Congressional act passed in 1789 to form the federal court system and to authorize writs of mandamus.

midnight appointment documents to Adams' appointees, and no appointee could legally take his position on the bench without showing an actual document proving his new authority. One of Adams' last-minute appointees, **William Marbury**, sued in court for the delivery of his commission. Specifically, he requested a **writ of mandamus**, which was the traditional instrument used for such a purpose, to compel Jefferson's Secretary of State (James Madison) to give him the document in question.

Marshall, a Federalist, instantly saw the beauty of the case. By deciding *against* Marbury, he would be thwarting the interests of his own party, and Jefferson would probably let the decision stand. Accordingly, Marshall, in his written opinion, argued that there was no explicit grant of authority in the Constitution for the courts to issue writs of mandamus, and therefore the **Judiciary Act of 1789**, which had set up the federal court system and specifically authorized writs of mandamus, had violated the Constitution.

It was a dubious argument. The Constitution had specified very few particulars pertaining to the federal court system, leaving Congress basically a free hand. The framers undoubtedly assumed that Congress would give the federal courts all powers that courts generally held, including that of mandamus. (Other such powers granted by the Judiciary Act went unchallenged.) However, finding no mention of mandamus in the constitutional text, the Chief Justice declared that clause of the Judiciary Act of 1789 null and void—in a word, unconstitutional.

Marshall went on to lay out four separate arguments justifying what he had done. None of the four was entirely convincing, perhaps, but judicial review did make constitutional sense. For one thing, it conclusively (and not unreasonably) answered the question posed earlier: *who decides?* For another, it added a whole new dimension of checks and balances to the federal system, very much in keeping with the framers' strategy. The judicial branch was now in a position to check both the legislative and executive, and they had to take that possibility into account.

Wisely, Marshall used the power of judicial review sparingly. What was important for him was to establish the precedent. Later, however, Supreme Court justices would have no such qualms. Like the Bill of Rights, judicial review would become something of a Pandora's box, especially when all those loose phrases found in the one came up for adjudication by the other. In due course, the U.S. Supreme Court would become an active player in the many-sided American political game.

By 1803, the year the *Marbury v. Madison* decision was handed down, it was clear that the American Constitution could indeed be put into action, and that constitutional government actually worked. Americans gained confidence in their new system, which they now spoke of as having been divinely inspired. At the same time, they increasingly thought of themselves *as* Americans rather than Virginians or New Yorkers. Proof of the Constitution's success was galvanizing a feeling of nationhood.

THE 4 MA'S OF JUDICIAL REVIEW

Marshall

Marbury

Madison

Mandamus

Figure 7.11 Contemplation of Justice *outside the Supreme Court Building.*

Timeline

Contextual Events

- Northwest Ordinance **1787**
- Constitutional Convention begins (May)
- Constitutional Convention ends (Sep)
- Delaware ratifies (Dec)
- New Jersey ratifies
- Georgia ratifies (Jan) **1788**
- Connecticut ratifies
- Maryland ratifies (Apr)
- South Carolina ratifies (May)
- Judiciary Act of 1789 **1789**
- North Carolina ratifies (Nov)
- 11th Amendment ratified **1795**
- Washington's Farewell Address **1796**
- John Adams wins Presidential Election

- Jefferson makes the Louisiana Purchase **1803**

1780 · 1790 · 1800

Chapter Specific Events

- **1787** (Dec) Pennsylvania ratifies
- **1788** (Feb) Massachusetts ratifies
  - (Jun) New Hampshire ratifies
  - (Jun) Virginia ratifies
  - (Jul) New York ratifies
  - Washington becomes President
- **1789** Judiciary Act of 1789
- **1790** (May) Rhode Island ratifies
- **1791** Bill of Rights ratified

- **1800** Thomas Jefferson wins Presidential Election
- **1801** John Adams makes his famous *Midnight Appointments* to the federal judiciary
  John Marshall becomes Chief Justice
- **1803** *Marbury v. Madison* decided—
  Justice John Marshall establishes Judicial Review

Key Terms

Federalists
natural rights
judicial review
Anti-Federalists
civil rights
The Federalist
"The Great Oughts"

Key People

Edmund Randolph
Samuel Adams
Alexander Hamilton
William Marbury
John Marshall
George Mason
Patrick Henry
John Jay
Thomas Jefferson
Elbridge Gerry
James Madison
George Washington
John Adams

Questions

1. What complaints did the Anti-Federalists have about the new constitution? Which of these complaints were remedied by the Federalists? Do we see the results of any of these concerns and subsequent changes in our government today?
2. As the first president of the United States, George Washington established precedents. What are some of the traditions instituted by Washington? How might our nation be different today had a more personally ambitious man been elected president?
3. Which amendments are drafted narrowly with more concrete language, and which are drafted broadly with more abstract language? What are the implications of each?
4. Looking at our society today, were the concerns of the Federalists regarding the inclusion of a Bill of Rights justified?
5. Explain the events leading to *Marbury v. Madison* and discuss the decision of the Supreme Court. What are the modern implications of judicial review?

Chapter 8

From Unity to Political Parties

Abstract

During the election of 1800, Alexander Hamilton and Thomas Jefferson presented the American people with two very different views of America, leading to the formation of two political parties vying for power. While parties are not explicitly provisioned in the Constitution, the prominence of two large parties is a result of the Founders' constitutional structure. This chapter explores how Representatives, Senators, the President and Supreme Court Justices are placed in office, and how these constitutional processes have influenced the present American two-party system.

Ratification of the Constitution brought home the reality that the new government, along with all of its intricacies, now had to be implemented. Fortunately there was one great unifying figure who could be relied upon—George Washington. For most of the population, he was the only acceptable choice for the first President of the United States. They had much more confidence in him than they did in their new constitution. Each of the electors selected for the Electoral College were to vote for two candidates without giving preference to one as President and the other as Vice President. Every elector voting listed George Washington on his ballot. John Adams received the next most votes and was selected as Vice President. Washington's status and prestige was unrivaled and no other political figure would dare directly challenge him. He assembled an extraordinary cabinet with Thomas Jefferson as Secretary of State, Alexander Hamilton as Secretary of Treasury, Henry Knox as Secretary of War, and Edmund Randolph as Attorney General. Those four, with Adams, brought an extraordinary collection of brilliance, legal skills, and statesmanship in the fledgling government of the new republic.

Loyal opposition

When losers in the political game continue to support the system, even when the system is against their ideology.

Washington proceeded with insight and caution to establish precedents in the day-to-day operations of government. He successfully kept most of the trappings of monarchy out of the new government. It was decided that he should be called "Mr. President" rather than "Your Majesty," although he would have found today's easy familiarity with political leaders appalling. Washington and his administration determined the practical meaning of phrases in the Constitution. At first, he assumed that the phrase "advice and consent of the Senate" meant that he should meet with the senators and receive their counsel. However, after listening to hours of contradictory advice from senators in love with the sound of their own voices, he wisely determined that "advice and consent" simply meant a vote of approval from the senate.

George Washington hoped for unity within his administration. Indeed, the founders did not really understand or accept the notion of a **loyal opposition**. They had been called traitors by loyalists committed to the King of England, and they saw those same loyalists or Tories as disloyal to the revolutionary cause. Political parties were seen as factions to be lamented and then controlled. They held the hope that the new government with popular sovereignty and filtered consent would avoid political parties and the worst of factional behavior. Their hope for the absence of political parties was soon shredded by the realities of political life.

Two American Ideologies

Two towering figures dominated government in Washington's first administration—Alexander Hamilton and Thomas Jefferson (see Figure 8.1). Jefferson, the nation's first Secretary of State, was devoted to the ideal of a society composed of free, self-reliant individuals with a small government to protect their rights. He used the term "yeoman farmers" to describe the people who would dominate his ideal society. Ideally, he thought, self-governing farmers would till the soil by day and read science and political philosophy by candlelight in the evening. This ideal, pastoral society would produce intelligent, virtuous people in little need of government. Should tyranny arise, they would support another revolution and put things aright again. Kings, with all their elegant trappings, held no philosophical attraction for Jefferson. He supported the bloody and

WASHINGTON'S CABINET

President	Vice President	Secretary of State	Secretary of Treasury	Attorney General	Secretary of War

George Washington	John Adams	Thomas Jefferson	Alexander Hamilton	Edmund Randolph	Henry Knox

Figure 8.1 Washington's cabinet contained some of the most talented political and legal minds of the time.

chaotic revolution then under way in France with an enthusiasm and hope not shared by many other Americans. Jefferson pretended to be uninterested in political power and intrigue. He claimed that he just wanted to retreat to his beloved Monticello to study, read, and farm. But he was always drawn back to politics and the challenges and intrigues of government.

At the other side of the personality spectrum, Alexander Hamilton, Secretary of the Treasury, did not disguise his ambition. He viewed the common people of Jefferson's vision with distrust and disdain. There must always be the rulers and the ruled. He intended to be the one who ruled. He envisioned the main task before the new government as being the creation of a great empire that would dominate the western hemisphere and compete successfully with the established powers of Europe. For Hamilton, it was only a question of how best to create a great and powerful nation. It was inevitable that Hamilton and Jefferson, two brilliant, strong-willed, and talented men with very different political ideologies, would clash. Indeed, they dominated America's political scene most of the time until 1804 when Vice President Aaron Burr fatally shot Hamilton in a duel on a ledge above the Hudson River.

The Duel of Hamilton and Burr

Alexander Hamilton and Aaron Burr had been political opponents for many years, but the primary cause of the duel appears to have been a letter from one Dr. Charles D. Cooper opposing Burr's candidacy that was published in the *Albany Register*. It referred to comments made by Hamilton against Burr at a dinner party. Burr pushed for an apology, but Hamilton refused because he could not recall the details of the conversation to which Cooper had alluded and would not be held responsible for Cooper's interpretation of his remarks.

On July 11, 1804, Burr and Hamilton with several others crossed the Hudson to New Jersey (dueling was against the law in New York). Different accounts state that Hamilton fired first into the air away from Burr, after which Burr fired and struck Hamilton in the lower abdomen. Hamilton died the next day in Manhattan. Burr was charged with murder for the duel several times. In most instances the case never went to trial. In the final indictment years later, Burr was acquitted.

In a letter written by Hamilton the night before the duel, he stated his opposition to the practice of dueling on both religious and practical grounds, and wrote, "I have resolved, if our interview is conducted in the usual manner, and it pleases God to give me the opportunity, to reserve and throw away my first fire, and I have thoughts even of reserving my second fire."

Figure 8.2 An artistic rendering of the duel by J. Mund.

As Secretary of Treasury, Hamilton proposed an ambitious economic program to begin to build a nation, if not an empire. He argued with skill that the national government should do the following:

1. Assume the revolutionary war debt of all thirteen states.
2. Pay off in full all the debt of the federal government, thereby establishing the financial solvency and reputation of the new nation.
3. Establish a Bank of the United States, patterned after the Bank of England, to manage the financial affairs of the country and to discipline and control private banks.
4. Negotiate a trade agreement with Great Britain.
5. Impose tariffs (taxes on imported goods) to encourage and protect domestic manufactures.

Hamilton's program was a U.S. version of the mercantilism of the British Empire. He wanted to use the power of the government to direct the economy in the path he thought best. He was certainly not committed to a free market economy that would follow its own course.

Jefferson was appalled by Hamilton's economic proposals. Paying off the states' debt in full would enrich Hamilton's speculator friends who had bought up much of the debt for pennies on the dollar. Jefferson felt that a Bank of the United States would put too much power in a few hands. Tariffs would hurt the yeoman farmers and create a merchant and manufacturing class that would be much like nobility. Jefferson believed in minimal government, not active management of the economy. He accepted Adam Smith's critique of mercantilism as well as Smith's advocacy of an unregulated market economy. Jefferson argued that the Constitution narrowly limited the powers of government. Jefferson viewed Hamilton's program as one that assumed dangerous powers not granted to the government by the Constitution. Moreover, Jefferson saw revolutionary France as the young country's natural ally rather than Great Britain, its former master.

The Birth of Political Parties

Out of this clash of personalities and ideologies, America's first two political parties were born. On one side were those who wanted a relatively powerful federal government capable of managing the economy and

JEFFERSON'S IDEOLOGY

Democratic-Republican Party

- Leaders: Jefferson and Madison
- Ideal Society: Self-reliant individuals with small government that protects rights
- Supported French Revolution and alliances with France
- Smaller federal government
- Narrow interpretation of constitutional powers

HAMILTON'S IDEOLOGY

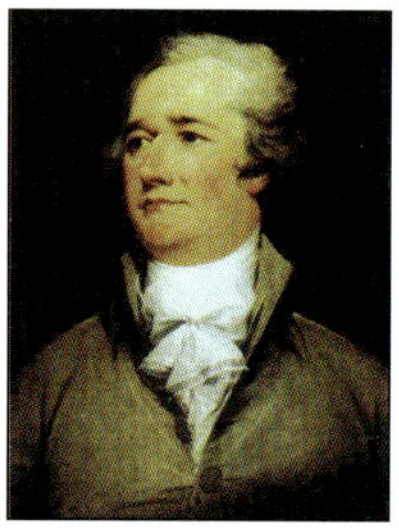

Federalist Party

- Leaders: Hamilton and John Adams
- Ideal Society: Great empire that would dominate the Western Hemisphere and compete with European powers
- Supported alliances with Great Britain
- Powerful federal government
- Broad interpretation of constitutional powers

Figure 8.3 Differences between Jeffersonian and Hamiltonian ideologies.

putting down any impulses of anarchy that might pop up among the common people. This government should align itself in Europe with the forces of stability (Great Britain), not the turbulent forces of revolution (France). This faction came to be known as the **Federalist Party** (not to be confused with the "Federalists" who championed the ratification of the Constitution). John Adams and Alexander Hamilton led the Federalist Party with the reluctant blessing of George Washington.

On the other side were those who wanted a narrow and strict interpretation of the Constitution. They pointed out that the Constitution said nothing about creating a bank or encouraging domestic manufacturing with tariffs. Government was best that governed least. Jefferson and James Madison led this group. But they had a problem. They couldn't be known as anti-federalists, the name reserved for opponents to the Constitution. So they adopted the awkward title of **Democratic-Republicans**.

By 1796, Washington had served two terms as President and was ready to retire to Mount Vernon. Adams, with Hamilton's influence behind the scenes, became the Federalist candidate, while Jefferson became the candidate of the Democratic-Republicans. A hard-fought campaign ended with Adams narrowly winning with 71 electoral votes to 68 votes for Jefferson. Because of the nature of the Electoral College, Adams became President and Jefferson, Vice-President. While Adams was occupied with the Presidency, Jefferson and Madison were free to build a Democratic-Republican Party that would dominate national politics for the next thirty years. Jefferson, with the aid of Madison, unified the Democratic-Republican Party, but the Federalist Party was soon split between Adams and Alexander Hamilton. Only a few years after the adoption of the Constitution, these two distinct political parties emerged to contest the elections for President and Congress as defined in the Constitution.

In his 1796 farewell address, a somewhat dispirited George Washington took the opportunity to "warn you in the most solemn manner against the baneful effects of the spirit of party generally." The new nation's politicians and common voters blithely ignored that warning.

The Constitution and Politics

The Constitution makes no mention of political parties, but it does outline a structure for selection of members of the House of Representatives, the Senate, the Supreme Court, and the President. This structure exerts powerful influences on the nature of American politics, elections, and campaigns. It is important to understand that the link between the constitutional structure and politics is the fundamental principle of a representative democracy and the rule of law—consent. The people gave their **original consent** when the Constitution was ratified and first implemented. With each election since then, citizens of the United States have renewed their consent or given **periodic consent** for the government to function. The Constitution requires these elections in order for the people to give the government legitimate authority.

The founders rejected the idea of direct democracy where the people vote directly on laws and directly manage the affairs of government. They knew that direct democracy could only work in a very small country or a city. They had some experience with direct democracy at the town or county level. But classical history and their own experiences clearly showed that even in a small country or state direct democracy could easily be corrupted by a mob mentality or by clever demagogues who promised anything to lead the

Federalist Party

Political party founded by Hamilton and John Adams that envisioned a great Western empire with a strong federal government and a broad interpretation of Constitutional powers.

Democratic-Republican Party

Political party led by Jefferson and Madison that championed a society of self-reliant individuals to protect rights, a smaller federal government, and a narrow and strict interpretation of the Constitution.

Original consent

Giving consent to a provision or law the first time, such as the ratification of the Constitution.

Periodic consent

Giving continuing consent at certain intervals (through means such as elections) to a provision or law to which original consent has already been given.

people astray. The founders were confronted with the fundamental problem of democracy: How to give the power of consent to the people without exposing the country to manifest dangers of mob rule?

The founders chose **filtered consent**, in which the citizenry selected representatives who then selected other representatives in the government. This process of election of representatives was to filter out the temporary feelings and mistakes of judgment that might dominate the direct vote of the people on matters of government. The more removed a decision was from the people, the more filtered the consent of the people. Of course, a government that is too removed from the people loses consent and legitimacy, and becomes more susceptible to tyranny.

The technical problem was to pick the right set of filters that preserved democratic control while preventing the problems associated with the temporary passions of the people. Consent becomes more filtered if:

- Elections are further apart in time.
- Each representative represents a larger population.
- The selection process is more indirect (i.e., the voters pick representatives who select the government officials.)

Filtered consent

When the selection of government officials is distanced from direct election by the people in order to protect against mob rule and public whim. Filters include indirect election, time between elections, and size of representative regions.

The Filters of Consent in the Constitution

The House of Representatives

The Constitution specifies direct election of members of the House of Representatives every two years in districts of more or less equal population. A census is taken every ten years to determine the changes in population and if necessary re-draw the boundaries of the House districts in light of population change. Each elected representative originally represented no more than 30,000 people (with the provision of slaves being counted as three-fifths of a person). As the population grew, more representatives were added, with each representing more and more people. After the apportionment based

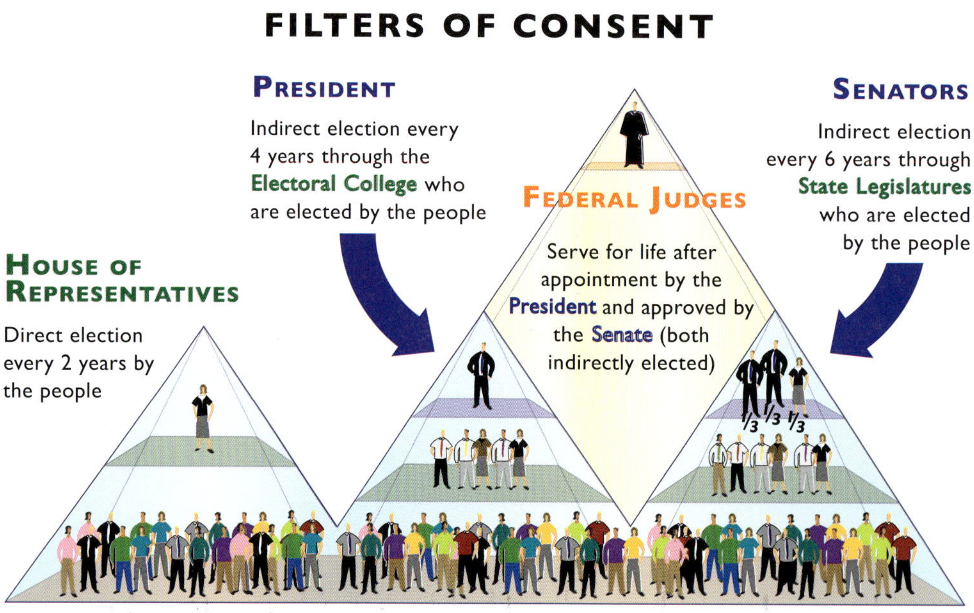

FILTERS OF CONSENT
PRESIDENT
Indirect election every 4 years through the Electoral College who are elected by the people
SENATORS
Indirect election every 6 years through State Legislatures who are elected by the people
FEDERAL JUDGES
Serve for life after appointment by the President and approved by the Senate (both indirectly elected)
HOUSE OF REPRESENTATIVES
Direct election every 2 years by the people
1/3 1/3 1/3

Figure 8.4 Filters of Consent within the United States government. The filters were instituted to prevent certain government offices from being subject to the whims of the general populace and the vulnerability of mob rule.

on the 2000 census, each Congressional District in the House of Representatives was composed of roughly 650,000 people, except in small states having only one representative. By using direct election every two years, the founders designed the House of Representatives as the part of government closest to the people. All other parts of the federal government were designed to be (and still are) more removed from the people because of more filters of consent.

The Senate

The Senate represents more filtered consent. The two senators from each state were originally selected by the legislatures of each state, with each senator serving a term of six years. The 17th Amendment passed in 1913 changed this practice, requiring that senators be elected to a sex-year term directly by the people. Thus, senators are more removed from the people than are members of the House of Representatives because they represent a larger group of people and are selected for a longer period of time.

The President

The Constitution provides an elaborate and somewhat confusing mechanism for the selection of the President. Each state selects a number of people called electors (equal to the number of senators plus the number of representatives from that state) to vote for the President of the United States. This group of electors is known as the **Electoral College**. (See Figure 8.5.)

The Constitution gives the state legislatures the power to determine how members of the Electoral College are selected with the restriction that no members of Congress or federal officials shall serve as electors. In practice, each political party (or Presidential candidate) chooses a slate of Presidential electors for each state. For example, in Utah, where there are five electoral votes, the Republican Party chooses five electors and the Democratic Party chooses five electors. As a citizen casts a vote for the Presidency, he or she is in reality voting for that candidate's slate of electors. Once the candidate with the highest popular vote has been determined by the people of a particular state, the electors for that candidate cast electoral votes on behalf of the state.

Electoral college

The group of electors selected by the people who are responsible for the selection of the president.

Electors originally voted for two individuals for the Presidency, so the person with the most electoral votes became President and the person with the second most

HOW PRESIDENTIAL ELECTIONS WORK

In 2004, Florida had two Senators and 25 Representatives. The state thus had 27 electoral votes, all of which went to George W. Bush because he won a majority of the vote.

	BUSH	KERRY
Before the Election	Chooses 27 electors	Chooses 27 electors
Election Results (Florida)	52.10%	47.09%
After Election	Bush's electors get all 27 electoral votes, and vote for him	Kerry's electors get 0 electoral votes

The Presidential candidate receiving the majority (270) of the 538 electoral votes from all states combined wins the election. If no candidate has a majority of electoral votes, the decision goes to the House, where each state, regardless of size, forms a delegation and casts one vote.

Figure 8.5 Illustration of how current presidential elections work.

became Vice President. This created a deadlock in the election of 1800 when Thomas Jefferson and Aaron Burr, both Democratic-Republicans, received the same number of electoral votes even though Jefferson was the presumed Presidential candidate and Burr, the Vice Presidential candidate.

The 12th Amendment prevented future such deadlocks by requiring that electors cast separate votes for President and Vice President. If a candidate receives a **majority** of the electoral vote, they become President of the United States. If no candidate receives a majority of the electoral vote, the selection of the President moves to the House of Representatives with a complex procedure specified in the Constitution. Fortunately, that method has only been used twice, in 1800 and again in 1824. As in the case with the Senate, consent for the President is more filtered than the consent for the House of Representatives because the President serves for four years, represents the whole nation, and is not directly chosen by popular vote.

Majority

Receiving more than 50% of the votes.

Plurality

Receiving the largest percentage of the votes.

The Supreme Court

The part of government most removed from the people is clearly the Supreme Court and the federal judiciary. Justices for the Supreme Court and other federal judges are nominated by the President and confirmed by the Senate. They serve as judges for life or until they choose to retire. Consequently, consent of the people for the judiciary flows through the consent given to the President and senators.

Removal from Office

Each branch of Congress (the House of Representatives and the Senate) is the judge of the qualifications of its members and may refuse entry, or remove a member deemed unworthy to be a representative or senator. The House of Representatives also has the power to impeach the President or any federal judge. Impeachment is similar to an indictment in a criminal case. Once impeached, the person is judged in a trial with all senators acting as jurors. If convicted, the official is removed from office. Two Presidents, Andrew Johnson and William Clinton, have been impeached, but neither was convicted by the Senate.

Figure 8.6 The strength of the filters of consent vary according to the area of government.

FILTERS OF CONSENT IN THE UNITED STATES GOVERNMENT

Governmental Unit	Scope of Representation	Method of Selection	Term of Office
House of Representatives	Small: Districts proportional to population	Plurality winner in the district	Two years
Senate	Medium: State	Originally selected by state legislatures Now plurality winner in the state	Six years
President	Large: U.S.	Majority of the electoral vote. If no candidate has a majority, the winner is determined by state votes in the House of Representatives	Four years with maximum of two terms
Supreme Court	Large: U.S.	Nominated by President with consent of the Senate	Lifetime or voluntary retirement

Effects of Structure of the U.S. Electoral System

For 150 years, the Republican and Democratic Parties have been dominant, sharing and jockeying for political power while being successful at either blocking or absorbing all third-party movements. From 1796 until 1860, numerous political parties were formed, but there were usually two dominant parties—first, Federalists and Democratic-Republicans, then Democrats and National Republicans, then Democrats and Whigs, and since 1856, Democrats and Republicans. The American political structure contains strong incentives for two political parties sharing the power of government to the exclusion of all smaller parties.

The Constitution sets out an elaborate structure of elections to implement the principle of popular consent required for the rule of law. These legal structure have four main characteristics:

- The President is elected separately from the legislature.
- A **single representative** is elected from each district or state.
- Plurality of votes being sufficient for election except in the Electoral College.
- Fixed intervals for elections.

When we talk about the structure of elections, we refer to the rules and laws by which elections are carried out. The implications of these four structural election devices the behavior of the voters and the candidates.

President Elected Separately from Legislature

In the United States we have elections for the President and separate elections for the legislature. In contrast to this system, parliamentary democracies like Italy and Spain have no separate election for their executive leader. (See Figure 8.7.) Instead, the majority party in the legislature (or a coalition of parties that makes a majority) selects a Prime Minister to act as the executive. If no single political party has a majority in the legislature, a number of small parties will organize themselves into a loose coalition, form a majority, and select a Prime Minister.

Single representative districts

Representational structure where each geographical region elects its one representative independent of outcomes in other regions.

PARLIAMENTARY ELECTORAL SYSTEM

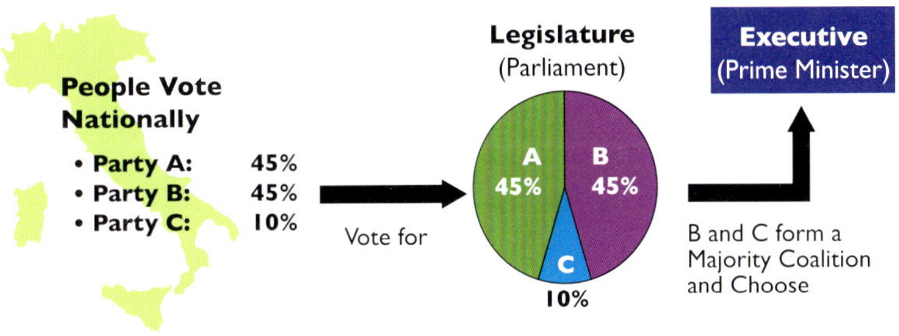

AMERICAN ELECTORAL SYSTEM

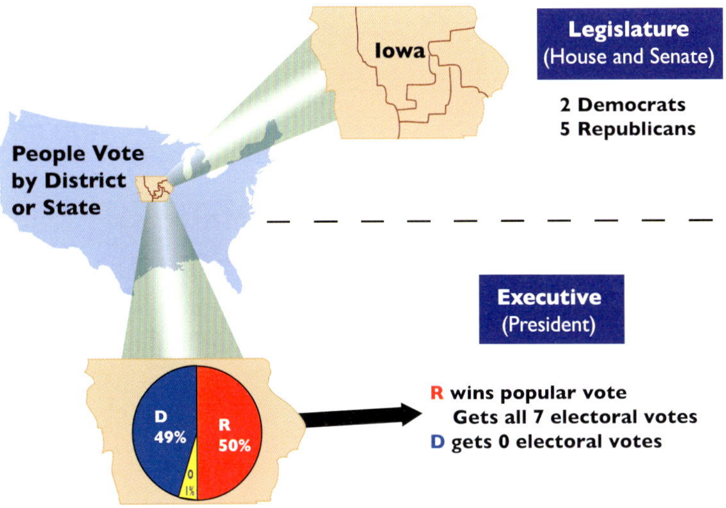

Figure 8.7 Unlike a parliamentary system where a majority coalition of the legislature selects the executive officer, the American electoral system has the executive officer selected (albeit through the filter of the electoral college) by the people. This practice allows for the executive officer to be from a different party than the majority of the legislature, and prevents small political parties from gaining too much power in the legislature.

These large coalitions in non-U.S. electoral systems may give small parties within the coalition a disproportionate amount of power. In reference to Figure 8.7, notice that no party in the parliamentary electoral system has a majority (over 50%) in the legislature. In order to form a majority coalition and appoint a prime minister, Party A or Party B must convince Party C, through promises and other incentives, to join its coalition. Because Party A and Party B both want control of the government (and because each needs Party C to be part of a majority coalition), they are each willing to give Party C more power and influence than Party C's size would normally warrant.

Unlike the parliamentary system (where parliament chooses the executive) the U.S. Presidential electoral system gives no incentives for larger parties to work with (let alone give concessions to) small parties. Because the American President is elected independently (structure), third-party candidates can only affect Presidential elections by capturing electoral votes and then throwing that support to one of the major candidates. Because the structure of the U.S. Presidential electoral system has limited the influence of third parties in Presidential elections, the other political parties are mainly "alternative voices," offering little more than a diversity of view points.

Single Representative Districts/States

American Congressional elections are conducted differently than many parliamentary elections. In parliamentary democracies, representatives in the legislature are often proportional to the national (not state or district) vote for each political party. A nationwide vote of 20 percent for the labor party translates into that party gaining approximately 20 percent of the representation in the parliament or legislature. Consequently, a party with some strength or popularity throughout the country will gain representation in the legislature, even though it may not receive the most votes in any part or district of the country.

In the United States, however, separate and distinct districts or states are the basis for the election of each representative and senator. This is a major structural element of the United States electoral system. Senators are elected by the citizens of the entire state, while Congressmen (representatives in the House) are elected by residents of their voting districts. (See how Iowa is divided into five districts in Figure 8.7.) When residents vote for senators or representatives, there is only one winner from each race. The losers get nothing. The collection of winners from the House and Senate races in each state across the nation make up the House and Senate for the nation.

Because of this structure, it is very difficult for a third party to elect even a few representatives in the U.S. To do so, that party must gain a plurality in one or more Congressional districts—a real challenge. Note again that it is much easier for smaller parties to gain representation in a legislature if there is representation based on national or regional voter percentages instead of single representative districts. Even if a third party were to prevail in one or two Congressional districts, the two major parties may exclude the third-party representatives from influential Congressional committee assignments, where legislation and government budgets are formulated. Voters who elect a third-party candidate soon realize that their choice may be the best man, but he is still ineffective in the day-to-day business of Congress. In the past 150 years, there have only been a handful of Congressional representatives not affiliated with one of the two major parties.

Plurality vs. Majority

Ordinarily, candidates in U.S. general elections have to receive a plurality (not a majority) of votes to be elected. This means that the winning candidate has to get the largest share of total votes, no matter the percentage. The requirement of a majority typically would imply some sort of run-off election between the top two vote-getters. States have rarely used run-off elections because of the added expense of an additional election and because most U.S. elections involve only two candidates. Primary elections, an election system not mentioned in the Constitution, are usually used today to pick the candidates for each political party.

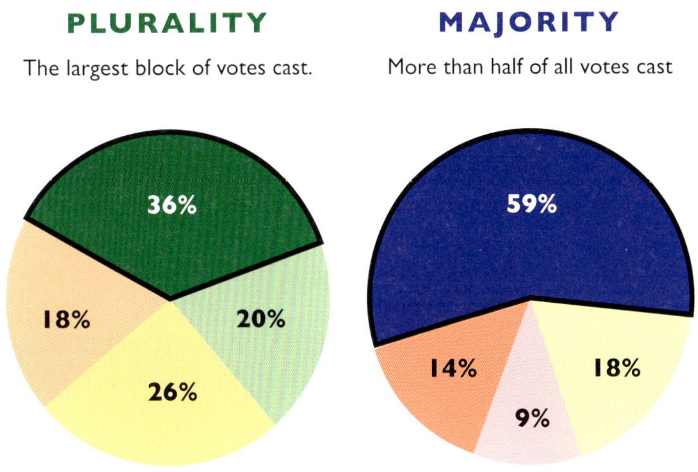

Figure 8.8 Most elections only require a plurality to win. If there are multiple candidates, a majority rule would require a run-off election (as the left circle demonstrates).

Fixed Intervals

U.S. elections are set at fixed intervals—every two years for the House of Representatives, every six years for the Senate (about one-third of the Senate is up for election every two years), and every four years for the Presidency. Many parliamentary systems of government do not have fixed intervals for elections. Instead, elections are called by the legislative body within certain time limits.

The Effect of Third Parties in the U.S.

This dominance by two parties does not mean that other parties do not exist or are not able to participate in elections. There have been many small parties throughout U.S. history, representing a variety of viewpoints. The colorful list of parties that have attracted some support over the years include Anti-Masonic, Free Soil, Greenback, Prohibition, Union-Labor, Populist, Socialist, Progressive, Farmer-Labor, Communist, States' Rights, American-Independent, Libertarian, Independent, Green, Constitution, and Reform, to name only the most prominent. However, none of these parties has gained enough ballot box support to enable them to share directly in government power. Even though the U.S. electoral system works against the formation of small third parties, third parties can affect elections in two ways.

First, a third party can act as a spoiler by siphoning enough votes away from one candidate to elect the other major candidate. One of the most successful third-party candidates in recent history was H. Ross Perot, who ran in the 1992 and 1996 Presidential elections. In 1992, Perot received no electoral votes but won 19 percent of the popular vote. Election researchers theorize that Perot acted as a spoiler because he may have garnered many votes that would otherwise have been cast for George H. W. Bush. As a result, William Clinton became President with less than a majority of votes. For the 1996 election, Perot garnered only eight percent of the popular vote, but even that was a respectable amount for a third-party candidate. A stronger case may be made that Ralph Nader and the Green Party were spoilers in the hotly contested election of 2000 by costing Vice President Al Gore the electoral votes of Florida. An election must be very close for a third party to have this effect on a Presidential election, and the third party must be willing to see the candidate less congenial to their policies win the election. For most Presidential elections, the effect of third-party candidates on the electoral vote and Presidential elections has been negligible.

Second, third parties affect the political positions and campaign efforts of the two major parties. If a third party becomes significant at all, one or both of the major parties will try to co-opt the third party by supporting issues important to the third party and inducing leaders of the third party to join their ranks. Through this effort, third parties or their members are usually absorbed into the two major parties. Suppose a "Green Party" were to become a significant third party in American politics. One of the two major parties would probably add some pro-environment planks to its official platform in order to attract voters and candidates away from the Green Party. Thus, American politics is almost completely dominated by the two large political parties.

Effects of Two Large Parties

Middle of the Road Politics

Competition between two parties for political power changes not only the nature of elections but also the nature of political campaigns. There is a simple strategy to win an election with only two candidates. Candidates will try to position themselves in the middle of voter sentiment and to portray their opponent as extreme in some way. Figure 8.9 illustrates this strategy for an election involving one issue—the size and power of the federal government. To win the election, a candidate will take position A in the middle of the distribution of voter sentiment and will portray the opponent at either position B or position C. In a two-candidate race, the politician who successfully occupies the middle of the political terrain wins the election. If one candidate is successful at portraying the other candidate as "out of the mainstream," the election may turn into a landslide for the middle-of-the-road candidate.

In the election of 1800 discussed earlier, Thomas Jefferson and his supporters tried to take the middle ground and convince voters that John Adams and the Federalists wanted too much governmental power and too much control over the people (position C). John Adams and his supporters tried to paint Jefferson as a lover of the French Revolution and disorder (position B). Both wanted the coveted middle ground in American elections (position A). Neither candidate was successful at convincing the voters that the opponent was an extremist, and the election is close; Jefferson prevailed over Adams with 53 percent of the electoral votes.

This general pattern of "I'm in the

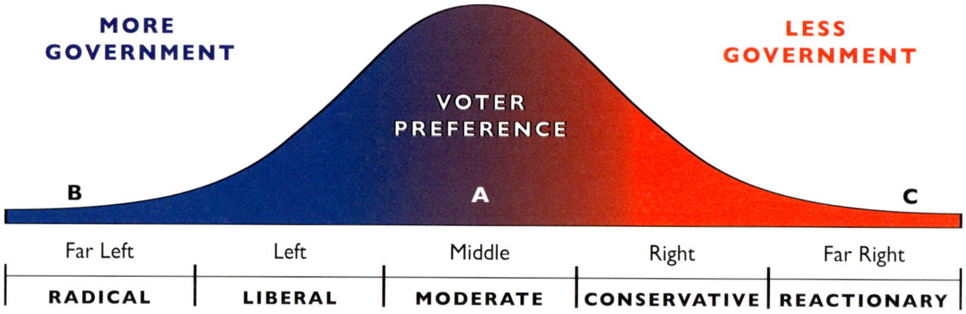

Figure 8.9 Because most voters are moderate, a candidate wants to be perceived by voters as being at position A (a moderate), whereas he would like his opponent to be viewed at position B or C (extremist). Consequently, both candidates try to position themselves close to A.

middle and my opponent is extreme" campaigning holds for Presidential campaigns as well as for campaigns for the Senate and the House of Representatives. Politicians concentrate on the distribution of voters relative to their campaign—at a district, state, or national level for representatives, senators, and Presidents respectively. If politicians know that one group, say the elderly, votes at a higher rate than another group, say young voters, they will tailor their appeal to the elderly and ignore the young, or try to mobilize a group of non-voters to participate in the election.

Superficial Campaigning

If both candidates are portraying themselves as middle of the road and have little chance of characterizing their opponent as extreme, campaigns are likely to be reduced to superficial issues and to personalities. Charisma, personal traits, campaign ads, and small differences in debates are likely to determine victory. (Election folklore has it that Richard Nixon lost the 1960 election to John Kennedy because he did not use make up to hide the five o'clock shadow on his face during a televised debate. Similarly, John Dewey, the Republican candidate in 1944 and 1948, supposedly lost votes because of his mustache. Hair or lack of hair can be an issue in a close election.) In modern elections, campaign consultants and pollsters try to measure voter sentiment and position their candidate accordingly. They also fashion attack ads to portray the opponent as unethical or extreme. Campaigns are often slick, superficial, and dirty when no big issues clearly separate the two candidates. Personality and real or rumored personal failings will receive undue emphasis.

Party affiliation may also play a role in a campaign. If the voters of a particular state or district view a national party as extreme, candidates from that party will have a difficult time trying to separate their own views from the views of the national party. In terms of Figure 8.9, a candidate may consider him or herself at position A, but his or her national political party is perceived by voters to be at position B or C. Such a candidate will probably lose the election, because voters believe the national party will have undue influence on the representative.

Figure 8.10 Senator John F. Kennedy debates Vice President Richard M. Nixon in the first televised debates in 1960.

Influence of Elections and Campaigns on Government

Many Americans are dissatisfied with elections and campaigns. They complain that election campaigns last too long, are too dirty and too expensive, and have too little discussion of important issues. But the more important issue might be to ask how elections affect government. Because U.S. politics is centrist, government policy changes little in response to victory by either party. Government budgets typically grow at about the same rate regardless of the party in power. (There have been important exceptions, however. Consider the enormous government spending increases under President Bush's two terms stemming from the Iraq

War and the War on Terror.) Legislation passed by one regime is rarely repealed when the other party comes to power. Just as campaigns concentrate on the middle range of voter sentiment, government concentrates on middle-of-the-road policies. Only on rare occasions is there a wide policy difference between the major parties. Only rarely does American government shift dramatically from one policy position to another. Changes are small and sporadic.

Many voters, especially voters without strong attachment to either party, respond to the times. If they believe the economy and society are going in the right direction, they will vote for incumbents (people in office) and reward the party in power. If they believe the direction is wrong, they will vote for a change in party. Voters often punish officeholders for corruption or personal failings by turning them out of office. Hence, U.S. elections tend to punish poor performance by political parties or individual politicians.

Finally, because both parties position themselves with their official platforms as moderate middle-of-the-road parties, either party can successfully mobilize the government and the people in times of crisis unless voters become severely polarized into two opposing camps. If there is truly a need for unity, such as World War II or the attacks of September 11, 2001, the government is able to command widespread support as long as the public perceives the need for unity.

Figure 8.11 Suffrage parade, New York City, May 6, 1912. There have been several amendments that have changed the electoral process since the Founding.

Electoral Changes in the Constitution since the Founding

The primary changes in elections have been increases in suffrage (those eligible to vote). The Fifteenth Amendment (1870) prohibited racial restrictions of the right to vote (although that amendment was often circumvented until the Voting Rights Act of 1965). The Seventeenth Amendment (1913) required direct election of senators rather than selection by the state legislature. The Nineteenth Amendment (1920) extended voting to women, though women had been voting in various local and state elections for some time. The Twenty-fourth Amendment (1964) prohibited the use of poll taxes (a payment to vote) in federal elections. The Twenty-sixth Amendment (1971) reduced the voting age from twenty-one to eighteen.

Several other constitutional amendments deal with changes in the Presidency. The Twelfth Amendment (1804) required Presidential electors to vote once for President and once for Vice President in order to avoid the problem encountered in the election of 1800. The Twentieth Amendment (1933) changed the starting date of the term of office from March 4 to January 20 following the election and clarified the line of succession to the Presidency under various scenarios. The Twenty-second Amendment (1951) restricted a President to two elected terms (or one term for a Vice President who served more than two years as President after assuming the presidency of someone else). Finally, the Twenty-fifth Amendment (1967) clarified the process to be followed if a President became incapacitated and unable to discharge his duties.

Why Do Voters Vote?

Between 50% and 65% of the population over 18 vote in Presidential elections. Typically, less than half of potential voters cast ballots in Congressional elections when the Presidency is not at stake. Newspaper editorials and public officials urge people to vote and express wonder that so many people do not exercise their right to vote. To many social scientists, especially those who believe individuals are self-interested and rational, the question is really the reverse. Why do voters vote at all? The voter must find out where to vote, go to the polls, often stand in line, cast their ballot, and travel home. Furthermore, to cast an informed vote, a voter must study some of the issues and the candidates' positions. This effort takes time that could be devoted to some other worthwhile activity. Why do voters incur the costs in time and effort to vote? The motive does not appear to be one of self-interest in the usual narrow sense of that term.

A rational voter realizes that it is very unlikely that his or her particular vote will determine the outcome of an election. The Presidential election of 2000 was an exception if you were an eligible voter in Florida. In that election, a shift of 269 votes in the official and disputed count would have tipped the electoral votes of Florida to Vice President Gore, making him the 42nd President of the United States instead of George W. Bush. Even in this election, which was closer than any election in history, one had to live in Florida and 268 other people had to vote differently in order for one person's vote to determine the election. So it is unlikely that people vote because they think their vote will tip the balance. Rational voters also realize that their vote is unlikely to have an economic effect on them and their family. It would be difficult to find a direct link between personal economic conditions and voting.

2000 ELECTION RESULTS

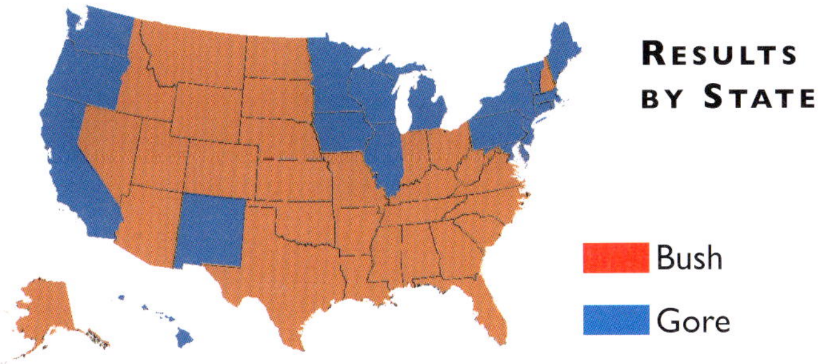

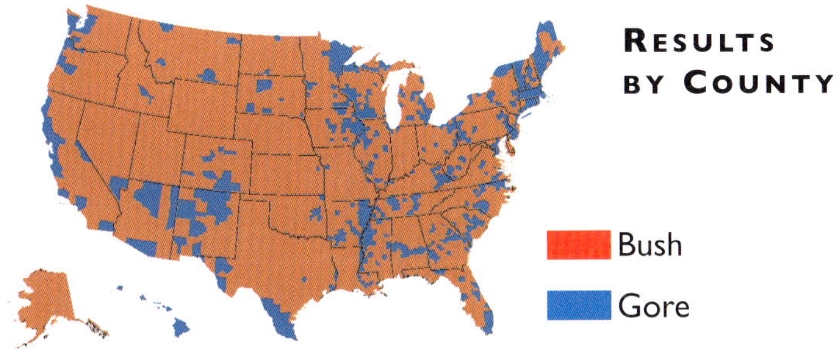

Figure 8.12 Why vote? Below are maps of the 2000 election results by state and county. The final outcome was decided by mere hundreds of votes after millions of voters went to the polls.

Voting appears to be motivated by personal non-economic considerations. Some may vote because they enjoy the process of studying the issues and going to the polls. Others may vote out of a sense of obligation or duty as a citizen. But it is hard to see how a rational person would vote out of self-interest. Perhaps the most compelling reason to vote is philosophical. Following the fundamental principles set forth by the Founding Fathers, voting represents the act of consent by the voters. By participating in the voting process, voters are empowering the selected representatives with their consent to be governed by the laws and actions of those representatives chosen. This essential relationship between voting and consent may be the strongest and certainly the most profound reason for participation in elections and voting.

Timeline

Contextual Events

1789 Judiciary Act of 1789

1791 Bill of Rights ratified

1795 11th Amendment ratified (judical powers)

1803 *Marbury v. Madison* decided—Justice John Marshall establishes Judicial Review

1860 Abraham Lincoln wins Presidential Election

1861 Civil War begins with the Battle of Fort Sumter

1865 Civil War ends

13th Amendment ratified (abolishes slavery)

1868 14th Amendment ratified (equal protection)

1913 16th Amendment ratified (income tax)

Chapter Specific Events

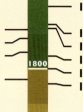

1783 Treaty of Paris signed (official end of Revolutionary War)

1787 Constitutional Convention

1788 Washington becomes President

1796 John Adams wins Presidential Election

1800 Thomas Jefferson wins Presidential Election

1804 12th Amendment ratified (electoral college)

1828 Andrew Jackson wins Presidential Election

1870 15th Amendment ratified (all male suffrage)

1913 17th Amendment ratified (direct senatorial election)

1920 19th Amendment ratified (female suffrage)

1933 20th Amendment ratified (changes when presidential and congressional electees take office)

1951 22nd Amendment ratified (limits presidential terms to two)

1964 24th Amendment ratified (forbids poll tax)

1967 25th Amendment ratified (presidential and vice-presidential succession)

1971 26th Amendment ratified (voting age set at 18)

Key Terms

Jeffersonians
Democratic-Republican Party
Electoral College
popular sovereignty
filtered consent
single representative districts
Hamiltonians
Whig Party
structure
original consent
majority
Federalist Party
Republican Party
loyal opposition
periodic consent
plurality

Key People

Thomas Jefferson
Alexander Hamilton
James Madison
George Washington
John Adams
Andrew Jackson

Questions

1. Why did the founders "reject the idea of direct democracy"?
2. How do the different aspects of government differ in the amount they are filtered from the people?
3. What four constitutional and legal structures affect elections? How does each of these affect campaigns and/or elections?
4. "In the United States, the president is elected separately and independently from the legislature." What are the effects of that election structure?
5. How does the structural attribute of single representative districts affect small third parties?
6. Why are campaigns between candidates in the US often "reduced to superficial issues and to personalities"?
7. Why is there "less party discipline in the US Congress compared to the legislatures of most other democracies"?

We the People of the
insure domestic Tranquility provide for the common

Reflections on the Founding

Although there is no single, specific date for the completion of the American Founding, we have chosen to discuss the Founding through the peaceful transfer of power from the Federalist party to the Democratic-Republican party after the election of 1800. By this time, the basic form and practices of American government and economic life were clearly recognizable even though later changes would modify the Constitution and institutions of government in small ways. Yet, a comparison of today's government and government in 1800 would clearly demonstrate that the basic structure has stayed the same.

To describe the Founding, we laid out the problem of government—the challenge of avoiding both tyranny and anarchy in order to build an orderly, free, and good society. Both the intellectual traditions in the colonies and colonial governments reflected the great political legacy of England. The bedrock of this English legacy is the rule of law and its key elements. We also described the impact of the cultural and religious foundations of the colonies. While there were certainly differences among the colonists, the colonists did share many values that united them at the end of the 18th century—a commitment to self-government, to liberty, and to a modest toleration of diverse views.

The Founding also relied implicitly on a free market economy. While Hamilton and others were willing to see government manage the economy a bit, there was clear consensus that markets would prevail in most circumstances. This general adoption of free markets reduced the scope of government, making freedom a realistic goal.

When we think of the Founding, we are naturally drawn to the principles of the

Declaration of Independence. The ideals of legal equality among individuals and the rights of life, liberty, and the pursuit of happiness animate and energize the Founding. They give purpose and motivation to government. American government has always been compared to these ideals, often found wanting, but always called to a higher level.

Most of all, the Founding is associated with the Constitution and the structure specified in its articles. The three co-equal branches of government with the intricate checks and balances inducing cooperation among the branches is the centerpiece of the actual government designed by the Founders.

A Founding based firmly on the sovereignty of the people could not be complete until the people gave their original consent to the Constitution through ratification. The political struggles over ratification were real with uncertainty as to the outcome. Eventually, ratification came with a clear mandate from the people that the Founding should include enumerated rights to protect the people from this new government and its powers. Thus the Bill of Rights became a part of the Founding.

As the country united around the presidency of George Washington, the structure of the Constitution and the nature of political debate seemed to lead naturally to two contesting political parties. As much as Washington and the other Founders worried about factions, two political parties—the most durable and public expression of factions—emerged as a constant, natural part of the American Founding. In the election of 1800, the country had its first bitterly contested presidential campaign, ending with a peaceful transfer of power.

The transfer of power in 1800 just happened to create the case that spawned the last great element of the American Founding—judicial review. *Marbury v. Madison* provided John Marshall with the opportunity to assert the primacy of the Supreme Court as the final interpreter of the Constitution. With that element in place, the Founding could be considered largely complete.

As you review and reflect on the American Founding, you might want to answer three basic questions:

1. What are the important ideas drawn from political philosophy, economics, and the historical experiences of the colonies that formed the intellectual foundation for the Founding?
2. What are the structural devices, economic, governmental, and political, that form the structure of the Founding?
3. What general outcomes were expected to be the fruits of the constitutional structure the Founders designed?

The American Founding should not be viewed like a great work of art—Michaelangelo's *David* or Rembrandt's paintings of the apostles—finished, complete, and perfect. Rather the Founding both affects American experience and is affected and often changed by that experience. The rest of this book examines key changes, important periods, and crucial issues that relate to the American Founding as we experience it today. In the coming chapters, we consider the massive changes that have occurred since 1800, historical events like the Civil War that changed the Founding and constitutional structure permanently, consistent trends like the growth of government and increased political participation by the masses, as well as recurring issues such as war and international conflict and the proper role for judicial review. By examining the Founding through time, we understand it better and we understand how its legacy changes. Viewing the Constitution and other aspects of the Founding under the intense pressure of events and crises divides the essential from the transitory in the American Founding.

Chapter 9

The Founding and Historical Change

Abstract

Since the Founding, America has been experiencing what might be called "deep change." This chapter describes some of the changes that have influenced the interpretation and operation of the the Constitution and the Founding. Each of these changes required the Founding and Constitution to adjust to new circumstances. Ensuing chapters examine the effect of these changes on the Founding and their effect on modern life.

By the time of Jefferson's election to the Presidency in 1800, the Founding was more or less complete. The American Revolution had succeeded. A national government was in place and operating. Americans were coming together and identifying themselves with a fixed set of ideals. Party politics, while still not wholly legitimized, were functioning well enough to transfer power peacefully to a loyal opposition. The Bill of Rights had been adopted. And while judicial review was not yet an official component of the system, it would be soon.

The events of the Founding occurred at a time of important historical changes in the Old World as well as in the new America. Many of the Founders saw these major social, political and economic changes as working in their favor, not against them. They pictured themselves as being on the cutting edge of a worldwide revolution that would bring their self-evident truths and popular government into full flower.

Deep change

Fundamental alteration in the way life is lived. Usually unnoticed because it is a slow process.

To some extent, history has justified their optimism. On the other hand, history has also brought changes of a magnitude and sweep that few could have foreseen. Many of these changes have impacted the Founding in ways the Founders could not have imagined.

Deep Change

The term "**deep change**" denotes some fundamental shift taking place in human society. Many societies have functioned for hundreds, even thousands of years without experiencing any major "deep" change. American society, on the other hand, began to undergo deep change almost from the beginning. It was brought about by many factors: wide open spaces, free land to the west, abundant opportunity, an eager and energetic populace, rapid technological change, and not least, a Founding that really worked.

Change from early colonization to the Founding was relatively slow. If Captain John Smith of Jamestown Colony fame stepped into a time machine and visited Virginia in 1807, 200 years after first setting foot on its shores in 1607, he would have found a world that was not all *that* different. He would encounter Englishmen living in roughly similar circumstances, performing roughly familiar tasks, transporting their goods, and communicating with each other more or less as they had done back in his time. Now consider the case of Thomas Jefferson, whom Smith might meet in 1807, borrowing Smith's time machine and going forward two centuries to visit the Virginia of today. How different would be Virginia—and the rest of the United States—in the year 2007! Such are the results of deep change.

In the following sections, we consider some of those deep changes and how they might conceivably impact the legacy of the Founding.

People

Population

The size of the United States, from 1776 to today, grew from 820,000 square miles at the time of the Founding to 3,537,000 square miles, with some of it beyond the Caribbean and South Pacific, and with Alaska separated from other states by Canada. Thirty-seven new states would be added to the Union, more than a dozen of which would have a larger population than the original thirteen combined, and one of which would exceed that population nearly tenfold.

Between 1790 and 2003 the U.S. population surged from just under four million to just over 290 million—an increase of seventy-three times. In a moment of exuberance Benjamin Franklin supposed that the American population might someday reach the astonishing total of 25 million. He undershot the mark by a factor of twelve.

This massive increase in population was generated through three forces:

- An increasing life expectancy
- A high birthrate, especially in the nation's first 100 years
- A constant flow of immigrants

Health and Longevity

The chart in Figure 9.1 shows the dramatic increase in life expectancy since the Founding. Infant mortality has fallen from rates as high as one out of five births to about one in 150 births. The average lifespan (life expectancy) has doubled from about 40 years to nearly 80 years. This change in lifespan affects nearly every aspect of life: birthrates, retirement, health care, size of government, living arrangements, recreation, and politics, to name just a few. In one sense, the increase in lifespan comes slowly, with increases of one or two-tenths a year, but the cumulative effect over time is profound. Such is the nature of deep change. It is so slow it is hardly noticeable, but its effect is powerful.

The increase in longevity is the result of increased income, better public health systems, better sanitation, cleaner water sup-

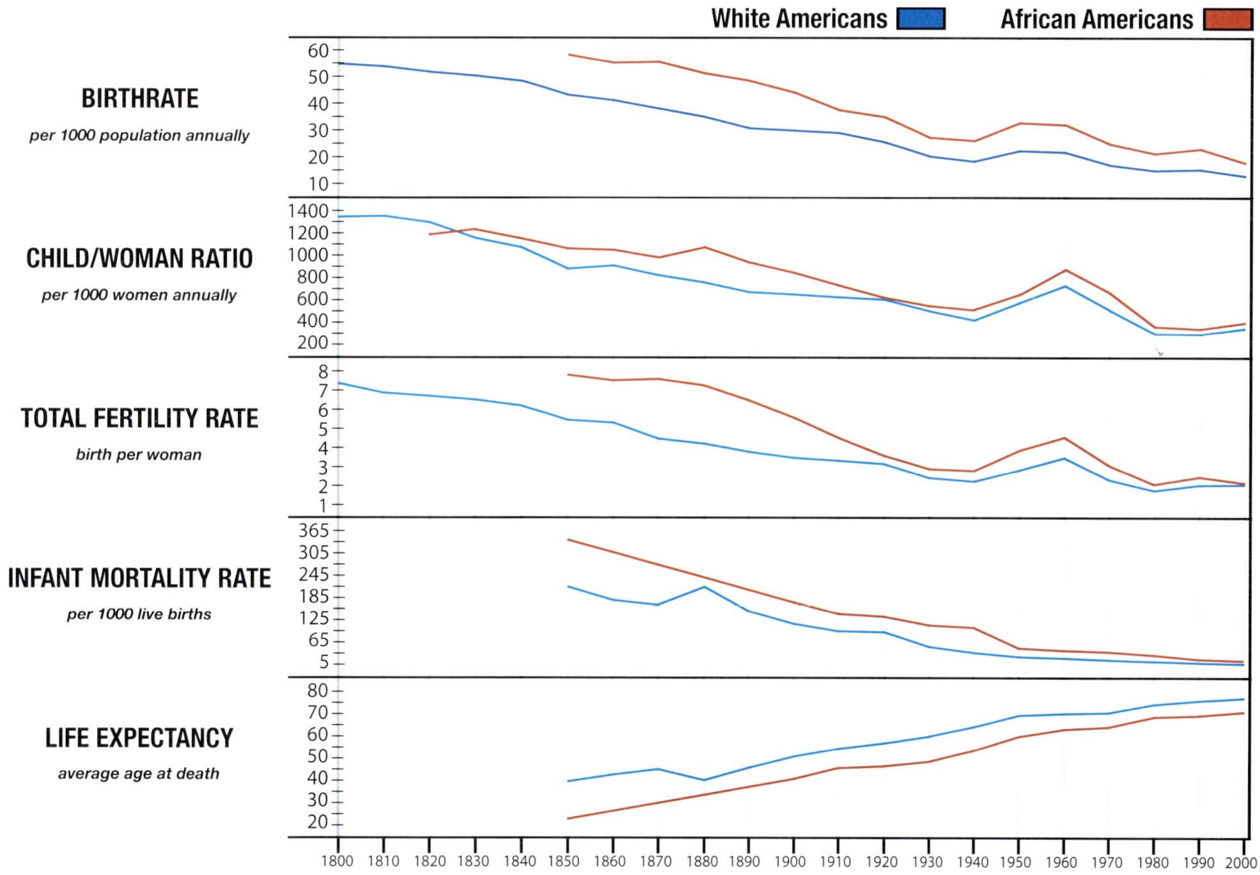

Figure 9.1 Fertility and mortality in the United States 1800–2000.

plies, and better health care. Early in our history better nutrition and better housing were the principal sources of longer life. Late in the 19th century, many cities started bringing clean water and milk to their inhabitants while transporting sewage away.

In the years just preceding the 20th century, medical practitioners began to accept the "germ theory" of disease, leading to a successful assault on infectious disease. Eight major diseases—tuberculosis, typhoid fever, scarlet fever, diphtheria, whopping cough, measles, small pox, and polio—were virtually wiped out. Their demise was affected by the steady march of medical research and discovery, which brought about antibiotics and sulfas, x-ray diagnostics, surgical anesthesia, insulin, aspirin, vaccination, open-heart and neurosurgery, organ transplants, hip replacements, ultrasound, electro-magnetic imaging, radiation and chemotherapy, and the use of highly sophisticated computers and innumerable drugs. Where there were only two medical schools in the United States in 1790, by 2001 there would be 125.

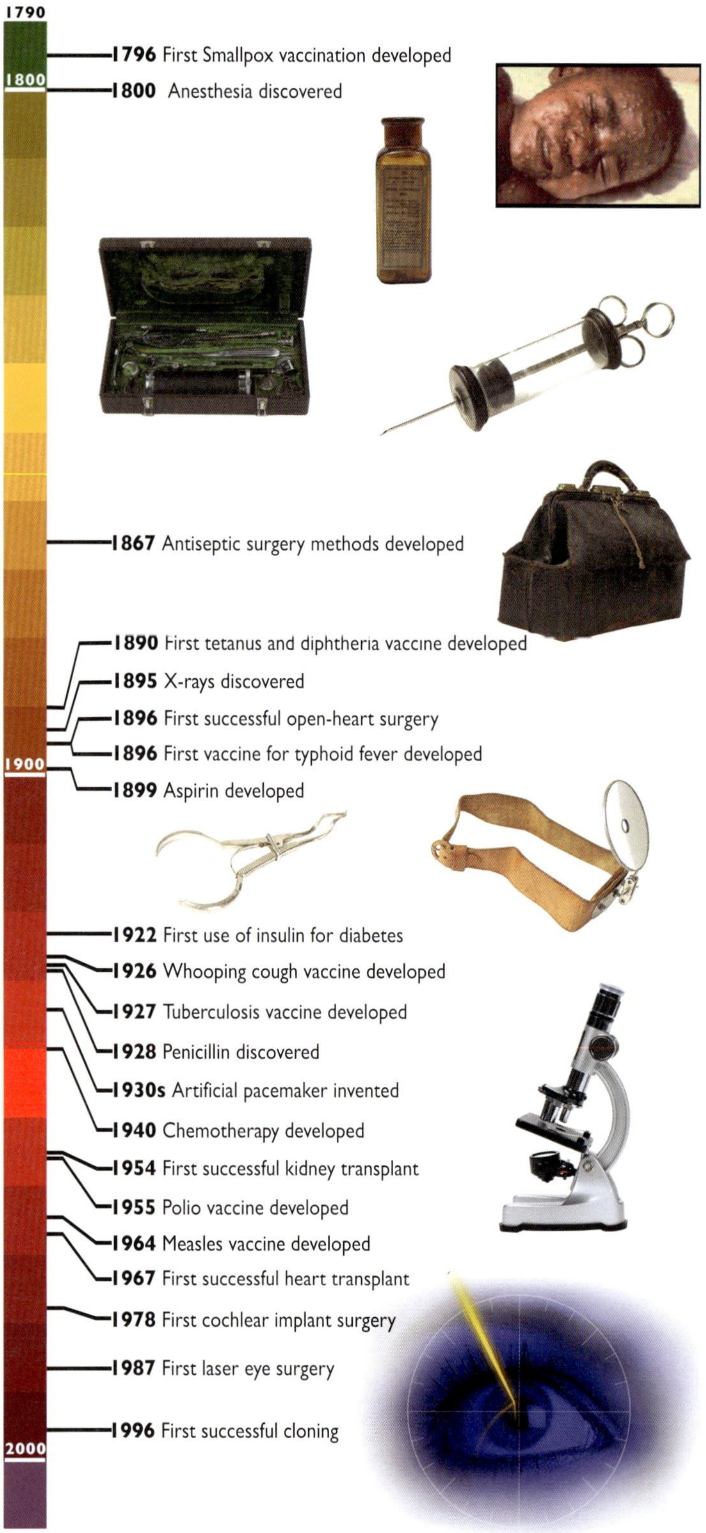

Figure 9.2 Timeline of medical advances.

Birthrate

As life expectancy increased, the birthrate began a steady decline from an average of 7 children per woman in 1800 to less than two children today (see Figure 9.1). This decline in the birthrate came partly

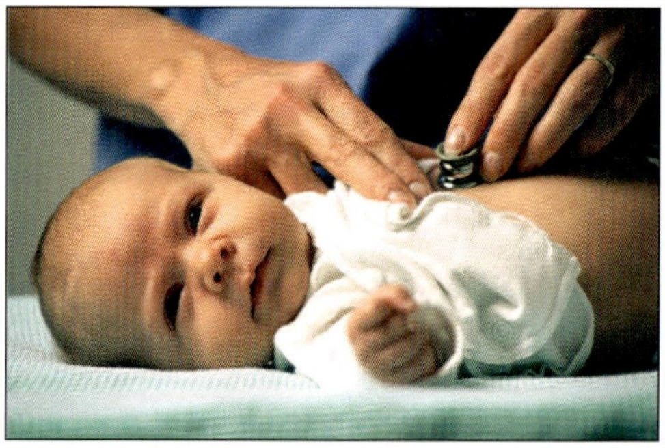

Figure 9.3 Advances in nutrition, hygiene, and medicine have greatly reduced the infant mortality rate.

because many families moved from farms to cities, where many children per family were not as necessary to help support the family. Other factors in the declining birth rate included the rise of inexpensive public education, with families emphasizing the quality, rather than quantity, of their children, the development of better means of contraception, and the increase in the percentage of women working outside the home. There were working women in 1900 but only 5.6% of them were married. By 2002 some 60% of American women worked outside the home—and 61% of those were married.

Families

The American family changed in response to the fall in fertility and the increase in longevity. In 1790 there had been nearly six people in the average family. By 1900 the family had shrunk to 4.6 persons, and by 2000 shrunk further to 2.59. In 1790, there were very few older couples living alone. By 2000, 26 percent of the population was living alone and 36 percent of households had no children under 18 years of age. The U.S. had shifted from a society with children in nearly every household and households with more than two generations to a society of small nuclear families and "empty nesters."

Moreover, mom and dad were less likely to be married in the twenty-first century. As recently as 1960, cohabitation affected less than one quarter of one percent of the population. By the turn of the current century, however, some five million Americans were "living in sin," as their parents might have put it, a stunning seven-fold increase. Equally stunning was the increase of illegitimate births. On the eve of World War II, there were 90,000 illegitimate births in the United States, constituting 3.8% of all births. By the year 2001, births to unwed mothers had skyrocketed to 1,349,249 and now account for 33.8% of all births. As noted by our change

Figure 9.4 Family size has decreased significantly over the last 100 years.

in terminology, such births are no longer called "illegitimate."

Before 1972, abortion was illegal throughout most of the United States. That year a Supreme Court decision, *Roe v. Wade*, legalized the practice and the number of abortions surged to 587,000 a year. That was only the beginning. By the turn of the century, abortions would climb to 1,313,000. The 42 million fetuses aborted since the *Roe* decision would more than populate Spain.

Immigration

At the time of the Founding, 92% of the American populace (not including the nearly 700,000 African-American slaves) was of British stock, and the remainder hailed either from Germany or the Netherlands.

Beginning in the 1820s, immigration became an important new theme in American life. Newcomers arrived by the tens of thousands in that decade and by the hundreds of thousands after the last great Irish

Figure 9.5 Four immigrants and their belongings, on a dock, looking out over the water, 1912.

potato famine of the 1840s. The steady flow of immigrants swelled into a torrent by the 1880s as the United States became an industrial nation. The flood crested in the decade between 1905 and 1915—during which no less than *ten million* immigrants arrived.

The "New Immigrants," as these latter were called, were no longer mainly from the British Isles or Northern Europe but from far to the south and east. Most of them did not speak English and very few were Protestant. Many were not even Christian. Jewish immigrants alone arrived in the millions.

Immigration slackened in the 1920s after government legislation nearly throttled it, but it recommenced after the Vietnam War. Between 1990 and 2000, the number of immigrants arriving in America was over forty-six million. Here is the breakdown by continent:

Europe	8,194,000
Africa	3,549,000
Asia	419,000
Americas	34,202,000

Figure 9.6 New York—Welcome to the land of freedom—*An ocean steamer passing the Statue of Liberty: Scene on the steerage deck. Staff sketch from Frank Leslie's illustrated newspaper, 1887 July 2.*

From the time of Emma Lazarus's famous poem found on the base of the Statue of Liberty—"Give me your tired, your poor, / your huddled masses yearning to breathe free"—the United States was known as a nation of immigrants. By the dawn of the twenty-first century, it was a nation of the entire world.

Figure 9.7 Pioneers crossing the plains. Based on an engraving by H. B. Hall, 1874.

Migration Westward

The sparsely settled frontier of Jefferson's day would transform dramatically into the "Heartland." Jefferson's world would shrink in the big picture to become just "the Eastern Seaboard." As shown in Figure 9.8, millions of Americans moved west to seek land and new opportunity. They tended to move more or less straight west. For example, there were just over 18,000 persons living in Louisiana, Texas, and Arkansas in 1850 who had migrated from New England or the large Middle Atlantic states like New York and Pennsylvania, but over 200,000 persons who had migrated from the Carolinas, Georgia, and Alabama. The same held for Northern states strengthening the North-South divisions and weakening any East-West divisions.

Settlers began making their way toward the west as soon as the best lands were taken up in the tidewater areas. They moved up into the Piedmont region, an area of low

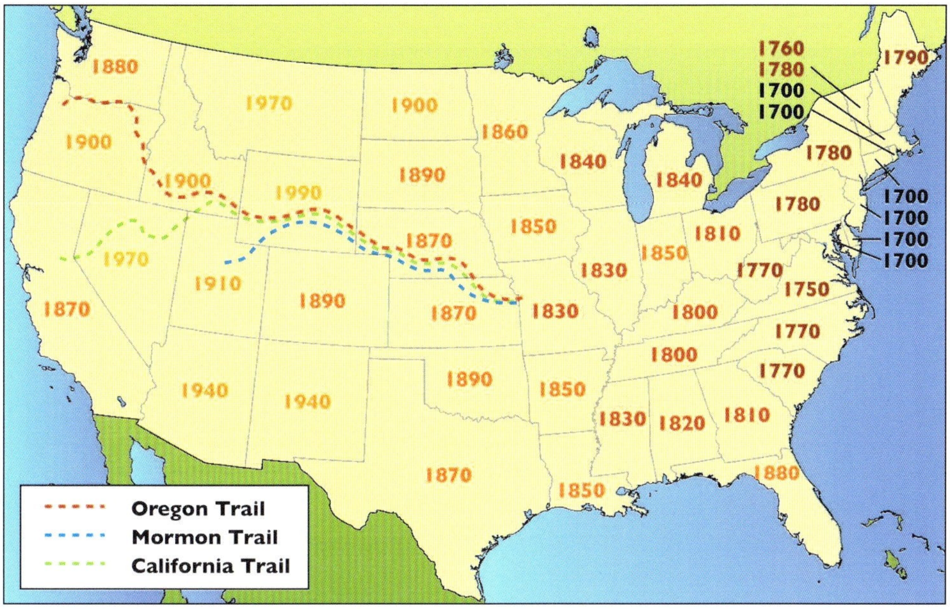

Figure 9.8 Westward expansion of the U.S. population. The dates indicate the decade when each state reached an average population density of five people per square mile.

rolling hills, and then into the mountain valleys beyond the reach of easy river navigation. They soon started to settle in areas that had never been surveyed, becoming "squatters" without formal ownership of the land. Easterners referred to the hinterland as the "backcountry," where life was much different from that of the seaboard. For one thing, settlers in the backcountry seemed much freer of social restraint, ignoring many of the affectations and mores of polite society. The "backcountry" settlers often dressed in buckskin, plaited their hair and oiled it with bear grease, and in the winter commonly sheltered with their livestock. Some drank to excess, brawled violently, engaged in blood feuds, and now and then touched off Indian wars. They wed at a younger age, had larger families, and sometimes dispensed with marriage altogether, "for want of a preacher." Social class was more fluid among them and social position of less account. In their world, actual accomplishments counted far more than birth or breeding. Such characters were the inspiration for author James Fenimore Cooper's backwoods heroes. A few decades later, the American cowboy replaced the coonskin-capped trapper as an American icon.

Eventually, western towns and small cities produced the trappings of civilized society and competed with other areas for population growth. There was a never-ending campaign of "**boosterism**" to make the west seem more civilized, more laden with opportunity, and more of a Garden of Eden than it could ever be. For 100 years, there was always a frontier somewhere in the West. The impoverished, the discontented, the restless could always pack up and move towards the sunset. This frontier, whether it was the reality of pioneer life or the dream of a new beginning, dominated American democracy and much of its economy until the frontier "closed" around 1890. Then the gravitational pull shifted from a movement westward to a movement to the city.

Boosterism

Promoting one's town or city, sometimes in an excessive or exaggerated manner, in order to increase both its quality and its public perception.

Urbanization

At the time of the Founding, the largest cities in America were Philadelphia (42,440), New York (33,130), and Boston (18,058) respectively. By the year 2000, New York had a population of 8,008,000. The eastern seaboard from Boston to Washington, D.C. had become one sprawling urban corridor of 40 million people by the year 2000.

Whereas in 1790 there were eight cities with a population of 7,000 or more, at the turn of the 21st century there would be 1,800 cities in that category, seventy-five of them with a population greater than 250,000. In two hundred years, America switched from 95% rural to 75% urban. Thomas Jefferson, for one, detested cities and everything about them, calling them "sores on the body politic." Such relentless urbanization would have taken him completely by surprise.

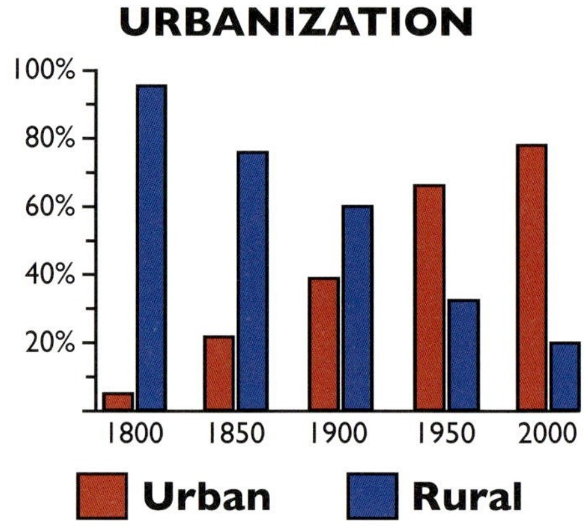

Figure 9.9 The percentage of the U.S. population living in urban versus rural areas from 1800 to 2000.

City life was very different from the rural life those early Americans had known. Cities were cramped and crowded, with little in the way of amenities, and everyone had to live within trolley distance of their jobs. City life was often squalid, too, especially for newly arrived immigrants, with jumbled tenements and cold-water flats by the square mile, and with few established public standards for citizens' health and safety. It was impersonal. Diverse ethnic and cultural groups were thrown together cheek by jowl, each group clinging to its native language and Old County culture as its members tried at the same time to assimilate and learn the ways of the New Country. City life was also dangerous if one lived in the "working class" parts of the city. Gangs ruled many neighborhoods, battling one another for turf and creating over time an underworld that spawned sophisticated criminal enterprises that haunt us even today. All manner of thugs and "bully boys" walked the streets, which also teemed with flim-flam artists, swindlers, and racketeers. Runaway vehicles, electrical hazards, open sewers, and dogs running in semi-wild packs added to the sense of menace.

Yet urban life was also exciting. The bustle of the streets filled people with a sense of motion and purpose. The city became the

Figure 9.10 There is a great difference between the pastoral life that most Americans were living in Colonial times and life in a modern city. Urbanization has certainly effected "deep change" in American life and culture. (Left: Winter in the country. A cold morning. *Lithograph by Currier & Ives, 1863.)*

cutting edge of American culture, with new fashions, new music and art, new idiomatic expressions, and new developments in lifestyle exported to the hinterland. What would be called a sense of "cool" emanated from places like Harlem in New York and the Chicago South Side. Cities offered attractions unavailable to those living out on the farm. For the blue-collar mechanics, factory workers, seamstresses, and other people who labored primarily with their muscles, there were saloons and taverns, cafes, restaurants, deli's, brothels, melodeons, burlesque and vaudeville theaters, penny arcades, midways, drugstores, ball parks, beer gardens, skating rinks, prize-fight arenas, gyms, playgrounds, and after the turn of the century, movie palaces. For the upper classes, those who owned the factories and businesses, there were luxury hotels, exclusive clubs, resorts, theaters, opera houses, race courses, museums and galleries, libraries, and educational establishments of several kinds. But there were costs to urbanization as well.

Pollution

As cities began to grow in the 19th century, the country faced new challenges. The cities, proud of their new trolley systems but still largely unaware of what caused disease, were one enormous breeding ground for infectious disease. Epidemics of typhoid, influenza, and cholera were common. As coal replaced wood as the primary method of heating homes and offices in the large cities, air quality plummeted. Scientists started accepting the germ theory of medicine, and became more sophisticated about the root causes and spread of disease. The question of who was responsible for public health assumed greater political and medical importance.

Crime

To some extent, Jefferson may have been right about the cities, since crime rates rose with urbanization. On the other hand, the sources of cultural pluralism, materialist values, and the tradition of violence that contributed to rising crime could be traced back to colonial times. Both metropolitan and rural homicide rates rose steadily through the early 20th century until the start of WWII, when the nation's Great Depression economy was jump-started by industry moving quickly into a war-equipment economy. Homicide rates declined through World War II and the 1950s, and then began to rise to record levels before falling during the 1990s as the population grew older. Rape and sexual assaults rose throughout the 20th century. During the year 2002 there were nearly twelve million crimes, including 1,426,000 violent crimes, committed in the United States. Even so, the crimes of choice during the latter part of the 20th century were increasingly white collar: forgery, embezzlement, fraud, and identity theft.

Figure 9.11 Not all change is good.

Figure 9.12 Increases in both population and crime rates have swelled the number of people in prisons.

With the burgeoning of crime came a swelling of the national prison population. On June 30, 2002, it topped the dubious two million mark—roughly the population of Utah in 2002. At the same time, the treat-

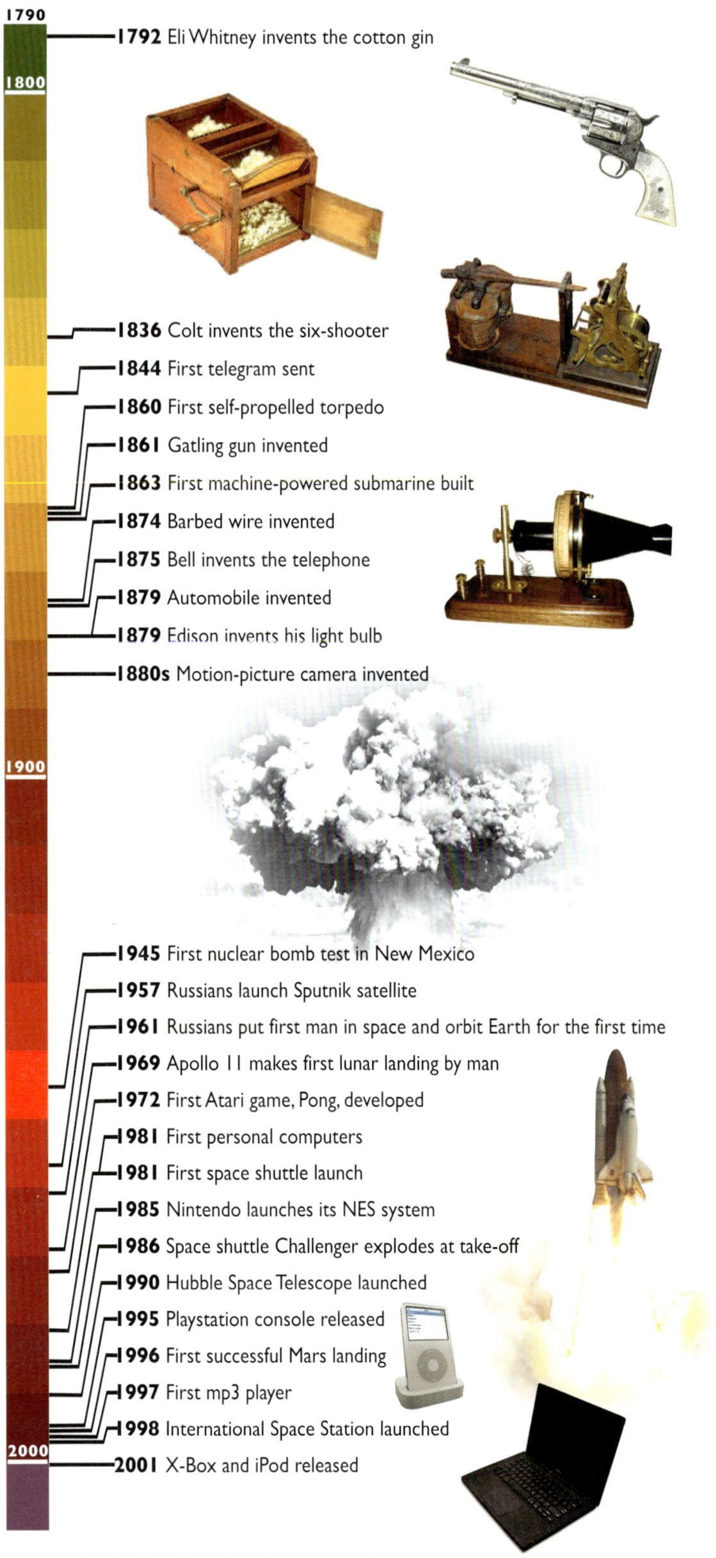

Figure 9.13 Timeline of technological advances.

ment of prisoners and those accused of crimes but not yet convicted came under a new constitutional scrutiny by civil rights groups. In the dark world of crime and criminals, the protections of the Bill of Rights would take on stark new meaning.

For the American people, the change from 1790 to today encompassed longer life, smaller families, migration for many and life in the city. Could a constitution formed and ratified by a rural society of large families who for the most part lived within 100 miles of the Atlantic Ocean withstand the demographic tides that moved the population over the next 200 years?

Technology and Economic Growth

Americans, observed Alexis de Tocqueville, took to science and new technology like ducks to water. It was the applied sciences that Tocqueville had in mind, not the theoretical sciences that had been so conspicuously advanced by Europeans. Americans simply wanted science to deliver more efficient machinery and higher crop yields.

Americans proved themselves adept at technological development. In the early years of the republic, they made astonishing breakthroughs in the design of textile machinery, woodworking machinery, steam boats, sailing vessels, locomotives, firearms, and a host of other devices that were faster, tougher, smaller, and more productive than the devices they replaced. Of particular note were innovations that made the production process faster and more efficient. The most important of these innovations was the concept of equipment production uniformity, i.e., interchangeable parts for manufacturing processes. When each close-fitting component of, say, a rifle no longer had to be handcrafted specifically to work smoothly with other parts of that same rifle, it was possible

Figure 9.14 This lithograph by Currier & Ives from 1876 depicts some of the major advancements developed during the 1800s. All would have tremendous impact on our standard of living.

to separate the rifle manufacturing process into discreet segments and arrange them in an orderly fashion. It was only a short step from this to the moving assembly line.

Some early American inventions literally changed the world. Eli Whitney's cotton gin helped entrench slavery and enhance Southern power. Samuel Morse's telegraph, Colt's six-shooter, and Glidden's barbed wire all had a revolutionary impact on economic development. Later, the electric light, the telephone, the radio, and the motion-picture camera had a similar influence on domestic life and introduced a host of technical jobs to the economy. By the turn of the century, the fragile horseless carriage was being perfected into the automobile, and only computers would have a more transforming effect on American culture in the 20th century. In the 1930s and 1940s, development of hybrid seeds increased agricultural yields, dramatically shifting even more people off small farms.

If technology brought blessings, it brought curses as well. The most conspicuous of these were improvements in the art and science of warfare. Spurred on by the Civil War, inventors began perfecting such weapons as the machine-gun, the self-propelled torpedo, the long-range howitzer, and the diesel-powered tank. By the mid-1920s, improved military aircraft were conclusively proving that they could not only scout enemy trenches, they could sink battleships. The massive aircraft carrier, designed to haul relatively short-range aircraft to within bombing range of the enemy, followed this discovery—and was itself eclipsed by missiles that could be fired from submarines. The ultimate military breakthrough of the 20th century, of course, occurred when a nuclear blast lit up the desert sky at

Alamogordo, New Mexico, on a summer's dawn in 1945. Nothing before or since more convincingly illustrated the problems and promises of change.

On the whole, change steadily improved the standard of living of the average American. By 1900, ordinary people had conveniences that only the very wealthy might have enjoyed a hundred years earlier. By 2000, the living standards of all had far outstripped the lifestyle of the wealthiest plutocrat of Jefferson's time.

Transportation

In 1790 a trip from New York to Boston required four hard days on rough roads by stagecoach. In 1844 the same trip took eleven, sometimes smoky but otherwise comfortable, hours by train. In 2005 that trip could be completed over a lunch hour on a comfortable, air-conditioned airplane. At the beginning of the 20th century, the dramatically more powerful and convenient gasoline-powered automobile began to replace Jefferson's horse and buggy. One hundred years later there would be 225 million faster, unimaginably more comfortable gas-powered vehicles on American roads.

Figure 9.15 Meeting of the rails at Promontory Summit. The completion of the intercontinental railroad on May 10, 1869, further opened the way for westward expansion and greatly reduced the cost of moving goods and people between the east and west coasts.

On the sea-lanes, dramatic advances in transportation lowered ocean shipping costs. Steamships were followed by ocean liners and freighters, which were supplanted by supertankers and containerized shipping. A container loaded with delicate electronics in Japan, Korea, or China now arrives in perfect condition in small Provo, Utah, having been efficiently transported by truck, train, and ship. Crude oil is routinely sucked from deep below a barren desert in Saudi Arabia, piped into a supertanker at anchor many miles away, and transferred to a refinery in Texas. The cost of shipping oil thousands of miles only adds a few pennies to the consumer's cost of a gallon of gasoline.

Communication

Technological change separated communication from transportation. Many business, personal, and government communications continue to be made by mail, the telegraph, transoceanic cable, and the telephone, but the development of computer networks in the last two decades has made communication faster and cheaper than ever before. By the beginning of the 20th century, all continents except Antarctica were tied together electronically so that communications took only minutes. By the end of the 20th century, satellites and the Internet had made the communication of voice, words, and pictures virtually instantaneous.

Today's cable-television news networks offer major competition for the former "big three" TV networks (ABC, CBS, and NBC), which had themselves earlier replaced the radio as the prime source of news and entertainment. Radio, a sensational new part of American culture when it arrived in the early part of the 20th century, offered faster news than even the most efficient metro newspaper, which had earlier replaced the town crier for the whole world. Now, a few years into

Figure 9.16 Technology has made instant global communication a reality.

the 21st century, cell phones the size of small candy bars instantaneously connect people in even the poorest parts of the globe.

Wealth and Income

Improvements in transportation, communication, production technology, healthcare, and education generated constant economic growth. Steady small improvements in production and transportation would ultimately have great consequences. Ever-increasing automation in the manufacturing process have caused manufacturing costs to be pushed lower and lower, and with ever-greater efficiency in the movement of goods, the area of profitable exchange has grown larger and larger. At the time of the Founding, most profitable exchange was limited to an area about the size of a modern county. Every improvement in transportation—canals, steamboats, railroads, internal combustion engines, airplanes—expanded that area of exchange. By the end of World War II, the area was nearly encompassing the entire globe. Today, it is a global economy.

Agriculture and Manufacturing

Americans made the most of mechanization in agriculture and new technology in manufacturing. The reason that just two percent of the population could raise an abundance of food for the remaining 98% is precisely because of machines. The iron plow, the mechanical seed drill, the combine harvester, and motorized tractor were only a few of the developments that made modern life and culture possible. In 1830 it took 300 labor hours to produce one hundred bushels of wheat, but by 1987 the same hundred bushels required just three hours. The same change was taking place in manufacturing. By the 1920s, the Ford Motor Company could comfortably produce a Model T every 45 seconds on its factory assembly lines.

The result was economic growth on an unprecedented scale. In 1790 Americans averaged $916 (expressed in year 2000 prices) in per capita income. By 1900 that amount, adjusted for inflation, had grown almost five fold to $4,943. And a century after that, per capita income would reach $34,800, a thirty-eight-fold increase. There were still rich and poor in the United States, to be sure, but both rich *and* poor were a lot wealthier. Most importantly in our capitalist economy, this rising income smoothly translated itself into rising consumption.

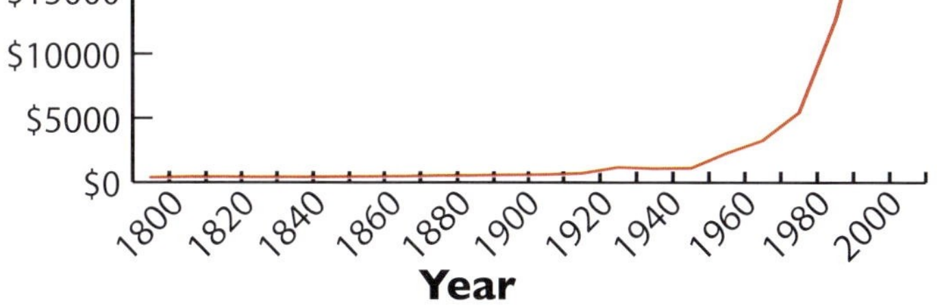

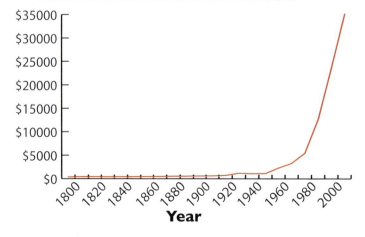

Figure 9.17 The increase in technology, communication, and transportation has also had a significant effect on the average income of Americans.

Food and Nutrition

The American diet was transformed over time as food processing improved. To the pioneer diet of dairy products, grains, and beef or pork, much of it salted or dried to last into the winter, were added fruit, vegetables, poultry, and fish—much of it canned, frozen, processed, pre-packaged, or consumed outside the home. Technology now makes it possible for a Kansas farmer to dine in the evening on Alaskan salmon caught that morning. And "traditional" American cooking that represented primarily English culture in early America has been tastefully supplemented by French, German, Italian, Mexican, Indian, Chinese, Japanese, Korean, Thai, Vietnamese, and other exotic cuisines. But cheaper, more abundant foods, made possible by the same technology that reduced both the number of work hours and the hard labor involved with most jobs, had its cost. Rates of obesity in the past half-century have grown alarmingly.

Housing

Part of the Americans' rising incomes was spent on more living space. A typical family of the founding generation lived in a log home of 500 to 1,000 square feet with a few windows to let in light and a fireplace for cooking and heating. By 1900, wood siding and bricks were common exteriors. Soon thereafter most houses had such amenities as electricity, indoor plumbing and central heating. The advent of convenient, low-cost air conditioning units after World War II enticed many to move south and west to hotter areas where homes could be made comfortable with the cooler indoor air. Transportation improvements made living further from one's work more feasible, and an increase in living space per person fol-

	Goods or Services	1900	1920	1940	1960	1980	2000
Household 	Running Water	24%		70%	93%	99%	99%
	Indoor Toilet	15%	20%	60%	87%	98%	99%
	Heating (oil, gas, electricity)	1%		22%	82%	95%	98%
	Electric Lighting	3%	35%	79%	99%	99%	99%
Appliances 	Refrigerator	Na	1%	44%	90%	100%	100%
	Washing Machine	Na	8%		73%	73%	79%
	Dishwasher						53%
	Microwave Oven						92%
Transportation	Automobile Ownership	0%	26%	60%	75%	84%	90%
Communication 	Telephone	Na	35 % *	37%	78%	93%	94%
	Cell Phone					5.2 mil	56%
	Computer						53%
Recreation 	Radio	Na	<10%	46%	95%	99%	99%
	Television			9% **	87%	98%	99%
	VCR					1%	85%

*1922 **1950

Figure 9.18 Distribution of consumer access to selected goods and services.

lowed. Segregation by income and race developed as cities spread out.

Consumer Goods

Improvements in manufacturing generated a never-ending stream of new products that became cheaper and more accessible within a few years of initial production. For example, a worker in an automobile plant in 1908 had to work nearly two years to purchase one of the plant's own products. But with the introduction of the continuous assembly line in 1910, a worker could buy a much better car in 1929 for about four months salary. As illustrated by Figure 9.18, new goods spread quickly through the whole society.

Leisure

In 1790 the vast majority of Americans had very little time and still less money for "frivolous" pursuits. Children commonly began adult labor in their early teens and worked until the day they died. Sixty to seventy hours of work a week was common for most factory laborers, while farmers chalked up ninety or a hundred. By the 1920s the workweek had declined to forty-five to fifty hours and by the end of the 20th century it was down to 35 to 40 hours. Retirement age declined as well. Americans not only had leisure time, they had the wherewithal to make use of it. In 2002 they spent:

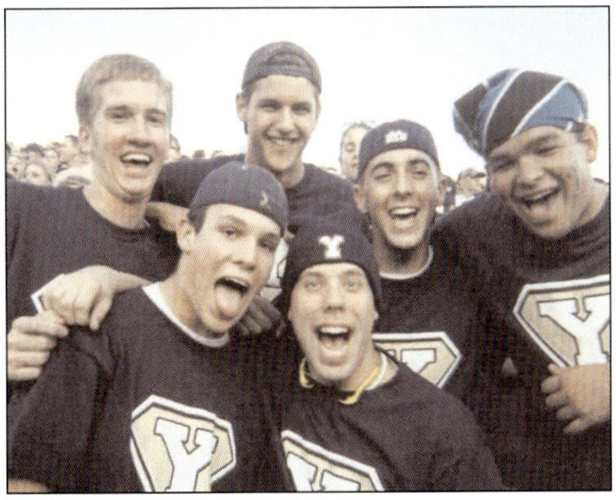

Figure 9.19 Even "impoverished" students enjoy leisure activities far beyond what their ancestors did 200 years ago.

- $9.6 billion on movies
- $35.8 billion on books
- $44.2 billion on computers
- $76.7 billion on audio and video equipment

Their total expenditure for arts, entertainment, and recreation came to a whopping $137 billion, including $20 billion for spectator sports, $8 billion for amusement parks, and $15 billion for fitness centers. They even scraped up $24 billion to feed into slot machines or lose at blackjack. Americans belonged to 3,700 country clubs, splashed in 5 million swimming pools, and sailed 12 million boats.

Education and Ideas

The rise in family income allowed families to invest more in the children's education. At the time of the Founding there was little formal schooling in the United States. The "common school," what we now call elementary school, was a neighborhood fixture in New England at the time of the Founding, while plantation owners in the South either hired live-in tutors or "home-schooled" their children. In frontier New York, Pennsylvania, and other areas on the edge of the wilderness, if the family was academically inclined, schooling might consist of reading the Bible and whatever newspapers or religious tracts might become available. Children might learn math by calculating cleared acreage or bushels of produce.

Once again, change was deep and dramatic. Beginning with Jackson-era initiatives, Americans started putting up one-room schoolhouses. In 1900, 6.3% of the U.S. population had completed high school—

considered progressive at the time. Then along came the tech-minded 20th century. In 2004, 85% of adults were high school graduates and 28% had received a college degree. Education had become big business. At the time of the Founding there were 14 colleges in the United States, graduating 170 students annually. By 2001 there would be 4,197 institutions of higher learning, with a total enrollment of 16.4 million. Half of all high school graduates would now attend college. Higher incomes and white-collar occupations were increasingly associated with college degrees.

The content of education changed as well. The Greek and Latin of Jefferson's day were replaced by math and science, and by an ever-increasing emphasis on technology. And with it came a new understanding of the world. The theory of evolution advanced by **Charles Darwin** and his many followers was to have broad social and cultural—not to mention religious—implications. In the mid-19th century, economics theorist **Karl Marx** developed what would become the most successful attack yet on the concept of a market economy. His ideas, centered on the immorality of the market distribution of income and the inevitability of historical forces, generated the powerful political movements of socialism and communism. **Sigmund Freud**, father of modern psychological sciences, argued that man was in the grip of psychological forces beyond his individual control. Indeed, all the social sciences' theories and practices weakened the premise of individual autonomy and responsibility assumed by the Founders. The work of **Albert Einstein** pointed to radically revised notions of time and space and matter.

Figure 9.20 How much has your own education been impacted by changes in economy and technology?

Charles Darwin

1809–1882
English naturalist known for writing *The Origin of Species*, in which he proposed the idea of natural selection as the primary means of species diversity.

Karl Marx

1818–1883
German philosopher who wrote the Communist Manifesto, championing communism and socialism and attacking market economies.

Sigmund Freud

1856–1939
The father of psychoanalytical psychology, Freud's theories were based on the idea that people were influenced in their behavior by subconscious and external factors beyond their control.

Albert Einstein

1879–1955
German-born theoretical physicist who is most known for his Special and General Theories of Relativity and the formula for mass-energy equivalence, $E=mc^2$.

Above all, science grew in overall power to rival, if not dominate, the attitudes and values of education, which in turn shaped the elites of society.

Growth of Government

All of these changes played into the growth of government. Government grew at all levels. It grew during good times and bad. It grew with the warm approval of some groups and against the dogged resistance of others. At times its growth became a political issue, and elections were tilted one way or the other by it. Whichever side won, government continued to grow. It grew in absolute terms, as part of a growing society, but it grew in relative terms as well, playing an ever larger role in American life.

Consider, for example, the number of civilian employees working for the federal government at three periods of time. They increase geometrically:

1816—4,837
1901—239,476
2003—2,743,063

Had the U.S. population grown as swiftly as the population of government workers, it would today exceed that of China and India combined!

Consider the price tag of government expenditures, which follows an even more fantastic upwards arc:

1791—$4,269,000
1900—$520,861,000
2006—$2,568,000,000,000

All those zeros are hard to come to grips with. Suffice it to say that the 1791 budget divides into the 2006 budget more than 600,000 times!

Deep Change and Constitutional Experience

Deep change (as characterized by rising income and consumption, longer life expectancy, lower fertility, urbanization, and increased education) played into America's constitutional experience in a variety of ways, some of them direct and obvious, others indirect and subtle. The following, framed as questions, give us some notion of the impact of deep change on how we view the United States Constitution.

Popular government

"Government of the people, by the people, for the people."

Who Participates?

The Founders consciously designed a **popular government**, but they shied away from democracy as such, which to them bespoke the frauds and tumults of ancient Athens. They assumed that political partic-

Figure 9.21 There have been many groups throughout history who were often not allowed a full voice in the political process but have since gained those rights in the U.S. Are there other groups today that are still disenfranchised? Are there groups that should NOT be given full right of participation?

ipation would remain more or less limited to those of property, learning, and accomplishment. Deep change made property ownership increasingly widespread, made learning ever more accessible, and defined accomplishment in new ways. The door of democracy began to creak open.

Democratic revolution

Change in political power by the voting of the people.

American constitutionalism would confront the **democratic revolution** in a variety of ways. Power would drift into the hands of highly organized groups and be wielded by professional brokers. Public opinion would be molded by the mass media. Finance and industry would forge new kinds of interest. And the phenomenon of "private power" would arise where political bosses or industrial magnates would become powers unto themselves.

Equally troublesome would be the issue of who could *not* participate, for in a democracy the disenfranchised are at society's mercy. In the actual practice of politics for much of the 19th and some of the 20th centuries, the disenfranchised would include slaves, ex-slaves freed by the Civil War, and even their children and grandchildren—often kept from political participation by the legal manipulations and racial fears of the white majority. Other ethnic groups were also not given full political participation—Native Americans, Hispanics, Chinese, and Japanese. Some groups were considered to have political views that were too radical and unacceptable for American politics, unions such as the International Workers of the World and political parties such as the Communists.

Democracy's stepchildren also included women until well into the 20th century. Throughout human history, women had rarely enjoyed full political participation. But the logic of the Founding, based on natural rights, seemed to apply to women as well as men. The campaign for women's suffrage would consciously echo the Declaration of Independence. The campaign for full participation of all groups would follow in the 20th century. But that does not mean that everyone felt comfortable with full equality. The debate would continue with new groups and new issues. What did equality before the law and at the ballot box mean?

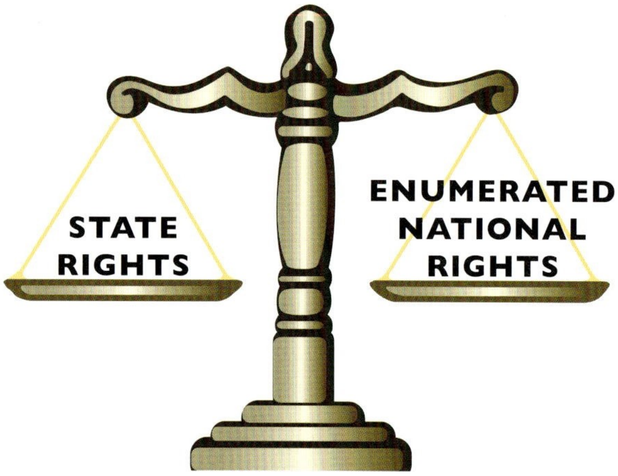

Figure 9.22 The federal government has certainly changed since the Founding. Do you think that the scales are still balanced? Should they be?

How Is Federalism to Be Kept in Balance?

In the minds of the Framers, the key to federalism lay in the concept of enumerated powers. If the national government exercised *only* the powers granted it under the Constitution, the federal balance would work. Once again, however, deep change would intervene.

To begin with, the growth and complexity of the modern world would give rise to growth and complexity in the national government. Changes in transportation and communication necessarily reduced the importance of the states as economic and cultural entities. Given such general wording as the "necessary and proper" clause, federal power could steadily expand and it often did. The government would take on dams, powerhouses, river and harbor projects, even (through subsidies) the building of intercontinental railroads. By the end of the 19th

ANDREW CARNEGIE

Owner of Carnegie Steel, a notable trust of the late 19th century.

JOHN D. ROCKEFELLER

Owner of Standard Oil, a major trust that was eventually broken up.

Figure 9.23 Anti-trust legislation is still an issue today. Does the government have the right to protect or regulate a free market economy? If so, is it truly a "free" market?

century it was ready for the largest single construction project on Earth—digging the Panama Canal.

It was the same with military power. Every time the United States would go to war the federal establishment would expand. At the beginning of the Civil War, for example, the national government was small and watchman-like. After four years of war it would be more than three times as large and a great deal more robust. In order to conduct a war on such scale, the government would have to mobilize every resource, conscript every combatant, enlist every ally, and ward off every enemy imaginable—utilizing powers the Founders never dreamed of.

The Civil War would affect federalism in another way as well. The conflict would result in no small measure from the inherent ambiguity of federalism: an uncertainty about where final authority lay in the federal balance. After General Lee surrendered to Union troops at Appomattox, there would be no further questions, north or south, to that question. The *final* authority lay in Washington.

As the American economy continued to expand, economic instability (and abuse) would be countered with ever larger doses of federal power. The rise of the "**trusts**," as they were called, would be answered by the rise of "progressive" administrations—**Theodore Roosevelt**'s and Woodrow Wilson's—capable of bringing the trust-masters to heel. In the Great Depression of the 1930s, there would be further demands that the government step in and do something, even as opponents of growing federal powers claimed that the Depression was *caused* by the government. **Franklin D. Roosevelt**'s "**New Deal**" gave the government blank check powers to do literally everything it could think of, from regulating industry, to inflating the money supply, to employing the jobless on work projects to improve the nation. The New Deal's alphabet soup of new agencies bespoke the burgeoning power of the federal government and the comparative impotence of the states.

World War II and the Cold War brought the United States to confront a world that was shaky, hostile, and downright lethal. It was a world symbolized by the duck-and-cover air-raid drill. The United States would go to war not once but repeatedly, in Europe, in Korea, in Vietnam, in Iraq, and in dozens of brushfire skirmishes all over the globe. American power would be the primary anchor on which free market democratic societies—and a few friendly dictatorships—would rely.

New conceptions of the Bill of Rights would create still another role for the federal government: that of civil rights watchdog. In case after case the FBI and federal marshals (at one point even the U.S. Army) would be summoned to disturbances involv-

Trust

A business entity created to monopolize and dominate a market.

Theodore Roosevelt

1858–1919
The 26th President of the United States, Roosevelt was known for his boisterous personality. He was known for trust-busting, championing environmental causes, and promoting his "big stick" foreign policy that called for American policing of the Western Hemisphere to protect its economic interests.

Franklin D. Roosevelt

1882–1945
The 32nd President of the United States, Roosevelt served four terms, the only U.S. President to serve more than two terms. His exuberant public personality helped bolster the nation's confidence as it struggled through the Depression and then entered World War II.

New Deal

Plan by Franklin D. Roosevelt involving the creation of various government agencies and programs designed to stimulate the economy and help the U.S. escape the Great Depression.

ing opposition to federally mandated desegregation in schools. The federal/state teeter-totter would tilt all the more toward federal power.

As modern life generated new hazards—pollution, contamination, pandemic contagion, resource depletion, global warming, drug abuse, identity theft, electronic eavesdropping, organized crime, computer viruses, internet fraud and, not least, terrorism—Americans would turn to the central government with ever greater insistence. For who else could fix what was wrong?

States would fight a rearguard action against the centripetal drift of power, but with no built-in mechanisms to maintain the balance, it was often a losing battle.

How Does the Founding Address the Economy?

The American Revolution was largely powered by protests against unfair mercantilist economics. Yet nothing in the founding documents specified any other economic model. Did the heaving of those tea chests into Boston Harbor mean that Americans were jettisoning mercantilism?

Early debates between Jefferson and Hamilton were often based on the issue of a free market versus an unfree mercantile economy. The election of Jefferson in 1800 seemed to mark a preference for the former. Still, Jefferson as President did not dismantle the Bank of the United States, undo Hamilton's tariff, or undermine the system of power and privilege that mercantilism supported. Something like mercantilist economics might have been taken by Americans as a given.

Still, political debates in the era of Jackson continued to focus on market themes. If, as the Jacksonians kept saying, "that government is best which governs least," it followed that government needed to keep its hands off of the economy. Indeed, this was increasingly the way that Americans would conceive freedom—the right to dig canals and build railroads and promote outrageous real estate schemes.

Yet the resulting megacorps could be scary, too, in a different way. By the end of the 19th century, Standard Oil would be virtually the only petroleum company in America, and Carnegie Steel would be virtually the sole producer of steel. There would be a copper trust, a silver trust, a sugar trust, a timber trust—the list went on and on. Here was the "free economy," reformers scoffed, and it left much to be desired.

Reform movements of the late 19th and early 20th centuries would address corporate concentration in a variety of ways. Some would seek to break up the trusts through court action and enforce competition. Others would prefer to supervise the trusts and regulate their activities. And still others would demand that the trusts be nationalized and run by the federal government. Utopian reformers like Edward Bellamy (*Looking Backward*) and Henry George (*Progress and Poverty*) would advance radical changes for combating economic injustice.

As American society continued to experience deep economic change, there would be ever more reasons—real or contrived—for the government to step in. Yet wherever the government moved in there would often be disturbing costs and consequences. In the airline industry, for example, government regulations and requirements for fair competition would ease the turbulence between the carriers competing for the same passengers, but the resulting ticket prices would be no more to the public's advantage than Rockefeller's oil or Carnegie's steel.

And so, to put the question in its simplest form: Was government intervention merely a natural and inevitable development, the price of living in a highly complex society, or was it a violation of America's free birthright?

How Does America's Relationship with the Rest of the World Affect Its Constitutional Government?

At the time of the Founding, the vast Atlantic Ocean made it possible for the U.S. to be isolated from foreign politics while still engaging in world trade. However, the French Revolution and ensuing European wars soon changed that. The American Republic found itself on treacherous footing among the contending belligerents, and not for the last time.

Deep change would shrink the size of the globe and land the U.S. in other entanglements. In 1812 it would be the Napoleonic Wars. In 1898 it would be the collapse of the Spanish Empire. And in 1917 it would be World War I. Increasingly, it would become clear that freedom itself was at stake. Yet was it possible (or even desirable) for the United States to defend freedom at every turn—in effect to police the world?

That prospect appealed indeed to some Americans. In the aftermath of the Spanish-American War, cartoons showed Theodore Roosevelt policing the Americas, sending marines here and gunboats there, his famous "**big stick**" at the ready. And in the aftermath of World War I, when **Woodrow Wilson** attended the Paris Peace Conference to implement his moralistic **Fourteen Points**, the political cartoonists would again have their day. Yet it would be no laughing matter for cheering crowds in London and Paris who hailed the American President as a political savior. If he wanted a **League of Nations** (his Fourteenth Point) to produce a lasting peace, they said, a League of Nations he should have. That Wilson's own countrymen should veto that dream would illustrate the difficulty of the American situation.

With the development of 20th-century weaponry, the peace and safety of the planet would increasingly be at risk, and the United States increasingly exposed. When deep change took the form of Japan's aircraft carriers with their state-of-the-art attack planes aimed directly at Pearl Harbor, it became clear to the dullest American that—for better or worse—this was one world.

America's participation in World War II would constitute a formal recognition of that fact. There could be no freedom for half of the world if the other half was in slavery. Yet while several bandit dictators came down with an ignominious thud at the war's end, other dictators would still be in place and thriving. When Joseph Stalin stretched an "iron curtain" across the eastern half of Europe and expanded his own reign of tyranny over several formerly free small nations, the United States would find itself at a major crossroads. Whichever fork it took, there could be no turning back.

The changes in science and technology would now manifest themselves in something called nuclear fission in the summer of 1945, heralding a nightmarish new world in which nuclear war might turn entire cities into radioactive soot in an instant. Science-based

Figure 9.24 The test of an 11-megaton nuclear weapon detonated by the U.S. from a barge near Bikini atoll on March 26, 1954. Deep change has made it dangerous if not impossible for the United States to remain aloof from the rest of the world.

Big stick

Part of the Theodore Roosevelt phrase: "Speak softly and carry a big stick," which represented the military might of the United States.

Woodrow Wilson

1856–1924
The 28th President of the United States, Wilson helped frame the Treaty of Versailles ending WWI and proposed Fourteen Points that included the formation of the League of Nations.

Fourteen Points

Moralistic ideals of Woodrow Wilson that were to be implemented after World War I in an attempt to have a lasting peace.

League of Nations

One of Woodrow Wilson's Fourteen Points at the end of WWI; it called for the creation of a group of nations to help ensure peace. The U.S. never joined because of a veto by Congress. After WWII, the United Nations was formed with similar goals.

weapons were difficult to keep secret. Nuclear technology quickly spread from the U.S. and Britain to the Soviet Union after World War II. France, China, India, Pakistan, Isreal, and, by the end of the century, Iran and North Korea had at least basic atomic bomb technology. And all the while mini-wars, each one capable of exploding into another world war, would continue to break out around the world.

With mortal dangers on all sides, the checks and balances set up by the Founders would come to seem quaint, if not counterproductive, to some. What good was it to refer questions of peace or war to Congress when any President could go to war with the push of a button? And why worry about constitutional liberties when spies, saboteurs, and turncoats were everywhere? The so-called Imperial Presidency would be one result of the Cold War mentality. The McCarthy-style witch-hunts would be another. And in time there would be many more.

Gibbons v. Ogden (1824)

Supreme Court case in which the power of the federal government was expanded by broad interpretation of the commerce clause.

McCullough v. Maryland (1819)

Supreme Court case in which greater federal power was established by maintaining the national bank.

So it was that Americans coming of age in the 21st century are facing the toughest questions of all. Is leadership of the free world a game that America can win? Is such leadership compatible with American institutions, consistent with long-held ideals? Or will it embroil the nation in a never-ending conflict that will ultimately compromise those institutions, tarnish those ideals and exhaust the political will of patriotic citizens? Has such deep change, in other words, robbed America of something precious and fundamental that can never be restored?

Judicial activism

When the courts use judicial power to achieve social goals.

What Are the Limits of Judicial Review?

A strong case could be—and was—made for judicial review. However, the actual practice of scrutinizing the Constitution by Supreme Court Justices would be subject to considerable latitude, ranging all the way from reactionary to radical. And as with federalism, there was no constitutional mechanism to keep the reviewers in narrow bounds.

As scrupulous as he was with his newfound power in the early years of this nation, Chief Justice John Marshall would hand down some momentous decisions affecting the way constitutional law would develop. In ***Gibbons v. Ogden***, for instance, he would argue for broad federal powers in the regulation of commerce, which could potentially mean the regulation of *everything*, and in ***McCullough v. Maryland*** he would significantly narrow the scope of state power. It was Marshall, in other words, who would give the federal/state teeter-totter its first big shove.

Some of Marshall's successors would be even less cautious. At the end of the 19th century, the Supreme Court, far from the passive referee that Hamilton and Madison had imagined in *The Federalist*, would begin to assume broader powers. The justices of that era would seize on a clause in the Fourteenth Amendment—the "due process clause"—and use it like a bludgeon to defeat state regulatory efforts. Never mind the will of the people in such matters, the Justices seemed to say, and never mind that the due process clause had been written for an entirely different purpose. The words were there on paper, and the Supreme Court could make of them what it wished.

There would eventually be a reckoning of sorts. In the 1930s, the Supreme Court would once again dive into the political fray, this time blocking the New Deal's efforts to cope with the Great Depression. President Roosevelt would hit back. In 1937 he proposed that the Court be reorganized and augmented in size, with the new appointments made by him, of course. There was an uproar across the land and Roosevelt's plan was defeated. But the Justices would gain at least a momentary respect for their limits.

Judicial activism would not die, however. We may think of it as one of those abiding temptations of constitutional government,

Figure 9.25 The United States Supreme Court, 2006. Seated left to right: Justice Anthony M. Kennedy, Justice John Paul Stevens, Chief Justice John G. Roberts, Jr., Justice Antonin Scalia, Justice David H. Souter. Standing left to right: Justice Stephen G. Breyer, Justice Clarence Thomas, Justice Ruth Bader Ginsburg, Justice Samuel Anthony Alito, Jr.

the exercise of power without the accountability of the ballot box. Congressmen, senators, even Presidents must come and go according to the popular will, but members of the Supreme Court sit behind their carved oaken bench for life, immune from all consequences. The urge to be involved with political affairs, to settle difficult or ambiguous matters once and for all, has constantly beset the high jurists.

Without surprise, then, we note that Supreme Court decisions in the areas of federalism, economic regulation, criminal process, civil rights, civil liberties, political participation, and the like would tip important scales in American life. To many it seemed that the Court had taken the place of Congress in formulating the law of the land.

Experience has made it clear that a written Constitution must be interpreted and applied to concrete cases. It has become equally clear that an activist court, unmindful of limitations, could essentially write its own Constitution.

Conclusion

By the end of the 20th century, some Americans stated or implied that their Founding as a nation, however successful in the past, was growing less relevant to the modern world. Deep change had simply changed the world too much.

On the other hand, an episode like the infamous Watergate scandal demonstrated that the machinery laid in place by the Founders worked as well in an age of high-tech espionage and imperial magistrates as it had back in the age of Jefferson. After all, the Constitution's various checks and balances operated with nearly flawless perfection, engaging legislative and judicial branches to root out corruption in the executive branch.

One thing was clear—the Founding had survived into a world appreciably different from that of the Founders, and that in order to do so it had developed and changed in important ways. What we must do now is take a closer look at those developments and alterations.

Timeline

1780 · 1800 · 1900 · 1980

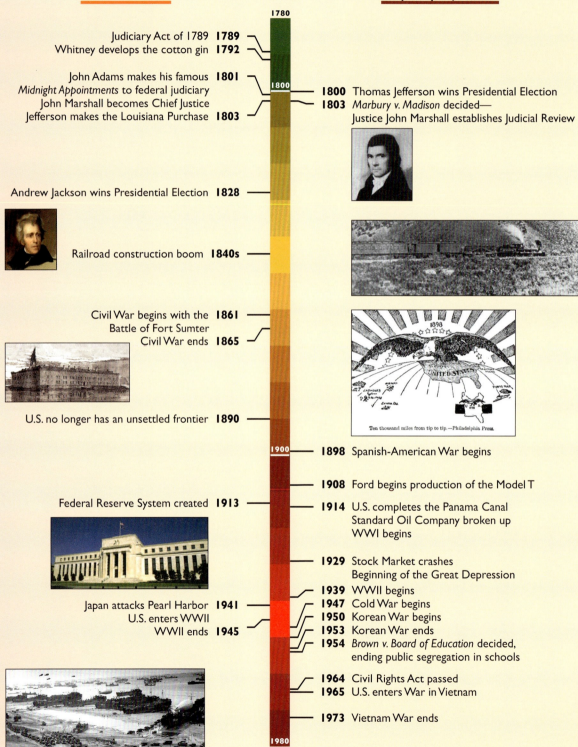

Contextual Events

- Judiciary Act of 1789 **1789**
- Whitney develops the cotton gin **1792**
- John Adams makes his famous *Midnight Appointments* to federal judiciary **1801**
- John Marshall becomes Chief Justice
- Jefferson makes the Louisiana Purchase **1803**
- Andrew Jackson wins Presidential Election **1828**

- Railroad construction boom **1840s**
- Civil War begins with the Battle of Fort Sumter **1861**
- Civil War ends **1865**

- U.S. no longer has an unsettled frontier **1890**
- Federal Reserve System created **1913**

- Japan attacks Pearl Harbor **1941**
- U.S. enters WWII
- WWII ends **1945**

Chapter Specific Events

- **1800** Thomas Jefferson wins Presidential Election
- **1803** *Marbury v. Madison* decided—Justice John Marshall establishes Judicial Review

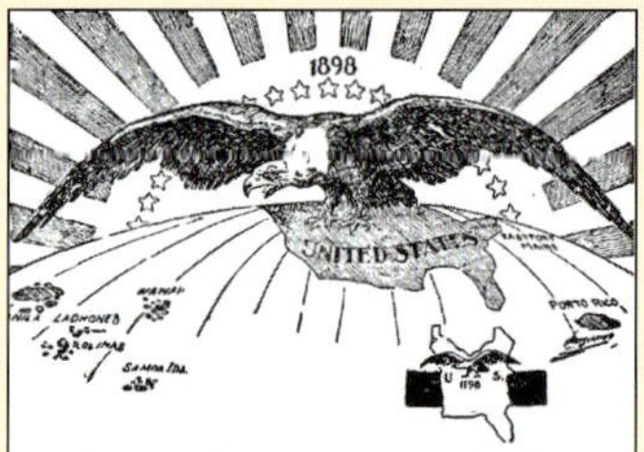

Ten thousand miles from tip to tip.—Philadelphia Press.

- **1898** Spanish-American War begins
- **1908** Ford begins production of the Model T
- **1914** U.S. completes the Panama Canal
  Standard Oil Company broken up
  WWI begins
- **1929** Stock Market crashes
  Beginning of the Great Depression
- **1939** WWII begins
- **1947** Cold War begins
- **1950** Korean War begins
- **1953** Korean War ends
- **1954** *Brown v. Board of Education* decided, ending public segregation in schools
- **1964** Civil Rights Act passed
- **1965** U.S. enters War in Vietnam
- **1973** Vietnam War ends

Key Terms

deep change
monopoly
League of Nations
federalism
judicial activism
New Deal
big stick

Key People

Charles Darwin
Albert Einstein
Franklin D. Roosevelt
Karl Marx
Theodore Roosevelt
John Marshall
Sigmund Freud
Woodrow Wilson

Questions

1. Give several examples of the "deep change" the U.S. has experienced over the last 200 years?
2. Analyze how one aspect of this deep change has put stress on the Constitution and Founding principles.
3. Do you think that the Founders would be proud of today's America? What would they want done differently?
4. Is "democracy" really fair when it excludes several less-organized groups and interests?
5. Has the government grown too large for its own good? Is there any way to reduce the size of government after it has grown?

Chapter 16

Empire of Liberty

Abstract

As innovative as American democracy was at the time of the Founding, it underwent many changes until it emerged as the democracy with which most Americans are now familiar. As the nation expanded westward, it also changed politically as government became more democratic and, some would argue, more corrupt. Thomas Jefferson and Andrew Jackson are given credit for initiating democratic change. This chapter examines the democratization of politics, the involvement of the masses in campaigns, political machines, the spoils system, and other government practices common to our day and how they developed as a result of the ensuing changes in American politics.

In Chapter 8 we saw how two-party politics came to be shaped by fundamental aspects of the Founding. In Chapter 9, we reviewed the changes that moved the American population westward and into cities. In this chapter we will examine how the American political system also tended to work its way toward democracy—not the democracy of the ancient world, where all citizens participated in the daily business of governance, but what we today call democracy, where all citizens may participate in the governing process by means of political activity. This development grew out of the Founders' general commitment to "popular government." Even more, it grew out of their decision to create a certain kind of polity, one that **Thomas Jefferson** referred to as an "empire of liberty."

Thomas Jefferson

1743–1826
Third President of the United States, Jefferson was the principal author of the Declaration of Independence and an influential Founding Father of the United States. He founded the Democratic-Republican Party and promoted the idea of a small federal government.

The Problem of the West

Colonial Americans did not think of their continent as an empire. Rather, they thought of each colony as a sort of outlying province of Great Britain and therefore as a part of some *other* empire. As far as the vast, unexplored wilderness was concerned, it was something that was just out there, like the deserts of Asia or the jungles of Africa, to be considered much later, if at all.

At the same time, however, several of the colonies gained claims to western lands, either because western boundaries had never been established in their charters or else because their charters specifically included certain vaguely described tracts to the west. In time, western lands became a hot-button issue for the colonies and played into the rivalries between them. The reason was simple. Land speculation was one of the approved routes to financial success in the 18th century. Anyone who could gain title to an area of wilderness was free to survey it, develop it, and ultimately resell it in parcels to others also seeking economic success. More than one of the Founding Fathers—including George Washington—entertained just such ambitions.

The colonies' British leadership viewed America's raw backcountry with skepticism, partly because opening it for settlement by Europeans complicated Indian affairs and partly because it aggravated the problem of control over groups moving into wild and generally uncharted territory. After all, how *did* one govern so vast an expanse of wilderness? In 1763, at the close of the French and Indian War, British authorities drew a line along the crest of the Appalachians and forbade settlement beyond it.

After the Revolution, settlers headed west once more. Many of them were veterans of the Continental Army who had been

Figure 10.1 The British government outlawed westward expansion in 1763.

Figure 10.2 The Last of the Mohicans, *by Newell Convers Wyeth, 1919. Inspired by James Fenimore Cooper's epic novel of the same name, which portrayed life in the backcountry of early America.*

granted homesteads in the Ohio country for their loyal service to the new nation. Speculators resumed their activities, too, trading in old colonial land claims in territories acquired from the Indians, or in lands simply there for the taking. If the area had been troublesome before, it was even more troublesome now. Indian hostility, British intrigues, Spanish possession of the lower Mississippi, competing land claims, and conflicting titles combined to make the West a fused powder keg.

Congress took a hand with the western problem long before the Constitutional Convention. The individual former colonies, now states, were asked to surrender their various western land claims to the Confederation government and allow it to assume jurisdiction. Then the government set about to determine how the West should be organized politically. Looking back, we might imagine that the solution to that problem would be obvious: simply create more states. It was by no means obvious at the time because of the government's main problem with the West. If the rugged western land claims of the states were fated to be a perennial backcountry, how could the people of such an area be trusted with self-governance?

Accordingly, many Congressmen favored a "colonial" approach to organizing the land westward. That is, the western part of the continent would always be kept subordinate to the East and given only limited sovereignty, just as the colonies had been subordinate to Great Britain. There were compelling reasons for such thinking. After all, the unsettled western territory was necessarily derivative in character. It must always be dependent on the East for capital, for know-how, for institutions. And the West would always run the risk of attracting losers, misfits, and malcontents who wanted to remain beyond the reach of law, order, and civilization.

There was one further reason for this colonial approach to the West: fear of democracy. When we think of it, the institutions of Old-World-style aristocracy—and there were still plenty of them in early America—depend on an ethos that didn't exist on the other side of that line drawn on the mountain crest. To maintain aristocracy,

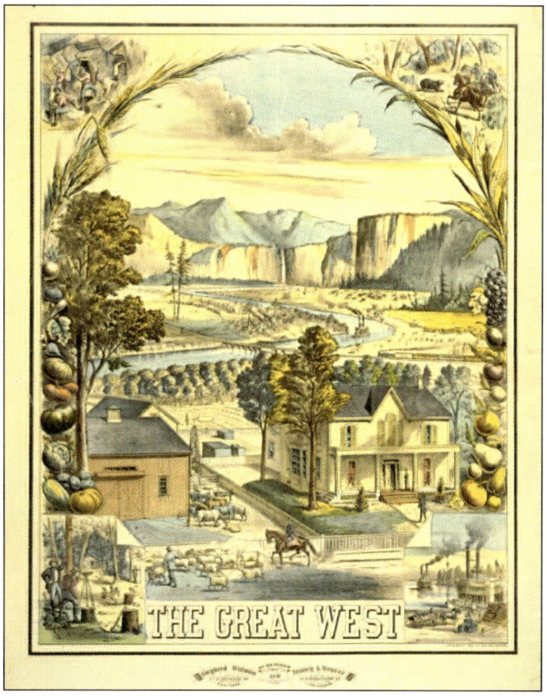

Figure 10.3 The Great West. *After the Revolution, determining how to handle the inevitable westward expansion became an issue even before the Constitutional Convention. Lithograph by Gaylord Watson, 1881.*

there must be a prescribed access to power, limited opportunities, and few available alternatives. Aristocracy seeks broad lands, it is true, but it seeks to bring them under centralized control and to place labor under tight restrictions. Without slavery (which was the wild card in the game), the aristocratic model was not going to work in the West.

Had the western territories been organized like the original colonies, a very different United States of America would have almost certainly resulted. One could easily imagine a western frontier in bondage to the East, carved into fiefdoms and bailiwicks, each of them beholden to powerful figures or prevailing interests. One could also imagine a tapestry of conflict as the expansive territory drifted politically, struggled for autonomy against the East (just as the colonies had struggled against the British Empire), conspired with foreign powers, fomented insurrections, or just generally ran amok. But young America's western wilderness was not to be organized in that way, nor was it to be left adrift from the newly established republic. A chance occurrence determined its destiny.

The Northwest Ordinance

In 1784, three years before the Constitutional Convention, Congress appointed a committee, chaired by Thomas Jefferson, to prepare a plan for the governance of western lands as the claims of the states were ceded to the national government. Jefferson dominated the committee's proceedings and personally drafted its report. This was to become the basis for two brilliant strokes of legislation that would solve the problem of the West once and for all.

Jefferson did not share the qualms of others. The western part of the continent, as he imagined it, was destined to become the domain of the common man, the yeoman farmer about whom he was steadily gaining enthusiasm. "Those who labor in the earth," he wrote, "are the chosen people of God, if ever he had a chosen people, whose breasts he has made his peculiar deposit for substantial and genuine virtue." Far from lapsing into barbarism, the wild and untamed west,

Figure 10.4 Left: Pioneers of America. Westward the star of empire takes its way *by Lithograph by Hunter & Co., 1870. Right: The Arkansas traveller. Scene in the back woods of Arkansas. Lithograph by Currier & Ives, 1870. Jefferson envisioned an enlightened rural citizen similar to the illustration on the left, a virtuous man of the earth. Some, however, felt that those who populated the West would be of a more unsavory stature and unfit to responsibly organize and govern themselves.*

as Jefferson saw it, would develop into what modern-day writers now call America's "heartland." He envisioned endless miles of rolling prairie covered with wheat and corn, tidy homestead farms, and white steepled churches. The real United States—as opposed to colonies torn away from the British Empire—would begin there.

Consequently, in his *Report of Government for Western Lands*, Jefferson envisaged as many as fourteen new states springing out of the frontier—*more* than the original thirteen—and all of them coming into the Union on the same footing and with the same privileges, as the existing states. It would be an empire, certainly, but nothing like the empires of old. There would be no London, no Fleet Street, no Windsor Castle or Westminster Abbey, no imperial metropolis upon which everyone must gaze in awe. Each state would be as good as every other, a republic unto itself, yet happily federated with its fellows.

Jefferson's ideal became the foundation of the proposal called the **Ordinance of 1784**. The Ordinance called for the organization of the nation's vast western domain into discrete territories, the naming of each of those territories, and a three-stage development of government institutions. At each stage, the territory in question would essentially govern itself with minimal supervision by Congress. Each territory would be responsible for improving the organization of its government while increasing its representation in the national legislature. When a territory's population reached that of the least populous of the existing states, the process would be complete and full statehood granted. Jefferson went even further. He insisted on fundamental rights for the new territories, including freedom of religion, right of habeas corpus, and trial by jury. He even insisted on the exclusion of slavery.

While the Ordinance of 1784 never took effect—a few states were still unwilling to surrender their western claims—it became the basis of two subsequent laws that did take effect. The first of them, the **Land Ordinance of 1785**, called for the systematic survey of the **Northwest Territory** (lands north

Figure 10.5 The Northwest Territory.

Ordinance of 1784

Plan of Thomas Jefferson to organize the national domain into discrete territories along with a three-stage development of government institutions.

Land Ordinance of 1785

Called for the systematic survey of the Northwest Territory and division into mile-square plots and organization into townships.

Northwest Territory

Lands north of the Ohio River.

Figure 10.6 A current overhead view of northwestern Ohio, showing the division into square plots. The plots were the result of the systematic survey of the Northwest Territory mandated by the Land Ordinance of 1785.

Figure 10.7 The Louisiana Purchase greatly increased the size of the United States from its original thirteen colonies, and created opportunities for later westward expansion.

of the Ohio River) and its meticulous division into mile-square plots, or sections, which can still be seen from the air today, marked off by ruler-straight boundary roads. Each grouping of 36 sections (six miles square) would constitute a township, and within each township one section would be reserved for the support of public education. America's western lands were to be nothing if not orderly.

In the second piece of legislation, the **Northwest Ordinance of 1787**, Jefferson's concept of the three stages of governmental development was fleshed out, made a bit more colonial and a bit less republican, and embellished with a few more refinements. But it preserved the main principle: the national domain was to be carved into self-governing republics, and these were to be systematically added to the Union. Anyone could do the math. The West was destined not only to join the East, it was destined to eclipse the East—and to redefine American nationhood.

During his own Presidency (1801–1809), Jefferson did much to bring his "Empire of Liberty" to pass. He adroitly negotiated the enormous **Louisiana Purchase** with France's Emperor **Napoleon**, who needed money more than distant land, thus adding to the new nation an area even larger than its original size. He dispatched the Lewis and Clark expedition to explore the new lands, record their natural resources, find the best route west, and show that the enormous area was ripe for settlement. And, as we will see, he both encouraged and inspired developments that could only be realized in the noble West of his imagination.

At the Constitutional Convention, the subject of the West came up time and again. After all, many interests were still on the line, not least of them the interests of the settlers. Convention Delegates were visited by one Manasseh Cutler, America's premier land speculator, who sought assurances that the Northwest Ordinance would still be valid under the Federal Constitution. The answer was yes. After much discussion, the Delegates decided that the procedures worked out in the old Congress would indeed guide expansion into the West.

Just how those procedures would work out in practice was reflected in one of the earliest settlements, Marietta, in what had been named Ohio Territory. It turned out to be a transplanted New England town, right down to the white church steeples, and thus an embodiment of Jefferson's vision. Further down the river, however, settlement was a little less orderly and a little more redolent of the frontier. That was the place where a young officer, fresh from the Indian wars, was about to mark out a suitable area for a plantation for himself, which he would name The Hermitage. His name was **Andrew Jackson**.

The Age of Jackson

In Chapter 8 we examined many of the dynamics of a two-party political system. Even though the two-party system made its first appearance in the 1790s, its develop-

Northwest Ordinance of 1787

Called for the governmental development of the west based on creating self-governing republics that would be systematically added to the Union.

Louisiana Purchase

Land purchased by Thomas Jefferson from France. Consists of much of the midwest United States.

Napoleon Bonaparte

1769–1821
French Emperor and European conqueror who sold France's North American holdings to the United States as the Louisiana Purchase.

ment into the political system we would recognize today dates from the Presidency of determined, opinionated, and iron-hard Andrew Jackson. And it was in Jackson's time, too, that the United States began to think of itself as a democracy.

We have seen, however, that processes which were strongly democratic in character date back to colonial times. Take voting, for example. In theory, colonial Americans had to qualify for the right to vote by owning a certain amount of property—a restriction that was common in Great Britain. In fact, suffrage requirements were set so low that most male heads of household could easily meet them. Voting was common in early America and a key part of consent. When the Founders spoke of "popular government," they had in mind the sort of government that the common people could and would support.

But politics was another matter entirely. Common people were not expected to run for office. They were not expected to deliberate policy, nominate candidates, or scrutinize the work of elected officials. Those activities were all tacitly reserved for gentlemen. It was generally accepted that the moneyed elite of the new nation were the only ones who possessed the stature, education, experience, and name recognition—not to mention financial resources—necessary to run for office.

Jefferson, himself one of those elite, was troubled by this. The more he thought about virtue and tillers of the soil, the more convinced he became that "pure republicanism"—an expression he often used—actually amounted to democracy. In other words, he came to believe that ordinary farmers and tradesmen and artisans ought to take a more active part in the political process. For him, consent meant more than simply casting ballots.

While Jefferson was President, a slow but steady process of democratization began in American public life. Democratization occurs when political power is brought closer to the people through extension of the right to vote, removal of filters of consent, and appeals to the masses during elections. Voting restrictions were gradually relaxed until most states had abandoned property requirements altogether and accepted full

Andrew Jackson

1767–1845
The seventh President of the United States, Jackson championed the U.S. as a democracy, pushing for more political involvement by the common man. He also vetoed the U.S. Bank's charter and made other reforms to keep the federal government small.

Figure 10.8 From The Brave Boy of the Waxhaws, *by Currier and Ives, 1876. In 1780 Andrew Jackson, when a boy of 13, enlisted in the cause of his country, and was taken prisoner by the British. When ordered by an officer to clean the officer's boots, he indignantly refused, and received a sword cut for his temerity.*

suffrage for all adult white males. Commoners, those who had become by their own talents and efforts successful, began running for office, too, although mostly on the state and local levels. They played more active roles in their political parties as well, especially in places where the two parties were evenly matched and a few extra workers could make a difference. Democratization developed as an outgrowth of the Founding, as a reflection of those memorable words in the Declaration of Independence, and as the destiny of popular government.

Significantly, it developed first and foremost in the West. Of course, there were no restrictions against democracy in the East, and in time it would make its way there as well. But the weight of old institutions and traditional practices was much heavier in the East. Aristocratic ways of thinking were rooted there as well, along with the fortunes and influence of great families. In the West, by contrast, it was much easier to break out of the traditional mold. There were far fewer gentry expecting deference, far fewer mansion houses, few well-groomed estates and monogrammed carriages. In the new and essentially raw western territories, few public offices were held as a matter of privilege.

Figure 10.9 The Battle of New Orleans, *by Herbert Morton Stoops, 1815. At the end of the War of 1812, American forces led by Andrew Jackson defeated British troops and gained control of New Orleans. The victory sparked Jackson's reputation as a hero, even though the battle was probably unnecessary. The treaty ending the war had been ratified two weeks earlier, but news had not yet reached the southern front.*

In the West, ordinary people—adventurous men with their own visions of the future, some with families, a few tools, and maybe a Bible and the Works of Shakespeare in their little cabins, gained economic and social power in a variety of practical ways. They bought and cleared the raw land for farming. They started businesses, tinkered with inventions, published newspapers, organized communities, and made their local world work. Political power came to them as a matter of course.

Jackson himself was typical of this new breed of western man. He was born in the Carolina backcountry and schooled only sporadically. As a teenager, though, he made an arrangement to read law with a practicing attorney and soon proved his abilities in the courtroom. Like many another frontiersman, he engaged in brawls, fiercely resented insults, and killed a man in a duel who had been rash enough to slander his wife. As he prospered, he began buying land and slaves to work it. He became a two-fisted border captain, venturing against the Indians on several occasions and lynching a pair of British subjects whom he suspected of inciting the Seminoles against the local American settlers.

Politics came to him as naturally as military service. He was the first Congressman elected from the state of Tennessee and served a brief tour of duty in the Senate as well. He became a hero in the War of 1812, defeating the British at the Battle of New Orleans (which took place after American and British diplomats had signed the official peace treaty) and sending the redcoats, in the words of a popular song, running "through places where a rabbit couldn't go." By 1824 he was a candidate for President.

Once the decision was made to carve

the nation's western lands into states that could be admitted to the Union on an equal footing with all other states, political democracy became all but inevitable. There was simply no way that the forms and manners of aristocracy could survive in a Jacksonian world.

As President, Jackson came to symbolize the democratic revolution, of which he clearly approved. He became best known for his dogged and ultimately triumphant battle against the Bank of the United States, which had been a prominent feature of Alexander Hamilton's economic program years before. The bank organized by Hamilton had prospered. It had accomplished precisely what Hamilton believed it should, tying the fortunes of the rich and the well born to the economic health of the federal government. And so it had become a symbol of the reemergence of both economic mercantilism and political aristocracy. Jackson believed he was speaking for the common people when he decided to veto the Bank's charter in 1832 when it came up for renewal. The veto was no simple affair, for the Bank of the United States was financially powerful and politically well connected. But Jackson persevered. "It is trying to kill me!" he fumed to an aide, "but I will kill *it!*" In the end, he not only terminated the Bank's charter, but also transferred federal deposits out of its vaults and redeposited them in state banks instead. He was bound and determined to win.

Rightly or wrongly, Jackson's war against the "monster bank" was understood by Americans to be a blow against aristocracy. Other blows followed. Many of the franchises and monopolies that had arisen from Hamilton's economic program were throttled as ignominiously as the Bank of the United States had been. Jackson vetoed the construction of a major federal highway because, he explained, it would more likely serve the special interests than the people. He applauded states rights and vowed to keep the federal government small and service-minded. His followers, the Jacksonians, saw all of these moves as opening doors of opportunity.

Figure 10.10 A Democratic cartoon showing Jackson taking on the monster bank. While Jackson succeeded in vetoing the Second Bank of the United States' charter in 1832, the resulting dispersion of funds and duties to numerous state and local banks fueled speculation and caused further economic problems.

But it was in the arena of party politics that the greatest changes took place. The restrictions on participation had continued to decline since the days of Jefferson, and ordinary voters were playing an ever-larger role in the political process. During Jackson's presidency, these developments were not only acknowledged, they were discussed, welcomed, and even celebrated. Jackson's lieutenants, of whom Martin Van Buren was the most important, began working out a whole new political style, geared not only to the acceptance of parties but to the participation of the rank and file. Here are a few of its features:

Public togetherness

Aspect of party politics in which groups of political party members would gather together in order to have more solidarity and support.

Public Togetherness

Politics came to include torchlight parades, public demonstrations, theatricals, barbecues, and even chummy affairs of state. Beginning in Jackson's administration, admirers took to sending enormous cheeses to the White House (one gifted cheese weighed some 1,400 pounds!)—so that the

President could invite the public to join him for cheese and crackers. Never before had there been so much public togetherness.

Party newspaper

A journal used by a political party for disseminating party information to and encouraging more active participation among the grass roots voters.

Party Newspapers

Political parties, once aloof from the rank and file, began communicating with them nonstop by means of party newspapers. These organizations expounded philosophy, explained tactics, justified controversial positions, and reported on the activities of candidates and officeholders alike. Their purpose was to encourage an ever more active participation among the grass roots citizenry.

Popular campaigning

Promoting candidates as being from (and therefore representing) the common masses, rather than as elite gentlemen-politicians.

Popular Campaigning

Where politics had once favored the elegant manners of gentlemen, it now embraced the manners of the crowd. Candidates were expected to speak from the stump, confound hecklers, and animate a sluggish or disinterested audience. Broadsides popped up in storefront windows. Slogans were painted on barns. Complex positions were reduced to comic jingles. And popular heroes of the day, such as "Tippecanoe" (General William Henry) Harrison and witty frontiersman Davy Crockett, were pressed into running for office. In one campaign, supporters of a Presidential candidate trundled a facsimile of the log cabin their man had been born in through the streets to prove their hero was truly a man of the people.

Figure 10.11 1840 Broadside for a Harrisonian rally. The banner in the illustration labels William Henry Harrison as "The Farmer of North Bend," a man of the people.

Political convention

Large meeting of party delegates for the purpose of nominating candidates, often held with much pomp and ballyhoo.

Political Conventions

In the old days a caucus of the party's elite had nominated candidates, but now delegates at a grand convention nominated them. Conventions became filled with ballyhoo: banners, bunting, painted signs, colorful costumes, band music, and side-shows, all swirling in a carnival atmosphere. It was accepted that supporters of a winning candidate were obliged to out-shout and out-cheer all rivals.

Get out the vote activity

Aspect of party politics in which voters are systematically rounded up and helped to get to the polling place.

Get Out the Vote Activities

These ran all the way from home visits by the party's local agent, or "ward heeler," to the systematic roundup of voters on Election Day. Polling places often featured free food, free liquor, abundant handshakes, and, yes, even baby-kissing by the hopeful candidates.

Costs and Benefits of the New Politics

Full participation meant that people could run for office themselves, not just cheer from the sidelines. In order to accommodate such a striking innovation, much new thinking was required. For example, now that office was no longer the domain of gentlemen (as it had been for a thousand years) it slipped perceptibly in public esteem. Similarly devalued was the idea of the officeholder as a civil servant with special background or training. Andrew Jackson believed that public callings ought to be simplified to the point where any citizen could fill them. That worked, conceivably, for the case of mail delivery or road repair, but it didn't work for accounting, civil engineering, or advanced crime detection. There was *some* public business that couldn't be performed by just anybody—and growing incompetence was the outcome.

It was only a short step from these developments to the idea of office as a political prize. Jackson was in the forefront of this trend as well. "To the victor go the spoils!" he proclaimed upon turning out a host of appointed officials after one election and putting his own followers in the newly vacated offices. The fact that the party faithful could be rewarded with government posts that required little effort, held a degree of prestige, and paid at least modest salaries altered the nature of political participation. Many friends and followers of a candidate and a particular political philosophy no longer took part in political campaigns for purely idealistic motives.

There were corresponding developments in the nature of political leadership. In Washington's day, leaders had to inspire their followers with the hope of honor or shared achievement. No longer. Now a party leader could simply pay his people off, as it were. Executives who could dispense plums to their loyal supporters acquired a new kind of political power, divorced from ideology or persuasion. *You look after me and I'll look after you* took American politics back to the world of British patronage, which under the name of "corruption" had become one of the burning issues of the Revolution.

Democracy had other drawbacks. For instance, many voters had no knowledge of

"THAT'S WHAT'S THE MATTER."

Boss Tweed. "As long as I count the Votes, what are you going to do about it? say?"

Figure 10.12 Boss Tweed, *by Thomas Nast. William M. Tweed was a notorious figure in the New York democratic political machine known as Tammany Hall. In addition to controlling nominations and gaining political offices for others through illegal means, Boss Tweed stole an estimated $200 million from New York through graft and fraud.*

the candidates' qualifications for office and little understanding of complex issues gave rise to the phenomenon of the **political machine**, a group of party loyalists whose sole purpose was to do one thing: deliver the vote on Election Day to their candidate. How they accomplished this task was up to them. Their tactics were developed early and have continued to be successful to the present day. Candidates' political machines haven't changed much from the 19th century. They still buy votes, rig ballot boxes, and intimidate the supporters of rival candidates—if and when they can get away with such illegal and unethical antics. Election day riots became common in the Jackson era. Several of them resulted in pitched battles, while the tongue-in-cheek advice to "vote early and often" revealed the yawning depth of public cynicism at the ballot box.

Machines went hand in hand with public stealing, too. Bribery, extortion, embezzle-

Political machine

Group of party loyalists organized to deliver the vote on election day. Historically they often used questionable or illegal means such as buying votes or intimidation at the polls.

John Quincy Adams

1767–1848
Sixth President of the United States, Adams is known for formulating the Monroe Doctrine.

Daniel Webster

1782–1852
A leading American statesman and senator during the Pre-Civil War era.

Henry Clay

1777–1852
American statesman and congressman who founded the Whig party.

ment, kickbacks from the bestowal of public contracts, and other forms of graft were routinely unearthed in machine-dominated administrations. Some machines became informal welfare agencies, seeing to it that newly arrived immigrants were given jobs, legal representation, bail if necessary, perhaps even a Christmas turkey or new shoes for the kids, with the tactic understanding that they voted "early and often" for their machine's candidate. Supporters had to be kept happy at all costs, with the funds coming from contract kickbacks and other income resulting from reaching into the public till.

Democratic politics brought forth its own style of leadership. Where the machines held power, the real leaders were the party bosses, many of them saloon keepers, hustlers, and other ambitious men on the make. Decision making about who should run for a particular public office took place out of public sight, in smoke-filled back rooms of saloons. But even where machines were absent, leadership proved to be elusive. There was great care on the part of elected officials to keep within the bounds set by public opinion. No city mayor or state governor wanted to be turned out of office for failing to be in step with public opinion, and so being in step came to mean everything. In Jacksonian America public opinion often left much to be desired, too. Racism, sexism, chauvinism, and xenophobia were only some of the worst offenses, along with religious bigotry so virulent that it often resulted in violence. Americans could have used more of the exalted style of leadership exhibited by the Founders.

To offset these weaknesses, democracy also displayed conspicuous strengths. On many occasions, ordinary voters showed a wisdom and foresight that had been lacking in the old days of political privilege. They managed to elect, along with the proverbial fools and knaves, statesmen of great accomplishment such as Andrew Jackson, **John Quincy Adams, Daniel Webster, Henry Clay**, Thomas Hart Benton, Stephen A. Douglas, William Seward, and Abraham Lincoln.

Common voters also exhibited stronger moral sensibilities, on the whole, than did political elites. For them right and wrong often had a spiritual dimension, not just a secular one, and they were unwilling to compromise on matters of salvation. It was gentlemen, remember, who worked out the means for accommodating slavery in the United States—and it was the common people who ultimately refused to go along with that accommodation.

Democracy made its own contribution to liberty. Those who had no voice in politics could never be truly free; as the Declaration of Independence had made clear, those who would be free must be heard. Even when the people misused their power, when they tolerated corruption or supported the political bosses, it was their free choice to do so and they had no one to blame but themselves. Freedom, of course, had never promised a favorable outcome—only clear accountability.

Democracy's greatest strength lay in the way it altered consent. Consent had meant one thing at the time of the Founding, when ordinary American stood on the political sidelines and were essentially spectators in the political process. It came to mean quite another when politics became the nation's premier participation sport. Now the people themselves could truly approve or disapprove of governmental action and go on to do something about it.

The American Character

In 1892 the American frontier was officially pronounced closed. Homesteaders could no longer strike out across the prairie and file claim to a quarter section of free land (as specified by the Homestead Act of 1862). The following year, a young historian by the name of **Frederick Jackson Turner** published what may have been the most important scholarly paper in American historiography. In this essay, Turner asked what it was that had made the American experience unique. He was fully aware, of course, of the significance of British institutions, Enlightenment thought, and much else that had played into the Founding. Yet he argued quite persuasively that it was the experience of the American frontier over the previous century that had made America what it was.

Democracy, as Turner saw it, was not just a political theory or a style of governance. It was a way of building and operating an entire society—and of shaping a national character. It grew out of human optimism, self-reliance, ingenuity, creativity, and common sense. And since life on the frontier encouraged and rewarded these qualities, democracy could have been spawned nowhere else. Consider the homesteader out on the plains or the settler in his backwoods cabin. They and they alone ran their own lives, solve their own problems and build their own communities. Who on earth could think of denying them political say?

While some historians have quarreled with Turner's claim, there are aspects of his insight that still seem compelling. Granted, America was founded before the West was truly open, and yet the Founding was transformed by the West—by Jefferson's Empire of Liberty—in ways that might be difficult to explain but not difficult to feel. If there is such a thing as "the spirit of democracy," it may well have become the central ingredient both in the American character and the American way of life. And it certainly grew out of *something* in the national experience that was not imported from Europe but had its roots firmly planted in the New World.

Once the spirit of democracy was fully manifest, it proved to be not only self-sustaining but also progressive and expanding.

Figure 10.13 Young Corn. *Painting by Grant Wood, 1931. According to Turner, it was this frontier setting that truly shaped and defined the spirit of American democracy.*

- By the outbreak of the Civil War, adult male suffrage was the rule throughout the United States, with few or no property restrictions.
- In 1848, at Seneca, New York, feminists held their first American convention, and in language inspired by the Declaration of Independence, they demanded full political rights. The campaign to achieve such rights would be long and arduous, but it would ultimately succeed.
- At the close of the Civil War, the so-called Radical Republicans argued that freed slaves could only achieve their rightful place in society if they too were given political rights. While the constitutional amendments designed to give them such rights (Fourteenth and Fifteenth) were thwarted by Southern states' hostility and legal subterfuge, they would eventually come fully into play.
- In 1920, the Nineteenth Amendment was approved, giving women the right to vote. Forty-four years later, the Voting

Frederick Jackson Turner

1861–1932
American historian who studied and wrote about the American experience and what made it unique.

Rights Act secured the same privileges for Blacks and other minorities. And in 1971, the Twenty-Sixth Amendment guaranteed the right of suffrage for all Americans above the age of 18. Those who were old enough to fight for their country were now considered old enough to help choose their leaders and laws.

- Further obstacles to voting were removed by the National Voter Registration Act in 1993, popularly known as Motor Voter, making registration a simple task with uniform procedures.

Just how these advances have served the spirit of democracy is a question that Americans still vigorously debate.

Politics Today

Politics today is characterized by a "permanent campaign." Elections for the House of Representatives and a third of the Senate occur every two years and require extensive financing. Consequently, politicians must raise money and communicate their messages to voters on a steady, uninterrupted basis, much to the chagrin of some voters. To illustrate, the *New York Times* ran twenty-three stories on the 2008 presidential campaign in the two months following the election of 2004. The political season has become continuous.

Presidential politics, and to a lesser extent other political races, have become increasingly dominated by Presidential primaries and caucuses. The early events, such as the primary in New Hampshire and the caucus in Iowa, assume major importance in the nomination process of the two main parties and are covered exhaustively by the press. In the spirit of the permanent campaign, potential candidates will visit those early-voting states two or three years before the next election.

Primaries add an interesting wrinkle to the political process. Whereas general elections push candidates to the middle of the political spectrum, primaries push them away from the middle toward the preferences of the truly committed partisans of their respective parties. Independents and other voters with uncertain views are not likely to vote in primary elections. Hence the candidates must appeal to the "base" of their party, which is generally to the right or the left of the polled opinions of the general population. The combination of primaries, followed by a general election with only two viable candidates, treats the public to an interesting migration by Presidential aspirants. In January through March of the Presidential election year, candidates take rather extreme positions to appeal to the party activists who dominate the primaries. The victorious primary candidate then tries to supervise the party convention, usually in late July or early August, to prevent his or her own partisans from alienating the general public. By September, the candidate has reinvented his candidacy in terms of the middle-of-the-road politics that can win general elections.

Then and Now

At the dawn of the twenty-first century, Americans are still assessing the costs and benefits of political democracy. These have not changed appreciably since the rough-and-tumble days of Andrew Jackson.

While we no longer haul log cabins through the streets at election time, it is commonplace for high-priced campaign consultants to coach candidates on the fine points of acceptable public image. Jackson-era slogans like "Tippecanoe and Tyler too!" were not much less profound than "I like Ike" (1952) or "Hasta la vista, Baby!" (1996). As for recruiting popular icons into politics, we need only think about film star Ronald Reagan, who was elected President in 1980;

TV wrestler Jesse Ventura, who was elected governor of Minnesota in 1998; or Hollywood he-man Arnold Schwarzenegger, who in his 2003 California gubernatorial campaign said: "I am so rich, you can be sure no one will try to buy *me!*" Every campaign is now adorned with actors, rock stars, television personalities, fashion models, race-car drivers, and professional athletes attempting to steer their own admirers into the fold of their political candidate.

Machine-style politics in Jacksonian America finds its modern-day equivalent in the role of pressure groups, political action committees (PACS), and under-the-table corporate donors with deep pockets and their own hopes for special treatment by the winner. Election campaigns have become so expensive that, ironically enough, an ordinary citizen can hardly hope to run for office. Reformers have grappled with the money-and-politics mess for more than twenty years, with schemes to limit campaign contributions, include dollar-sized donations with tax payments, and punishment for politicos who cozy up to secret money sources. Even with official rules about campaign funding, and numerous official and unofficial watchdogs scrutinizing every candidate's actions and finances, new scandals come to light every year.

Listen to campaign ads in the next election, and you will hear phrases coined in the Age of Jackson. "He is a prisoner of the far left/right!" "She speaks out of both sides of her mouth!" "He is a tool of the XYZ interest!" "She represents *you!*" This is because the dynamics of popular campaigning haven't really changed very much. Then as now, the broad mass of middle-of-the-road voters wants more or less the same things: personal security, increasing prosperity, continuity with the past, and competent and trustworthy governance.

Just as Jacksonian voters agonized over moral issues such as slavery, voters in our

Figure 10.14 Ronald Reagan was a well-known movie actor before being elected as President of the United States.

own time agonize over abortion and gay rights. Democratic politics always mirrors such turmoil and perhaps even inflames it, owing to a connection that the Founders saw between politics and conscience. While we rarely use the term *virtue* to describe the duties of citizenship anymore, what we expect of political leaders (and are willing to punish them for if it is absent) is essentially just that. In their hearts, most voters seek to do what they think is right.

On a deeper level, America's politics-of-the-whole-nation continues to exemplify Jefferson's Empire of Liberty. The excitement

Figure 10.15 Governor Arnold Schwarzenegger announcing a state budget plan that includes increased funds for education, May 12, 2006. Schwarzenegger was also a well-known movie actor before entering politics.

of election campaigns helps overcome local attachments and cement the bonds of nationhood. Canadians, who share our language and culture, seem far away at election time, mere spectators to this ritual of togetherness. States that were created out of free land and coached into being by the federal government have become the most "American" part of the nation and the part most determined to remain free.

In America's Empire of Liberty, public opinion has become an active shaping force, filling every corner of American life. The techniques of advertising and public relations have been absorbed into the political process, with the result that politics is everywhere and election campaigns are in perpetual motion. Accompanying the never-ending modern political campaign is the assumption of many Americans that our government can (and ought to) tackle every problem that comes along. Andrew Jackson would find this notion appalling, for he believed with Jefferson that "that government is best which governs least." However, on reflection, he might understand that once the common people are given power, that power will be turned toward common ends.

Through it all, Americans' active, involved, and ongoing process of consent means that the Founding, though materially altered, continues to operate in their lives.

Timeline

Contextual Events

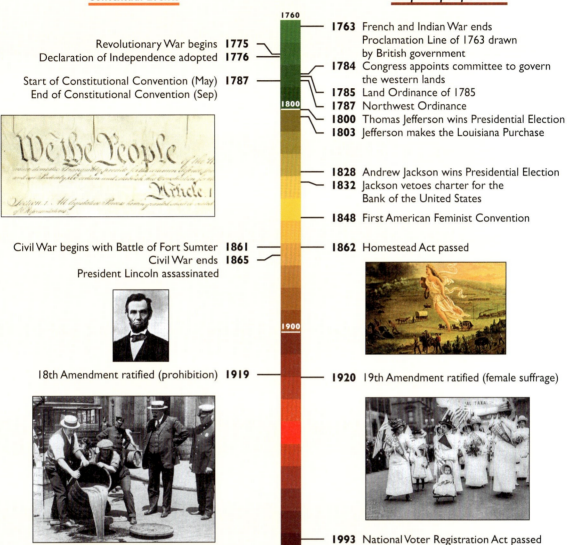

- **1775** Revolutionary War begins
- **1776** Declaration of Independence adopted
- **1787** Start of Constitutional Convention (May)
  End of Constitutional Convention (Sep)

- **1861** Civil War begins with Battle of Fort Sumter
- **1865** Civil War ends
  President Lincoln assassinated

- **1919** 18th Amendment ratified (prohibition)

Chapter Specific Events

1760

- **1763** French and Indian War ends
  Proclamation Line of 1763 drawn by British government
- **1784** Congress appoints committee to govern the western lands
- **1785** Land Ordinance of 1785
- **1787** Northwest Ordinance

1800

- **1800** Thomas Jefferson wins Presidential Election
- **1803** Jefferson makes the Louisiana Purchase
- **1828** Andrew Jackson wins Presidential Election
- **1832** Jackson vetoes charter for the Bank of the United States
- **1848** First American Feminist Convention
- **1862** Homestead Act passed

1900

- **1920** 19th Amendment ratified (female suffrage)

- **1993** National Voter Registration Act passed

2000

Key Terms

popular government
political convention
aristocracy
political machine
popular campaigning

Key People

Thomas Jefferson
Andrew Jackson

Questions

1. What did the British government do in an attempt to control the problem of western expansion in the colonies?
2. Why might the states' treatment of the settlement of western areas be described as an "empire of liberty?"
3. How did Andrew Jackson's views on the common citizen's participation in government differ from that of the Founders?
4. In what ways has the political structure of the United States changed since the Founding? In what ways has it stayed the same?
5. What tactics did political machines use to "deliver the vote on election day?"
6. How does the continual process of consent signify that the Founding continues to operate in our lives?

Chapter 11

FINISHING THE FOUNDING

Abstract

The Civil War was primarily a response to two features of the Constitution—acceptance of slavery and federalism. As moral and political sentiment in the North turned against slavery, the South saw their way of life threatened. The vague constitutional definitions of federalism led the southern states to believe they held ultimate sovereignty, and they seceded from the Union to preserve slavery. This chapter explains how the Civil War solved the ambiguities of slavery and federalism, making former slaves legal equals with the passing of the post-Civil War amendments, and solidifying supreme sovereign power in the hands of the national government.

National Flag of the 4th (West) Virginia Infantry, mustered in August 22, 1861, at Mason City.

The bloodiest day in American history was in September—not September 11, 2001, but September 17, 1862, early in the Civil War between the nation's northern states and eleven southern states. The Southern Confederacy's brilliant General **Robert E. Lee**, fresh from victory at the second battle at Manassas, Virginia, decided to take his Confederate army of Northern Virginia, 40,000 strong, north into Maryland to force U.S. President Abraham Lincoln to seek peace with the **Confederacy**. Confederate troops crossed the Potomac to Frederick, Maryland. Lee sent General Thomas "Stonewall" Jackson to capture the federal armory and garrison at Harper's Ferry, while Lee took the rest of his troops west to prepare for an attack in Pennsylvania. Near Sharpsburg, Maryland, Lee's group encountered the Union Army of the Potomac under General **George McClellan**, with some 60,000 troops, many of them newly recruited and untested in battle. Lee's troops dug in on high ground near a small church.

At dawn on September 17, Union artillery opened fire to soften Confederate positions for infantry attacks. The first large-scale battle was in local farmer Daniel Miller's twenty-acre cornfield. Confederate troops, positioned in the corn, waited for the Union troops to begin climbing over a small rail fence. Confederate soldiers then opened murderous fire; Union forces returned fire. The battle line surged from one side of the cornfield to the other. By 10:00 am, about 13,000 men had been wounded or killed in that cornfield.

Later in the morning and into the noon hour, the battle shifted to a nearby dirt lane used by farmers to haul their produce. Through years of use, the lane gradually had been worn down to below the level of the surrounding fields, making it an ideal place for the Confederate troops to set up a defensive position. Time and again, Union troops marched in ranks against the Confederates in the lane, but the Confederate troops held what appeared to be an ideal position to repel any attack. Eventually, that ideal defensive position was turned into a death trap by a successful Union flank attack. Two New York regiments captured the eastern end of the sunken lane and began to fire straight down the length of the lane, killing hundreds of soldiers and forcing the Confederate army into a hasty retreat.

Elsewhere, 12,000 Union soldiers attempted to cross a bridge over Antietam Creek to cut off Lee's tactical retreat. A few Confederate sharpshooters, positioned in rocks and trees above the bridge, held off the Union troops until about 1:00 pm, when the Union army finally gained control of the bridge. With Union troops now nearing Sharpsburg, the Confederate army was in great peril. However, new Confederate troops arrived from Harper's Ferry about 3:00 p.m., stopping the Union advance. Years later, General James Longstreet, Lee's chief com-

Confederacy

Alliance of southern states that seceded from the Union over slavery.

George B. McClellan

1826–1885
Union General who failed to press his advantage at the Battle of Antietam, and was later relieved of his command by President Lincoln.

Robert E. Lee

1807–1870
Confederate general and commander of the Army of Northern Virginia during the Civil War. After surrendering at Appomattox on April 9, 1865, Lee urged reconciliation with the North.

Figure 11.1 Ditch on the right wing of the Antietam battlefield filled with war dead, where Kimball's brigade fought so desperately, 1862. Stereograph by Alexander Gardner.

mander of infantry, claimed that 10,000 fresh Union troops could have captured the whole Army of Northern Virginia, thereby defeating the South. But General McClellan, not one to take risks, had kept 20,000 troops in reserve, losing a precious opportunity to end the war. By 5:30 p.m., the worst single-day battle in American history was over. The Army of Northern Virginia made its retreat back to Virginia, giving up the plan to attack Pennsylvania. They would come back to Pennsylvania the next summer and meet Union forces again at another small town—Gettysburg. Again, they would be driven back. And again, the larger, better-equipped Union army, hampered by inferior leadership, would fail to press its advantage.

September 17, 1862, was known as the battle of **Antietam** in the North and the battle of Sharpsburg in the South. Neither side really won that day, but Lincoln felt good enough about the result to issue his **Emancipation Proclamation**, freeing slaves in the areas of insurrection. Other battles fought over several days have had more casualties, but no single-day battle in American history has ever concluded with such devastating results—an estimated 3,650 dead and over 22,000 wounded, many of whom would die later from their wounds. In comparison, the all-out World War II attack by Allied forces against German-held French beaches on D-Day, June 6, 1944 resulted in only 2,510 American dead and wounded on Omaha and Utah beaches.

The Civil War battle of Antietam was just one of many in a war stretching over four years and costing over 600,000 American lives. In every city, town, and village in the North and South after the war, there were maimed veterans trying to rebuild their lives with an arm, a leg, or an eye missing. This was the worst kind of war—a civil war with horrific losses. A century and a half later, this terrible conflict still seizes America's imagination and raises significant questions. What caused this bloody conflict? How could the American nation have come to this just seventy-five years after a Founding with so much promise?

Figure 11.2 Allan Pinkerton, President Lincoln, and Maj. Gen. John A. McClellan at Antietam, MD, October 3, 1862. Photograph by Alexander Gardner.

Unfinished Business—Slavery and Federalism

Every political act in a representative democracy involves compromise, and the Constitution was a political act. The Founding Fathers, in an effort to reach agreement, left some issues unresolved, allowing each sectional interest to read its own comforting interpretation into the document. Two vexing issues were left to be resolved by future generations: slavery and federalism. Between 1787 and 1860, many attempts were made to grapple with these tough problems. But the attempts failed and the nation split into two geographical as well as political

Antietam

A severe Civil War battle that took place on September 17, 1862. It was the bloodiest day in American history. After the battle Abraham Lincoln issued the Emancipation Proclamation.

Emancipation Proclamation

Presidential order issued by Abraham Lincoln on January 1, 1863 that freed slaves in the areas of insurrection.

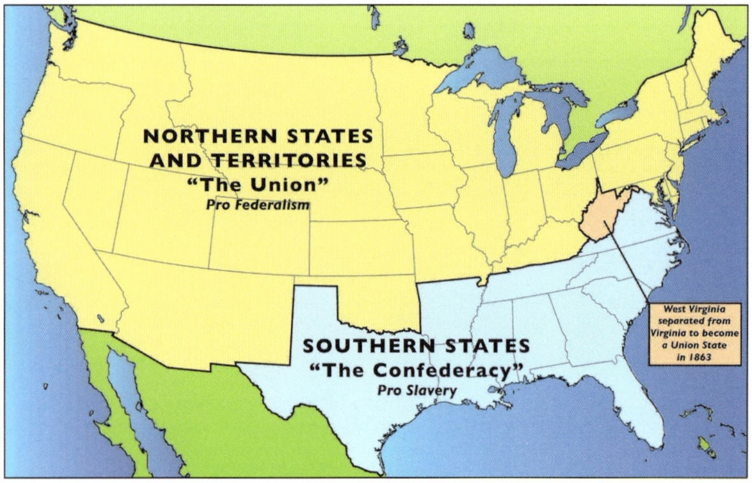

Figure 11.3 The issues of Federalism and Slavery divided the country into two primary geographical areas.

parts with the bloody conflict between them called the "Civil War" in the North and the "War Between the States" in the South.

The Founders did not confront the troubling issue of slavery despite its clear conflict with the ideals embodied in the Declaration of Independence and the new government. The issue was simply too volatile and too difficult to solve at that time, considering how entrenched slavery was in the South. Any direct attack on slavery would have been a "deal breaker" for the proposed Constitution. Instead, most of the Founders hoped that slavery would just fade away as the new nation developed its industry and commerce.

The hope that time would solve this challenging problem shriveled with invention of the cotton gin, simplifying the cleaning of raw cotton. This encouraged increased production of cotton in the South, making slavery even more profitable. Southerners moving west to farm new land in the Mississippi delta found the rich river-bottom land ideally suited to the plantation crops of sugar and cotton. Instead of fading away, slavery became more entrenched and economically important than ever in the early decades of the nineteenth century.

The Founders had also kept the shared sovereignty between the national government and state governments intentionally ambiguous. Federalism, this brilliant compromise between state sovereignty and national power, was at the heart of the movement from the Articles of Confederation to the Constitution. A precise definition of the relationship of the states to the national government was not conducive to ratification and acceptance of the new Constitution. In particular, a clear statement that either the states or the national government held ultimate sovereignty would have provoked intense opposition to the Constitution. Better to get it ratified, and leave it ambiguous for later generations to work it out.

These two issues—slavery and the precise nature of federalism—were at the heart of the political and economic conflicts leading to the Civil War. In some ways, the Civil War and its immediate aftermath may be viewed as the last chapter of the Founding, because both issues had to be resolved to make the Founding and the Constitution complete. This chapter will focus on the failure of the Founders' Constitution and political system to solve the immense social dilemma of slavery in a free society and its related political expression of shared sovereignty, or states' rights. It is a story of the failure of some of our most respected institutions. Ultimately the nation was saved and refounded through the virtue of ordinary people and the profound contribution of President Abraham Lincoln.

Slavery

The institution of slavery predates recorded history. The practice of one person actually owning another person—unthinkable as it is in our modern free society—was common throughout the ancient world. Europeans first conquered and then colonized the Western Hemisphere, looking for ways to exploit their vast and wild new dominions. Many of the first Europeans coming to the Western Hemisphere were opportunists—adventurers seeking riches who had limited skills and little intention of dirtying their hands with simple farm work. Slavery was the obvious solution to the demand for labor to work the land. Thus, Indians became the Europeans' first slaves, but their captors soon discovered that the original inhabitants of the New World were not suited to forced hard work and brutal treatment. Most of the Indian slaves simply died or disappeared instead of adapting.

Sugar cane, brought to the West Indies by Columbus, thrived in the mild, moist climate of the Caribbean. Europe's appetite for sweets and the devastation of the native population of the Caribbean (by disease, war, and slavery) caused Europeans to look to Africa for slaves to do the tough work of growing and processing sugar cane. About nine million African natives were forcibly brought to the Western Hemisphere as slaves. Most were sent to the Caribbean or Brazil, but over 400,000 slaves were eventually shipped into the area that would become the United States. All of the colonies had slaves, but slaves proved to be especially productive in growing four crops: tobacco (grown primarily in Virginia and North Carolina), rice (grown in South Carolina), cotton (grown in the Deep South from Georgia to Texas), and sugar (grown in Louisiana). These four plantation crops absorbed the productive efforts of most slaves, although many slaves would be trained for a variety of occupations, such as blacksmiths, carpenters, masons, sailors, and even cultured, well-dressed domestic servants for the wealthy.

Figure 11.4 This photo taken in 1864 shows a shop on Whitehall Street in Atlanta, Georgia, that sold slaves.

Figure 11.5 Photograph of restored slave quarters at George Washington's Mt. Vernon estate. These slave quarters were some of the nicer for the time, but share the common features of the cabins used to house slaves, such as multiple rows of bunks.

Slaves proved to be particularly productive on the large plantations that grew rice, sugar, and cotton. Because slaves did not have a choice about their working conditions, they could be organized to work in slave "gangs" that would move through a field of cotton or sugar cane like a giant machine performing the necessary tasks of planting, maintaining, and harvesting the crops. Older, less agile slaves directed young children and teenagers in caring for animals or gardening, while still other slaves were trained as craftsmen or skilled workers.

Economists studying slavery have concluded that slavery could be highly productive especially in plantation agriculture . Owners of slaves, which were an expensive investment, could expect a profitable return on their investment and a substantial income from the work of slaves.

The living conditions of slaves, though unacceptable by modern standards, were similar to the living conditions of unskilled free laborers of the time. Slaves usually lived as families in cabins. The plantation owner issued them clothing and many were allowed to keep a small garden near their cabin. Their diet, based on cornmeal, pork, and sweet potatoes, was relatively nutritious. During the slave years, the life expectancy of slaves in the United States was not quite as high as that of whites in the U.S., but surprisingly, they had a higher life expectancy than most Europeans and people living in cities like Philadelphia and New York of the time. Despite the unceasing hard work and near hopelessness of gaining freedom, the comparatively favorable living conditions caused the slave population to grow rapidly, even though slave importation had been outlawed in 1808. By 1860 the slave population had grown to nearly four million people.

Figure 11.6 Negro family representing five generations on Smith's Plantation, Beaufort, S.C. *A group portrait of ten slaves in front of the doorway to their quarters. Photograph by Timothy O'Sullivan.*

The burdens of slavery were more psychological than material for most slaves. Slaves were subject to the arbitrary, and at times, unpredictable will of their masters. The threat of separation and family breakup always hung over slaves because the master could sell slaves at will. Whipping was a common form of punishment on most plantations. Very few slaves were literate, and there were often laws against teaching them to read and write. Slaves could look forward to little more than a life of tedious, hard labor, where basic cheap food, clothing, and shelter were provided, but little else. Slaves could not expect to improve their conditions or their lives with any degree of certainty. Even the most talented and determined slave could seldom move above the position of a skilled craftsman, servant, or foreman.

Slavery had been an accepted part of society throughout most of human history. Even John Locke had defended the institution of slavery. By the time of the American Revolution, however, society was beginning to question the morality of this familiar institution. In 1776 the Society of Friends, or Quakers, passed a resolution condemning slavery and requiring its members to free their slaves. John Wesley, one of the founders of the Methodist Church, condemned the slave trade and preached against slavery. Other religious leaders took up the anti-slavery cause. By the 1830s the movement against slavery had a well-known name, "Abolitionism," and had become a crusade, mostly in the northern states. At first, abolitionists concentrated on the immorality of slavery and urged Southern slave owners to repent and free their slaves. Failing at this effort, some abolitionist leaders considered dissolving the union of states to be free of the stain of slavery.

The abolitionists did not attract wide support until the 1850s, when they succeeded in attaching the antislavery cause to the concerns of common white laborers fearful of slave competition and farmers interested in keeping the lands opening in the west free from slavery. The abolitionists reinforced their moral position with economic arguments by convincing many Northerners that slavery was also an outdated, outmoded system that was actually draining plantation profits. Southerners viewed these developments with increasing alarm, since about a third of the Southern states' total income was derived from slavery. They saw slavery and their cherished Southern culture as being gradually strangled by the elimination of the slave trade and the prohibition of slavery in some western territories. Southerners saw their "peculiar institution," the basis of their own romanticized view of Southern life, in great peril. The confrontation between the proponents of slavery and their critics represented the greatest crisis the United States had faced to that time.

The political expression of this conflict centered on interpretation of the rights of states and the rights of slaveholders. Southerners had long held the view that America's states, which predated the Constitution, held sway over the federal government.

South Carolina passed a law in 1832 declaring the national tariff law of 1828 null and void within that state. Even more radical measures were considered—withdrawing, or seceding, from the Union. The argument for **secession** rested on the belief that the states held sovereignty over the union they had formed. Because the delegates from the states wrote the Constitution and because the states had ratified it, states could unratify and withdraw from the Union. Slavery was a main cause for the conflict and the impetus for secession, but the more complicated political dispute about the sovereignty of the states in relation to the sovereignty of the national government was an underlying issue.

Figure 11.7 Arlington, Virginia, Band of 107th U.S. Colored Infantry at Fort Corcoran, November, 1865.

Constitutional Structure and the Slavery Crisis

Eight decades earlier, the Founders had sought to appease the Southern states' delegates and the interests of slave owners in three direct ways. First, for the purposes of representation in Congress, slaves were to be counted as three-fifths of a person, even though they were entitled to none of the privileges of citizenship. This provision gave Southern states increased representation in the House of Representatives. Second, the Constitution explicitly prevented Congress from passing any law prohibiting importation of slaves before 1808. Third, the Constitution explicitly required states to return slaves or indentured servants who had run away from their owners to another state. These requirements, as well as the deference given to states' sovereignty in the Constitution, caused some abolitionists to condemn the Constitution as a proslavery bulwark.

Secession

Formal withdrawal of states or regions from a nation.

Faction

A group of individuals who share the same specific political agenda.

Factionalism

When a city-state or nation has multiple factions that compete against each other. Madison felt that an extended republic would prevent factionalism from leading to tyranny because no faction could be large enough to dominate.

Sectionalism

Factionalism on a larger, more regional scale, with fewer but larger factions. Sectionalism during the 1800s over the slavery issue nullified the benefits of Madison's extended republic and led to the Civil War.

More moderate Northerners saw these three clauses as the price supporters of the new Constitution had to pay to secure Southern support for ratification. With good fortune, slavery would disappear when the slave trade was abolished. If not, later generations could solve this problem.

But did the Constitution provide the avenues and processes to resolve the slavery crisis?

Madison's large-republic argument is a good place to begin to see why the Constitution was not effective in the slavery crisis. Madison assumed that a large republic would have many different interests scattered among the various states. If power were kept at the state or local level, one interest or **faction** might become dominant. If power were transferred to the national level, many interests or factions would compete with one another for the attention and favor of the Congress and the Presidency. In the period before the Civil War, a different pattern of interests emerged. Interests became concentrated regionally rather than being dispersed nationally. Manufacturing and commerce dominated New England, New York City, and Philadelphia; family farms and their support dominated the Midwest; and slavery and plantation agriculture dominated the South.

This regionalization of interests created sectional or regional politics. Many political battles pitted one or two regions against the other. For example, economic conflicts between the increasingly industrialized Northern states and the primarily raw-material-producing Southern states left the states bickering not just over slavery, but also over tariffs. The North wanted a high tariff to make their textiles, shoes, and machinery more valuable, and the Southern states, not having much of an industrial base strongly opposed tariffs, which seemed to reduce foreign demand for their cotton. The West (which at that time was still Indiana, Ohio, Kansas, and other states and territories east of the Mississippi River) and the Northeast wanted the national government to subsidize transportation projects, such as railroads and canals. The South, blessed with an effective river system, opposed government involvement in transportation.

Powerful factions dominated each section of the country and often frustrated each other's wishes. On no issue was this sectional/factional rivalry more apparent than in the possible expansion of slavery to territories in the West. The explosive issue of slavery in the territories was first manifest when Missouri sought statehood status in

EXTENDED REPUBLIC **SECTIONALISM**

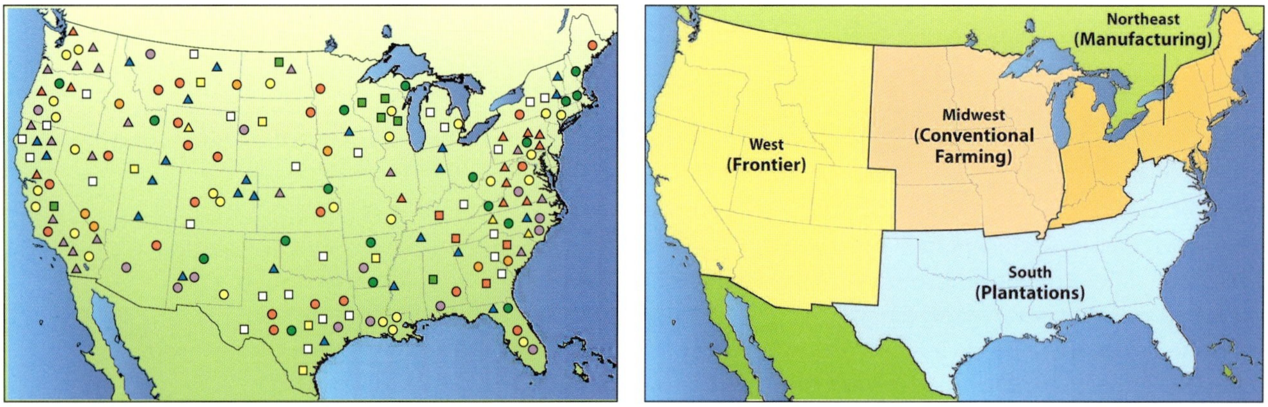

Figure 11.8 Madison's extended republic relied on interests being scattered throughout the nation, creating a diverse conglomerate. Prior to the Civil War, interests had become sectionalized based on economic and other differences between large regions.

1819. The Missouri Territory, whose settlers were bringing in large numbers of slaves, was to be admitted to the Union as a slave state, upsetting the previous balance of eleven slave states and eleven free states. The House of Representatives, with a Northern majority, approved a bill that would gradually turn Missouri into a free state by prohibiting further migration of slaves to Missouri after statehood, as well as freeing young slaves when they reached the age of twenty-five. The Senate rejected the bill from the House and passed a bill admitting Missouri as a state without any restriction on slavery.

The idea of Missouri as a slave state was disconcerting to the North. Parts of the northern border of Missouri reached as far north as Illinois, Indiana and Ohio. To the South, the thought of prohibiting the introduction of new slaves to Missouri and freeing young slaves was alarming. If these actions could be made legal in Missouri, what would keep those same laws from leaking over into other slave states? The issue appeared beyond compromise. An aging Thomas Jefferson, fearing for the Union, pronounced, "This momentous question, like a fire bell in the night, awakened and filled me with dread." The most ominous aspect of the Missouri crisis was the sectional nature of the vote in the House and Senate. Finally, a compromise was fashioned in 1820 that admitted Missouri as a slave state but restricted further expansion of slavery to the territory that is now Arkansas and Oklahoma. The **Missouri Compromise** did little to solve the long-term problem. It simply put it off and set the stage for the next regional crisis. Politics became increasingly regionalized. Madison's hopes for the benefits of controlling factions through a large republic simply did not apply here.

The mechanical devices of the Constitution were intended to promote compromise and cooperation among the branches of government and between the two parts of Congress. With block voting by regions, however, the system produced gridlock rather than compromise. The North tended to control the House, while the South held at least equal power in the Senate. Presidential candidates in the two-party race between the Whig Party and the Democratic Party had to appeal to the South, or at least to the border states, to win elections. In short, Presidential candidates had to avoid creating or supporting any solutions to the core issue of slavery and its expansion. Many other countries in the Western Hemisphere had successfully resolved the problem of slavery through "gradual emancipation," a process that freed slaves at some date in the future and prohibited additions to slavery. Gradual emancipation could not even be considered seriously in the United States because of the carefully constructed balance of power between North and South.

The potential for slavery's westward expansion reached its final crisis in the 1850s with the possible inclusion of Kansas and states mapped from the large Nebraska Territory. In 1854, **Stephen Douglas**, a senator from Illinois with Presidential ambitions, engineered passage of a bill that repealed the Missouri Compromise under cover of a requirement that each territory would vote on the issue of slavery. This requirement, known as "popular sovereignty," was an incentive for proslavery and antislavery factions to subvert the voting process by every method possible in order to have the vote on slavery go their way. Proslavery Missourians temporarily moved to Kansas to be counted as proslavery members of the territory. **John Brown**, an eccentric abolitionist of dubious background, led his sons and others on a search for proslavery settlers, which ended in the killing of five people at Pottawatomie Creek.

Popular sovereignty was leading to anarchy on the western frontier. Successive Presidents in the 1850s tried to appease the

Missouri Compromise

1820 agreement between slavery and anti-slavery factions in the United States that regulated slavery in western territories, prohibiting slavery above the border of Arkansas (except Missouri) and permitting it south of that border.

Stephen A. Douglas

1813–1861
An Illinois statesman who ran against Lincoln, Bell, and Breckinridge in the 1860 Presidential election on a popular sovereignty platform for slavery. Douglas also authored the Kansas–Nebraska Act, which repealed the Missouri Compromise and heightened the slavery debate.

John Brown

1800–1859
A controversial abolitionist who tried to start a slave rebellion and used sometimes violent guerrilla tactics in fighting against the institution of slavery.

Republican Party

Political party that stems from the controversy over slavery. It was dedicated to keeping future territories and states free from slavery.

South by pushing Kansas into the slavery column, with two unexpected results. First, Kansas settlers, tired of being pushed around by national politics, rejected statehood altogether. Second, political parties realigned, with sentiments about slavery and its expansion westward leading the way. Out of that realignment was born the **Republican Party**, committed to the idea of all future U.S. territories and states being free from slavery. This regional political party was completely unacceptable to the South.

The conflict over slavery foreshadowed the manner in which serious political and ethical conflicts are dealt with even today. Douglas's concept of popular sovereignty was attractive to many national politicians because they hoped it would remove pressure from them to solve a deeply troubling, passionately felt moral issue. Notice that politicians are quick to gravitate toward popular sovereignty when they do not want to deal

Figure 11.9 During a campaign for the senate in 1858, Illinois Republican Abraham Lincoln challenged Senator Stephen Douglas to a series of debates. The debates were mainly prompted by the Supreme Court's Dred Scott ruling. Although Lincoln lost the election, the debates on slavery attracted widespread attention.

Stephen Douglas' KANSAS-NEBRASKA ACT 1854

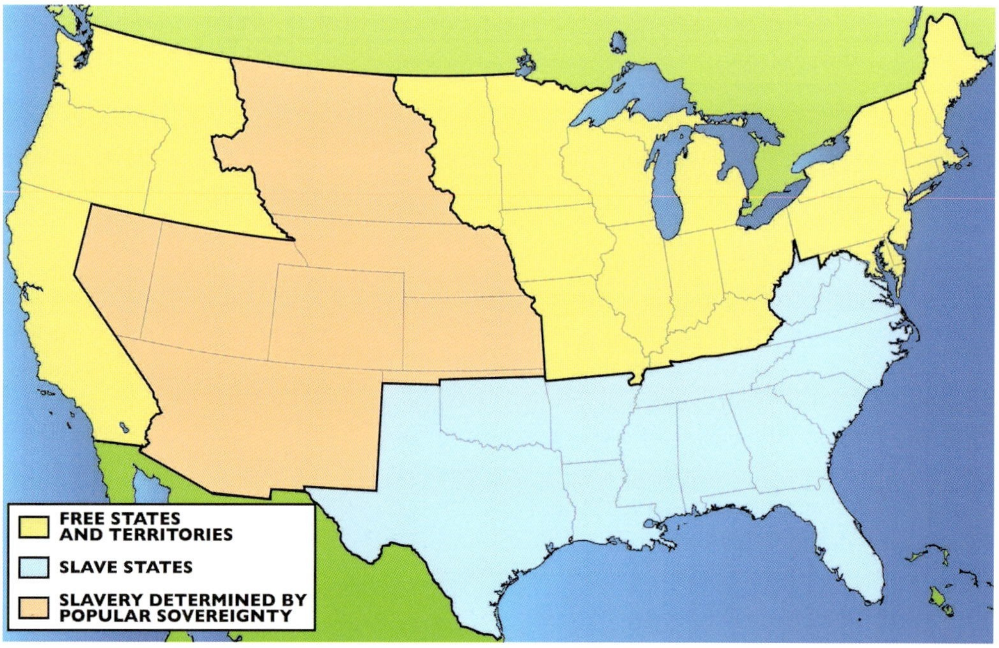

Figure 11.10 The Kansas-Nebraska Act essentially repealed the Missouri Compromise, allowing each territory within the area shown to vote on the issue of slavery. The Act spurred opposing Northerners to form the Republican Party in opposition and was a major factor in the political divisiveness prior to the Civil War.

with the issue in front of them. Modern social conservatives, having lost the battle on abortion in the *Roe vs. Wade* Supreme Court decision, now argue that abortion should be left up to the states. Twenty-first century liberals, caught between support for gay rights and concerns of their more traditional supporters, argue that the issue of gay marriage should also be left to the states. On the other hand, those who see a particular issue in stark moral terms do not usually accept voting as an appropriate resolution, arguing that morality cannot be legislated

Back in the mid 1800s, many reasoned that if Congress and the Presidency could not suggest an agreeable long-term solution to the problem of slavery, perhaps the Supreme Court could step in and resolve the issue. This led to the famous, or infamous, **Dred Scott** case. Scott, a slave owned by an army surgeon, had lived a number of years in Illinois and the Wisconsin Territory, where his owner had been posted. Illinois was a free state, and the Missouri Compromise had made Wisconsin a free territory. In 1847, Scott sued for his freedom. The case finally came to the Supreme Court in 1856. All the justices issued opinions about the case, but the opinion of Chief Justice **Roger B. Taney**, with which seven of the justices concurred, carried the most weight. Taney tried to end the controversy over slavery and the territories by declaring that the Missouri Compromise, which had prohibited slavery from some territories, was unconstitutional. Taney argued that Congress did not have the power to prohibit slavery in a territory or state. Taney's decision was a stunning victory for the South, but it infuriated the North.

A series of compromises delayed the ultimate military confrontation between North and South, but did nothing to solve the basic problem. By the 1850s, many Northerners had become committed to a "free-soil" position, that there should be no slavery in western settlements, a position that was intolerable to the South. Madison's brilliant and generally correct analysis of factions that he summarized in *Federalist 10* (see Appendix C) proved to be inappropriate for the pre-Civil War period. The power of sectional factions controlled politics and pitted regions against one another.

Politics and the Civil War

In the extreme and explosive political climate of the 1850s, with strong, unyielding regional factions and diametrically opposed moral views of slavery, meaningful compromise was impossible. Instead, the mechanical devices of the Constitution produced gridlock, with each group checking the other, making movement toward a solution impossible. In effect, each region held a veto, preventing a real solution to the problem of slavery—a solution that would have to be the emancipation of the slaves with minimal economic and cultural damage imposed on the South. Just as the mechanical devices of the Constitution were to promote compromise, the political structure of the Founding was to promote middle-of-the-road politics, helping the country find a solution acceptable to both North and South. Following the political turmoil and armed conflict between militias in Kansas and Missouri, the election of 1860 provided an opportunity for the political structure to use the will of the people to solve the most serious crisis facing the nation since the Revolutionary War.

The election of 1860 was the most divided and least centrist election in American history. Instead of two candidates with middle-of-the-road positions, the election brought forward four viable candidates representing markedly different points of view about slavery and its expansion westward. Abraham Lincoln was the nominee of the new Republican Party. In debate with

Dred Scott

c. 1795–1858
Slave who sued unsuccessfully for his freedom in 1857 because he had lived with his owner in several states where slavery was illegal. The ruling of *Dred Scott v. Sandford* determined that slaves were property and could not be freed by state laws. The ruling essentially nullified the Missouri Compromise and was a major factor contributing to the Civil War.

Roger B. Taney

1777–1864
Fifth Chief Justice of the Supreme Court, Taney ruled in *Dred Scott v. Sandford* that the Missouri Compromise was unconstitutional.

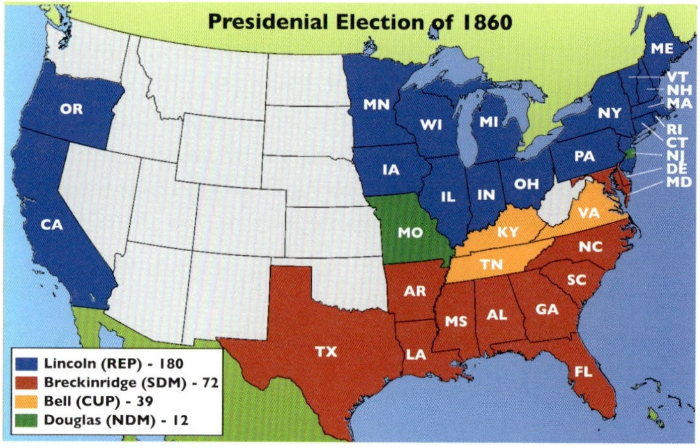

Figure 11.11 The candidates for President in 1860 and their respective views on slavery.

John Breckinridge

1821–1875
A Senator from Kentucky and the fourteenth Vice President of the United States, Breckinridge ran against Lincoln, Bell, and Douglas in the 1860 Presidential election on an extreme pro-slavery platform.

John Bell

1797–1869
A wealthy slaveowner from Tennessee who served in both the House and the Senate, Bell ran for U.S. President against Lincoln, Breckinridge, and Douglas in 1860 with the Constitutional Union Party on a moderate pro-slavery platform.

Stephen Douglas, Lincoln said that he did not believe the nation could exist half slave and half free. While that statement was undoubtedly true, it was also repugnant to Southern slaveholders. The Democratic Party could not agree on a national candidate. Southern Democrats supported **John Breckinridge** of Kentucky, running on an extreme proslavery platform. Northern Democrats ran on a platform of popular sovereignty, with Stephen Douglas as their candidate.

Some former Whigs who were dissatisfied with both the Democratic and Republican Party's platforms, formed the Constitutional Union Party, offering a platform of moderation on slavery and **John Bell** of Tennessee as their candidate. Figure 11.12 illustrates the basic positions of the candidates on the fundamental issue of slavery and its extension to the western territories. Though most people today would strongly support the position of Lincoln, his position and the position of Breckinridge represented the extremes of the day. Bell and Douglas represented centrist positions. The voting public was also divided between ardent supporters of slavery and equally ardent abolitionists, who were vowing to destroy slavery.

The 1860 election revealed the diversity

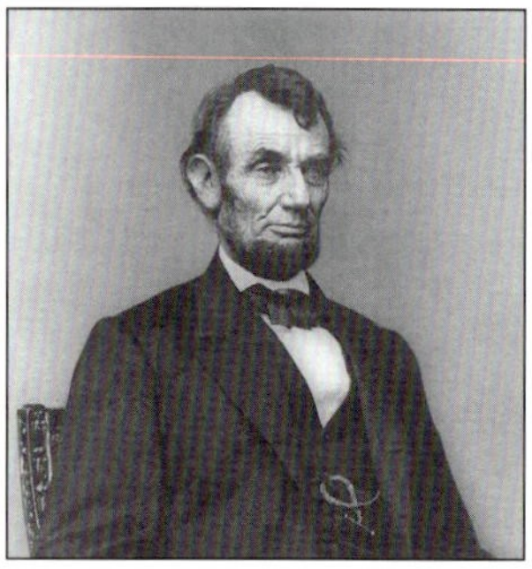

Abraham Lincoln
REPUBLICAN
Nation can not have both free and slave states and survive

Stephen A. Douglas
DEMOCRAT (NORTH)
Slave and free state status decided by popular sovereignty

John Bell
CONSTITUTIONAL UNION
No firm stand on slavery, must keep Union together

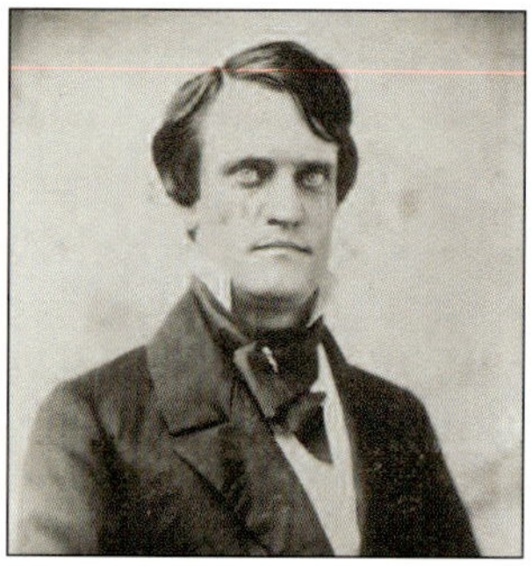

John C. Breckinridge
DEMOCRAT (SOUTH)
Pro-Slavery

Figure 11.12 The candidates for President in 1860 and their respective views on slavery.

of opinion among the voters, as well as the effect on the ballots of four, rather than two, viable candidates. Lincoln received about forty percent of the popular vote, but almost no votes for him came from the South and he won less than five percent of the popular vote counted in the border states. However, Lincoln received fifty-nine percent of the electoral vote because his votes were concentrated in the populous North. John Breckinridge received only eighteen percent of the popular vote, but twenty-four percent of the electoral vote because his votes were concentrated in the South. Stephen Douglas received nearly thirty percent of the popular vote, but only four percent of the electoral votes. He came in second in most states, but a strong second translates to no electoral votes. John Bell took the border states, winning nearly thirteen percent of both the popular vote and the electoral vote. Because votes were so regionalized, the candidates at the ends of the spectrum garnered the most electoral votes. Lincoln was the regional candidate of the more populous North, so he became President.

Can States Withdraw from the Union?

Ever since the adoption of the Constitution, various states, both in the North and the South, had considered withdrawing from the Union. No state had actually taken that step. With the election of **Abraham Lincoln**, Southern states were ready to test the nature of federalism and the actual binding power of the Union created by the Constitution. In December 1860, South Carolina held a special convention and repealed that state's ratification of the Constitution. By February 1861, six other states, Alabama, Florida, Georgia, Louisiana, Mississippi, and Texas, had also seceded.

Efforts at compromise failed. Lincoln, desiring to limit the secession to as few states as possible, tried to keep Virginia, Maryland, Delaware, and Kentucky in the Union. To do that, he needed the seceding states to appear extreme and anxious for war. Consequently, he was intent on making the Confederacy fire the first shots. He chose to send food and supplies to the most threatened federal facility—Fort Sumter in Charleston Harbor, South Carolina. That resupplying effort was considered by the South to be an act of provocation, and the Confederacy issued the fateful order to fire on the fort. Fort Sumter fell and Lincoln immediately issued a call for troops. The border states now had to choose a side. Virginia, at least the eastern two-thirds of it, eventually seceded, as did Arkansas, Tennessee, and North Carolina. The remaining states Delaware, Maryland, Kentucky, and Missouri—chose to stay with the Union.

Even with the smell of gun smoke in the air and calls for volunteers going out throughout the nation, the question was still asked: What was the nature of the Union formed by the Constitution? Did the states, many of which had existed before the Constitution, have the right to dissolve their rela-

Abraham Lincoln

1809–1865
The 16th President of the United States, Lincoln sought to end slavery and preserve the Union. He signed the Emancipation Proclamation and delivered his famous "Gettysburg Address."

Figure 11.13 The January 26, 1861 edition of Harper's Weekly *featured this illustration, showing Fort Sumter on the Eve of the Civil War. The caption reads: "Fort Sumter from the rear, at low water—From a sketch by an officer of Major Anderson's Command."*

The Confederacy vs. The Union

Jefferson Davis

1808–1869

President of the Confederate States

Abraham Lincoln

1809–1865

President of the United States

Robert E. Lee

1807–1870

General of Confederate forces

Ulysses S. Grant

1822–1885

Final general of Union forces

tionship with the Union? Or did the people, by ratifying the Constitution in state conventions, transfer sovereign power from the states to the national government? Shared sovereignty, which had seemed such an ingenious invention eighty years earlier, now seemed a political mirage. Shared sovereignty was serviceable when the nation dealt with ordinary issues and minor problems. In the end, however, only one entity—either the states or the national government—could have the final say.

Unfortunately, the Constitution was completely silent about the issue of state versus federal sovereignty or the right of secession. Each side was free to interpret federalism and the relationships between states and the national government as it wished. The states of the Confederacy asserted the right to secede from the Union. Lincoln, as President of the United States, rejected the right of the states to dissolve the bonds that brought them together under the Constitution. In his first inaugural address, he said:

> I hold that, in contemplation of universal law and of the Constitution, the Union of these States is perpetual. Perpetuity is implied, if not expressed, in the fundamental law of all national governments. It is safe to assert that no government proper ever had a provision in its organic law for its own termination. Continue to execute all the express provisions of our National Constitution, and the Union will endure forever—it being impossible to destroy it except by some action not provided for in the instrument itself.

If neither side acquiesced, then war was inevitable. Each side underestimated the resolve of the other side, and neither side could foresee the unimaginable catastrophe for which they were preparing.

Saving the Union

The structures and devices so brilliantly crafted during the Founding could not help the country resolve the issues of slavery or federalism, and the country reluctantly marched to the Civil War. Checks and balances made the gradual elimination of slavery impossible. No proposal for gradual emancipation of slaves was ever seriously considered in the U.S. Congress, in spite of the example of many other countries that had struggled with it. Supreme Court pronouncements were unenforceable and therefore ineffective. Politics produced no will toward compromise. The Constitution left open the legitimacy of secession. What then saved the Union? At the core, the Union was saved by the virtue of a few leaders and by the virtue and sacrifice of countless ordinary citizens.

Abraham Lincoln was the most extraordinary man to assume the Presidency since George Washington. Lincoln overcame widespread northern opposition to war, paper-shuffling generals unwilling to fight, foreign support for the South, and pervasive northern racism, to manage a terrible war, free the slaves, and maintain the Union. It is difficult to measure the impact of a single individual on a nation's history, but it is not difficult to imagine the breakup of the Union with someone other than Lincoln as President. Lincoln's perseverance and eventual success in the face of so many challenges was nothing short of miraculous.

Beyond the saving graces of a most extraordinary President, the virtue of the ordinary soldiers who volunteered to fight was also key to overcoming the crisis of secession. For whatever motives, millions left home and family to help save the Union with the high ethical corollary of freeing the slaves. Why did they volunteer? There were bounties paid to appeal to their self-interest, but by 1862 it was clear to all that this was a dangerous war, where hardship, pain, and actual loss of life was a significant possibility. Yet they continued to come to fight. Eventually, a draft was instituted, but most of the soldiers were volunteers who fought in part because the Union and freedom meant something to them.

Finally, there was virtue in the aftermath of the war. Southern leaders, defeated on the

Figure 11.14 Abraham Lincoln at the Gettysburg Address. Lincoln's speech was so short the photographer, Matthew Brady, missed Lincoln speaking. In this 1863 photograph, Lincoln (center of the small red box) is sitting down after delivering the Gettysburg address. (See Appendix C.)

Figure 11.15 Battle of Gettysburg, *July 1-3, 1863. Lithograph by L. Prang & Co., 1887.*

battlefield, could have conducted a protracted guerrilla war in which bands of southern soldiers harassed and impeded the lawful government. There were isolated incidents of this kind, but almost all Confederate leaders counseled cooperation with the federal government. The South had lost. It was time to move on and put the country back together. Without some good will on both sides, a guerrilla war or widespread resistance could have gone on for years and made reconciliation of the North and a devastated South impossible.

Rebuilding the South without slavery was very difficult, especially following the assassination of Lincoln just five days after the formal end of the war. Freed slaves needed to be housed and employed. The challenges of race relations and racism needed to be addressed. Political institutions needed to be remodeled to include freed slaves. The twelve-year-long Reconstruction Era following the war ended without any adequate and lasting solution to these problems. When state control was given back to southern whites, they immediately enacted state laws to maintain segregation and to disenfranchise freed slaves. There was occasional violence in the form of beatings and lynching to maintain race control. The Civil War had ended slavery—but not racial antipathy and discrimination.

Structural Changes after the Civil War

The Civil War led to important changes with long-lasting effects. No event of the 19th century changed political and social relationships as much as the Civil War did.

Every community, North and South, had disabled veterans to remind them of the cost of the war. American Political leadership was drawn from the veterans of the Union Army for the next half century. Only one (Grover Cleveland) of the next seven Presidents elected after Lincoln did not serve in the Union Army. The pensions of Union Army veterans and their widows became a major expenditure of the government, foreshadowing Social Security and other federal assistance programs. The Republican Party started each campaign reminding voters that they had prosecuted and won the war. Indeed, the Civil War was the defining memory and defining influence of American society for a half century.

The full effect of structural changes that followed the War had a lasting impact on the nature of the Constitutionally-driven Union. The victory of the North ended the ambiguity surrounding federalism. Four years of war had resolved the question of sovereignty in favor of the national government. States did not have the right to secede from the Union. Federal law and the federal government were supreme over state law. The war changed the character of federal/state relationships. States still handled important governmental functions, such as supervising the police and pro-

AMERICAN WAR CASUALTIES

Civil War

618,000

Union

360,222

Confederacy

258,000

OTHER AMERICAN LOSSES

Revolutionary War	25,324
World War I	116,516
World War II	405,399
Korean War	33,741
Vietnam War	58,148

Figure 11.16 About the same number of Americans were killed during the Civil War as all other American Wars combined.

viding education, but the federal government now had a different character and a different level of power.

Three amendments were adopted to eliminate slavery and to clarify the relationships between the national government and the states. The **Thirteenth Amendment** (1865) abolished slavery immediately throughout the country. The **Fourteenth Amendment** (1868) eliminated the constitutional clause counting slaves as three-fifths of a person for the purposes of representation, repudiated the debts of the Confederacy, and prohibited individuals active in the Confederacy's top tier of leadership from holding public office. The Fourteenth Amendment also applied the Bill of Rights and other rights mandated in the federal Constitution to freed slaves. Furthermore, this Amendment guaranteed all individuals equal protection under the law, including state law. Finally, the **Fifteenth Amendment** (1870) guaranteed the right of all adult male citizens to vote regardless of race, color, or previous condition of servitude. Unfortunately, southern states later found ways around this Amendment by applying literacy requirements unequally.

The Civil War was both tragic and heroic. It inflicted suffering and hardship on the whole country, yet it renewed the dedication of the United States to the ideals of the Declaration of Independence. To Lincoln and like-minded people, the war initiated a "new birth of freedom," which reaffirmed the fundamental propositions of equality before the law and government by the people.

Figure 11.17 This cartoon makes fun of a carpetbagger, a Northerner who moved to the South during reconstruction between 1865 and 1877. Because of their political connections they were often able to gain political office and participate in controlling various ex-Confederate states. Carpetbagger was a derogatory term, suggesting an exploiter who does not plan to stay.

Inconsistencies between the ideals of the Founding and the practice of government would continue to beset the country, but racial slavery, the greatest gap between ideals and reality, had finally been eliminated.

Just as the Founders had left vexing issues to be resolved by the Civil War generation, the leaders of the Civil War and Reconstruction left to others the challenge of bringing race relationships into conformity with the ideals of the Founding. Neither North nor South was ready to accept their African American populations as full and equal participants in the American enterprise. The problem of inequality among races would not be addressed until the birth of the civil rights movement a full one hundred years later.

Thirteenth Amendment

Abolished slavery in the United States.

Fourteenth Amendment

Defined citizenship and overturned the three-fifths compromise for slaves when determining representation, repudiated Confederate debts, and prohibited Confederate leaders from holding public office.

Fifteenth Amendment

All male citizens are granted the right to vote regardless of race, color, or previous condition of servitude.

Timeline

Contextual Events

Dred Scott v. Sanford decided **1857**
South Carolina secedes (Dec) from the Union **1860**
Mississippi secedes (Jan) **1861**
Florida secedes
Alabama secedes
Georgia secedes
Louisiana secedes
Texas secedes (Feb)
Virginia secedes
Arkansas secedes (May)
Tennessee secedes
North Carolina secedes

1820–1870

Chapter Specific Events

1820 Missouri Compromise

1860 Abraham Lincoln wins Presidential Election
1861 (Apr) Civil War begins with the Battle of Fort Sumter
1862 Battle of Antietam
(July) Battle of Gettysburg
1863 Emancipation Proclamation issued
1865 Civil War ends
President Lincoln assassinated
13th Amendment ratified (abolition of slavery)
1868 14th Amendment ratified (equal protection)
1870 15th Amendment ratified (all male suffrage)

Key Terms

Confederacy
secession
Missouri Compromise
Fifteenth Amendment
Emancipation Proclamation
factionalism/sectionalism
Thirteenth Amendment
federalism
state sovereignty
Fourteenth Amendment

Key People

Robert E. Lee
Stephen A. Douglas
John Breckinridge
George B. McClellan
Chief Justice Roger B. Taney
John Bell
Abraham Lincoln
Dred Scott

Questions

1. What two "issues" did the Founders leave to "be resolved by future generations?"
2. In what three ways had the Founders "sought to appease the interests of slave owners?"
3. How did Madison's theory of the large (or extended) republic fail to stop the Civil War?
4. Explain voter sentiment in the election of 1860. How was this election different from previous elections?
5. What three major structural changes took place after the Civil War?

Chapter 12

The Rise of Government

Abstract

With boat loads of immigrants coming ashore and rapid technological change, America enjoyed rapid economic growth at the turn of the 20th century. Along with this growing economy came growth in government. This chapter details the actions of the Progressives of the early 20th century who called on government to counter the growing power of private entities through economic and social reform. Government responded to these economic and social problems but also significantly changed several aspects of the Founding. The New Deal response to the Great Depression also expanded government as a response to the influence of impersonal market forces.

When Abraham Lincoln assumed the Presidency in 1861, he faced the greatest national crisis since the adoption of the Constitution. His office was in a rather large room on the second floor of the White House. In the center of the room was a large table where he held meetings with his cabinet and others while his desk was off in a corner of the room facing a wall. Two young personal secretaries—John Nicolay and John Hay—helped him conduct the affairs of the presidency. His cabinet consisted of seven men who were responsible for 36,000 civilian employees of the federal government. Of that number, 30,000 of these public servants were postmasters and mailmen sprinkled throughout the country. There were just over 2,000 government workers in the Washington D.C. area as Lincoln began his service as President.

Some eighty years later on the eve of World War II, Franklin Delano Roosevelt presided over a much different government. The civilian employees of the Federal government now numbered over a million with about 140,000 employees in Washington D.C. Besides the ten cabinet offices that managed much of the executive branch, many independent government agencies had been created. For ease in written communication, the agencies were reduced to three- or four-letter acronyms—ICC, IRS, etc. The White House of Lincoln's time had been continuously remodeled and expanded to meet the needs of a growing government. After the Civil War, the White House soon became too small to contain the Presidency. Shortly after Theodore Roosevelt took office in 1901, the building's west wing was constructed, followed by the east wing in 1942. By the time FDR faced a looming world war, the immediate staff of the President had grown into the hundreds.

Figure 12.1 The federal government under Abraham Lincoln was much smaller than the one under Franklin D. Roosevelt, as can be seen by the relative size of each President's White House.

The United States, a country founded on a rebellion against a king whose excesses now seem rather mild (taxes on tea and documents, failure to call colonial legislatures into session, and commercial regulations requiring goods headed to Europe to first be sent to England) now had a government that was large, impersonal, and powerful. The nation's Founders mistrusted government power, yet they recognized the need for a limited government to protect the nation and its citizens from invasion and each other. However, the word "limited" had faded into obscurity, and the Americans of 1940 were now under the control of a government that was taking nearly 20 cents of every dollar they earned. What had happened? Had the American Revolution been betrayed? Was a new revolution needed every so often, as Thomas Jefferson had prescribed? Or was the government simply meeting the needs of its citizens? How did we go from Abraham Lincoln and his two secretaries to FDR and a staff of hundreds? Why did government grow so much faster than the economy or the population of the country?

Much of the growth of government may be seen as a response to growth in private power. We will examine the circumstances that led the Progressives of the late 19th and early 20th centuries to call for an increased government power to counter the private power of trusts and political machines. We will then examine how FDR and his New Deal used government power to counter the impersonal forces of the market economy. Finally, we will study the implications of government growth on the Constitution and the Founding.

Public vs. Private Power

Americans have typically mistrusted large organizations and concentrated power. After all, in colonial times their own organizations—families, businesses, town or colony governments—were small and easy to supervise, while the only large organization they knew, the British Empire, was distant, impersonal, and oppressive. The problem of large organizations was like the problem of government: both wielded a great deal of power over people's lives and many citizens firmly believed that power—regardless of its honorable origins—eventually corrupted both societies and individuals. After the colonial Americans had fought their Revolution, they wanted no more gigantic bureaucracies. Indeed, it was only with the greatest difficulty that they could be persuaded to adopt the federal Constitution. In spite of this suspicion of government, government soon began to grow. To understand the roots of government growth, we must also understand the growth of private institutions and the fears it generated in ordinary people in the nation's early days following the Civil War.

Government exercised *public* power but what about *private* power? Americans had few worries on that score before the Civil War. But after Appomattox, which marked the end of the Civil War, large corporations amassing large amounts of capital began to dominate the economic scene. Among the deep changes the war-battered nation now had to contend with was the increasing reality of private power, which Americans encountered in all manner of organizations capable of shaping their lives. Private power would be exercised by fledgling labor unions and employer associations, by political parties, by organized interest groups, by railroads and steamship companies, by hospitals, by banks, by colleges and universities, by turnpike builders and power companies and the

Figure 12.2 Directors of the Union Pacific Railroad gather on the 100th meridian, approximately 250 miles west of Omaha, Nebraska Territory, in October 1866. The train in the background awaits the party of Eastern capitalists, newspapermen, and other prominent figures invited by the railroad executives.

Jay Gould

1836–1892

Often regarded as the most unethical of the Robber Barons, Gould was involved with Tammany Hall and Boss Tweed early in his career. After damaging his reputation in a gold speculation that instigated the panic of Black Friday in 1869, Gould went on to gain control of western railroads and by 1882 had controlling interest in 15% of the country's tracks. Although mistrusted by many of his contemporaries, Gould was recognized as a skilled businessman.

Leland Stanford

1824–1893
Stanford was the 8th governor of California and President of the Central Pacific Railroad. He hammered in the famous golden spike on May 10, 1869. A large portion of his fortune went into the founding of Leland Stanford Junior University, named after his teenage son who had died of typhoid while in Italy.

Figure 12.3 The Jupiter, which carried Leland Stanford (one of the "Big Four" owners of the Central Pacific) and other railway officials to the Golden Spike Ceremony.

owners of telephone exchanges, by ferry boats and trolley lines, and by hundreds of other services in the increasingly complex and sophisticated urban environment that was replacing rural America's simpler, season-governed way of life. Americans would be bothered by the demands, the regulations, and problems caused by all of these special-interest groups. Most of all, however, they would be bothered by the speedy rise of political machines and industrial monopolies—the so-called "trusts."

Cornelius Vanderbilt

1794–1877
Vanderbilt made his fortune in shipping and then later in railroads. A ruthless businessman, he did little in the way of philanthropy; it was his second wife's nephew that convinced him to help fund Vanderbilt University.

Figure 12.4 Cornelius Vanderbilt and James Fisk are shown in a race for control of New York's rails. Vanderbilt straddles his two railroads, the Hudson River and the New York Central.

The Rise of Big Business

Before the Civil War, most business enterprises were small. Family farms, craftsmen working with an apprentice and journeyman, small owner-supervised manufacturing and small commercial partnerships or banks were the lifeblood of the nation. There were a few exceptions, such as the large sugar, rice, and cotton plantations of the deep South, textile mills in New England employing hundreds of workers, and railroads slowly extending westward. Still, the vast majority of American workers were self-employed. Those who had a boss usually worked right along side their employer.

Growth in government started with the Civil War. In order to oppose the "slave power," which itself represented private power run amok, it was necessary to create a bigger government to supply and direct armed forces of over a million men. In the course of the struggle to bring the South back into the Union, the federal government tripled in size and started assuming and exercising powers well beyond the Founders' imagination. The pensions given to the veterans of the Union Army expanded the government still further and created the first large-scale financial transfer program administered by the federal government.

Something similar was happening to American industry. After the Civil War the scale of the economy grew rapidly, with immigrants flocking to America to take advantage of the opportunities they saw here. New technologies and new sources of energy attracted large-scale business and ambitious, visionary men. The Iron foundries of antebellum America that had employed a few dozen workers, all of whom had probably grown up together in the same community, were replaced by steel mills with thousands of workers, many of them new to the country and barely able to speak English. Whale oil that had been supplied by partners outfitting a ship or two was replaced by kerosene produced by companies working with millions of dollars of capital.

Most ominous to most Americans were the railroads, which spread rapidly over the whole country after the Civil War. Many railroad magnates were not shy about making financial gifts to influential elected officials for writing and passing bills that gave the railroad companies vast "rights of way," up to twenty miles of free land on each side of the tracks they were laying. The railroad companies' influence in Congress sometimes seemed to allow unfair control over Midwestern farm communities and anyone else whose livelihood depended on shipping and receiving goods over the rail lines.

This growth of corporations in American commerce during the second half of the 19th century had a significant impact on employer-employee relations. In early America, as in Europe for a thousand years earlier, such relations were generally characterized by a certain amount of *paternalism*. The factory owner was usually more than just a boss to his employees. In early America's relatively small communities, the factory owner was dominant, but he was also a friend, a father figure, even a neighbor. Workers went to him with their personal problems, sought his advice on financial matters, made him godfather to their children. This employer-employee fraternity disappeared with the growth of the large corporation. The impersonal corporate owner whom the average employee might never meet became something of an adversary. Workers began to form trade unions to protect themselves against what they perceived as impersonal, money-driven corporate power.

Figure 12.5 The USS Birmingham *is launched from Newport News shipyards on March 20, 1942. Huntington's shipyard has built many of the U.S. Navy's ships, including the first nuclear-powered aircraft carrier. It is also the only shipyard capable of building* Nimitz *class supercarriers.*

Monopolies

Large corporations often compete in the marketplace with different attitudes and different strategies. Small businesses generally assume their individual actions are too small to determine or influence prices. The wheat

Figure 12.6 The Frick Collection in Manhattan. Featured artists include Rembrandt, Titian, Goya, Vermeer, and Gainsborough.

Collis P. Huntington

1821–1900
One of the Big Four with Leland Stanford, Huntington was involved in both railroads and shipping. He founded Newport News Shipping, the largest privately owned shipyard in the United States.

Henry Clay Frick

1849–1919
Frick partnered with Andrew Carnegie and later helped form the United States Steel Corporation. He was a patron of the arts and amassed an extensive art collection that exhibits at a mansion he purchased on 5th Avenue in New York.

John D. Rockefeller

1839–1937
Rockefeller was the founder of the Standard Oil company. Known for his practice of buying out his competitors, Rockefeller was a favorite target of muckrakers. He was also a generous philanthropist. His name has become synonymous with massive wealth.

farmer and the shoemaker, for instance, assume that they have no influence over the retail price or wheat or shoes. They produce their products as efficiently as possible and sell at the prevailing, hopefully competitive price. A large corporation or a group of corporations acting together may dominate the market and set a high price for a product simply by withholding that product from the market (or influencing a government agency to set regulations that would force smaller competitors out of the market). Market power is often a goal of large corporate interests, and efforts to dominate a market have often encouraged unethical or illegal tactics. The path to market dominance often involved destroying competitors, buying them out, or in the case of trusts, making behind-the-scenes agreements to dominate a market. The sheer size of a corporation often caused people to assume it had more market "clout" than it actually had. Businesses taking advantage of cheap labor, new technology, expanding markets, and other post-Civil War conditions grew to keep pace with the nation's rapid population and economic growth. However, bigness does not necessarily imply monopoly power. Economic power depends on the extent of competition and the control a company or corporation can exercise over prices of its product. (See appendix A2 for a more complete discussion of monopoly as a market weakness.)

A few of the nation's new industries were inherently monopolistic. The New York Central Railroad and the Pennsylvania Railroad might compete between Chicago and New York. But each had monopoly power in the small towns on their particular routes through Indiana, Ohio, and Pennsylvania. Farmers and other shippers soon noticed that shipping rates from small towns to cities were much higher than rates between the major hubs where railroads competed with one another. When the builders of the Central Pacific Railroad succeeded in laying their rail lines across the Sierra Nevada Mountains, they conveniently monopolized all rail traffic in California—and soon all other forms of long-distance transport as well.

New technology with exclusive rights granted through government patents made its own contribution to several early monopolies. Frank Sprague's electric traction motor, which made the urban trolley possible, led to the creation of a giant corporation holding the Sprague patents. General Electric came about through Thomas Edison's patents, and the Westinghouse Corporation resulted from George Westinghouse's invention of the air brake.

How did late 19th century Americans feel about monopolies? Certainly, there were feelings of ambivalence. The rags to riches stories of some of the reigning corporate leaders were inspiring. In the story of **Andrew Carnegie**, for instance, they saw real-life reflections of the American Dream. Carnegie came to the United States from Scotland as a poor immigrant boy. He lived the life of an *Horatio Alger* storybook hero, studying nights to learn accounting, honing skills on the job, and winning the attention of powerful fig-

ORGANIZED UNDER THE MANUFACTURING LAWS OF OHIO
No. 19
250 Shares
STANDARD OIL COMPANY.
CAPITAL STOCK $3,500,000. ALL PAID.
35,000 SHARES. $100 EACH.
This is to Certify that William Rockefeller is entitled to Two Hundred and fifty Shares of One Hundred Dollars each in the Capital Stock of the Standard Oil Company, transferable on the Books of the Company in person or by Attorney only on the surrender of this Certificate and due payment of all liabilities on the part of the holder to the Company, subject to the provisions of Law and the By Laws of the Company. This Certificate is valid only when signed by the President and Secretary.
Cleveland O. April 3 1875
Secy
Prest

Figure 12.7 Standard Oil Company stock certificate signed by J.D.Rockefeller. Although criticized for forcing competitors into bankruptcy, he would frequently make a fair offer to buy them out first, sometimes offering Standard Oil stock as payment.

ures who could foster his career. He got into the steel-making business at the right time and in the right place and made the most of his opportunities. When financier J. P. Morgan bought Carnegie Steel, making it the foundation of another corporate giant, United States Steel, Carnegie retired to spend his enormous fortune on charities and endowments for hundreds of public libraries across the nation.

Journalists often portrayed the leaders of corporations in dark and sinister terms in the news media, labeling them "**Robber Barons**" to reflect their unscrupulous powers and grasping natures. In turn, journalist-reformers were often called "**muckrakers**" and their news stories were characterized as "yellow journalism." **John D. Rockefeller**, another rags-to-riches success story, represented the kind of businessman who was the prime target of muck-raking journalists, and with ample reasons. The Standard Oil Company founded by Rockefeller came to dominate the oil industry from drilling to refining to retail sales. Rockefeller bought out many of his competitors and ruthlessly competed with others in ways both fair and foul. He amassed a fortune which was valued at about $1 billion in 1912, perhaps $20 billion in contemporary terms. To ordinary Americans, he was considered to be much more powerful than even the Presidents of the United States. He and his family have shared much of the Rockefeller fortune with the nation and later the world through the Rockefeller foundation.

Figure 12.8 Skibo Castle in Scotland. Andrew Carnegie purchased it in 1898 and spent £2 million renovating it and adding a lake and an 18-hole golf course. It is currently operated as a luxury hotel.

The well-paid lawyers serving the "Robber Barons" found legal loopholes (and sometimes compliant judges who would reinterpret established laws), enabling their patrons to escape the constraints of the law. Indeed, corporate leaders often used local courts to issue rulings and injunctions to tilt the business field in their favor. Once in

Andrew Carnegie

1835–1919
Carnegie made his fortune in the steel industry. In his later years, he donated most of his money to establish schools, libraries, and universities around the world.

Robber baron

Muckraker term used for leaders of large corporations and trusts to reflect their power and unscrupulous natures.

Muckraker

Journalists that portrayed the leaders of corporations and the actions of their companies in unfavorable circumstances, writing "yellow journalism."

HORATIO ALGER, JR.

Horatio Alger, Jr. (1832–1899) was the author of over 130 dime novels. Most of his books were rags-to-riches stories of young boys who overcame poverty and adversity through hard work, determination, courage, and concern for others. Although his works tended to be formulaic, they were extremely popular, and helped to shape the modern idea of the American dream.

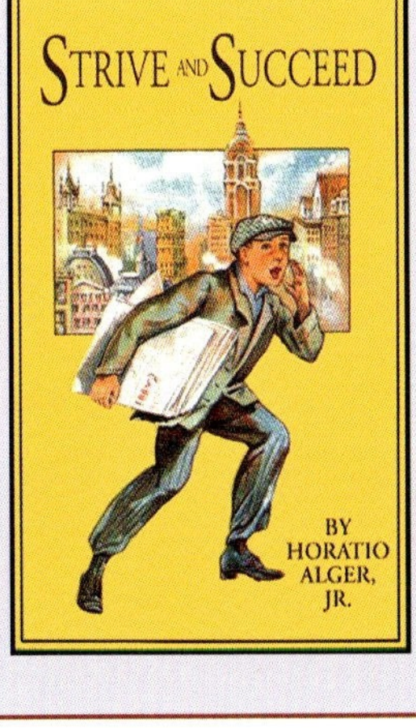

J. P. Morgan

1837–1913
Morgan was a banker and financier, and his firm, J. P. Morgan & Co., was one of the most powerful banking houses in the world. It financed the formation of the United States Steel Corporation, the world's first billion-dollar corporation. Morgan was also a benefactor of the Metropolitan Museum of Art, the American Museum of Natural History, and Harvard University, among others.

The Octopus

Frank Norris's novel that recounted the depredations of California railroads.

The Jungle

Upton Sinclair's muckraker book that exposed the practices of Chicago meat-packing plants.

Figure 12.9 The Metropolitan Museum of Art. J. P. Morgan served as its president from 1904 until his death in 1913. He loaned many works from his own collection to the Museum, and they were donated to the Met along with other of his pieces (about 7,000 in all) after his death.

power, monopolists used their considerable political clout to push for high tariffs, which shielded them from competition with foreign suppliers.

Beginning in the 1880s, Americans reacted against the presumption and power of such corporations. Many believed that railroads exercised unfair power over people and businesses dependent on the railroad. America saw bloody strikes—among them one at Carnegie's Homestead Plant—dramatizing Big Business's callous treatment of its workers. They had seen firsthand what happened to streetcar fares and heating oil prices when a trust of monopoly exerted their power over the market. Best-selling novels of the day included Frank Norris's ***The Octopus***, recounting the depredations of the California railroads, and Upton Sinclair's ***The Jungle***, exposing sins of the Chicago meat-packing plants.

Socialism

One proposed political remedy to the rapacity of trusts was socialism. It first appeared as a viable political movement in 19th century Europe and was brought to the United States by political refugees and others seeking to organize a utopian society within the new nation. Although socialism was a far cry from the ideals of the Constitution, long-time socialist Presidential candidate **Eugene V. Debs** garnered nearly a million votes—a breathtaking showing—in the 1912 General Election.

Other voices, speaking out against the trusts' greed and disregard for workers and consumers alike, made Debs claims sound positively moderate. Groups calling themselves Anarchists, Syndicalists, or the International Workers of the World prophesied an industrial Armageddon in the United States. They said workers would rise up and take charge of the factories once and for all, and all forms of government would melt away. Theodore Roosevelt and others would quietly incorporate some aspects of Debs's socialism into the expanding government bureaucracy, but the entire concept of socialism and its more radical variants never received as much power in the United States as they gained in Europe. America at the turn of the 20th century was still seen as a land of economic opportunity where you or your children could climb the ladder to economic success.

Figure 12.10 A 1904 campaign banner for the socialist party showing Eugene V. Debs for President. Socialism became a popular platform for many who wanted to disempower big business.

Despite the anti-trust stand of many major newspapers and the efforts of a few Congressmen and others, the problem of the trusts and corporate power did not go away. Individual consumers, workers, farmers, and small shopkeepers all felt powerless when confronted with corporate power. Correctly and incorrectly, they saw the corporations as economic tyranny—setting wages, determining working conditions, setting prices, and manipulating the political process for their own gain.

In principle, competitive markets were supposed to eliminate economic power. Businesses competed to obtain quality workers. If corporation X was making large profits while paying its workers 10 cents an hour, then competing corporation Y would attract X's workers by offering eleven cents an hour. Most workers weren't sure how a market economy actually worked, or even if it worked at all. Besides, markets that substituted the impersonal forces of demand and supply just put workers and consumers at the mercy of a capricious market rather than the personal prerogative of corporate leaders. In short, individuals often felt powerless in this new economy.

At the same time Americans also feared a tyranny of private power. The economy had been offering a rising standard of living to most citizens. Between 1870 and 1910, even as the fear of corporate power grew, income per person doubled, 40 million immigrants were effortlessly absorbed into the economy, and new consumer products such as electric lights and automobiles were becoming common in many American neighborhoods. Moreover, the work week was slightly shorter for most industrial workers, more young people were graduating from high school, infant mortality was falling, and life expectancy was increasing. While individuals distrusted and feared the economic power of corporations, the nation's standard of living was clearly rising.

Figure 12.11 Industrial Workers of the World (I.W.W.) demonstration, New York City, April 11, 1914.

Political Machines and Their Bosses

The big corporation wasn't the only form of organization that Americans found threatening. The era of big business also launched a new kind of "political machine," so named by a writer who saw the ever-more efficient and increasingly open way politicians and entire political parties were buying votes for candidates. The concept of trading favors, giving money or promises of government jobs in return for a person's vote at the ballot box has been around since before Andrew Jackson's time, but America had entered a prosperous era of semi-legal ("The results justify the means.") and outright corrupt trusts. The law-making influence of a Congressman in silent partnership with a trust could result in enormous trust profits and under-the-table wealth for the elected official. The political machines presented a new threat to the Founders' concept of representative democracy.

Eugene V. Debs

1855–1926

Debs ran for U.S. President five times as a socialist. The last attempt (1920) was made while he was serving time in prison for obstructing the draft of World War I.

Political machines of the early 1800s usually relied on free liquor passed out near the polling places on Election Day, often adding vague promises of government jobs and other favors for votes cast for their candidates. By the late 1800s advertising and other promotional advances made the development of more sophisticated political "machinery" possible. Early political bosses made use of paid-off newspaper editors to secure favorable comment about their candidates, and agents of various trusts and other organizations paid off political bosses and gave them funds to pass out to individual voters to vote "the right way" in coming elections. Thus, if the political machine was well-oiled by the trust, and was so corruptibly efficient that it could guarantee a candidate a healthy majority of votes from ward, precinct, even neighborhood blocks, the trust could tell its "purchased" elected officials what legislation to pass or kill. Such quietly purchased influence also enabled trusts to wield considerable control over the other branches of government.

Figure 12.12 The Tammany Lords and Their Constituents: The Bed of Roses and the Bed of Thorns. *Thomas Nast's political cartoon highlights the plight of those subject to the injustices of political machines.*

After the Civil War, cities grew rapidly and city services such as sanitation, streets, and transportation also became much more important. Contracts serving the public needs of local government became much more lucrative and political machines became the order of the day in many American cities. New York, Chicago, Kansas City, San Francisco, and many other urban centers found themselves under the thumb of political bosses. The bosses normally did not hold political office, but they ran the political machine that brought others to office under their control. Some bosses were clever, street-smart saloon brawlers, at home in a smoke-filled back room, making deals with other politicians, trust insiders, and other thugs. Others were distinguished citizens, polished and respected, usually considered heroes by the ethnic group they were part of and which received in one way or another much of the trust's bribe and vote-purchase money. What all political bosses had in common then, as now, was their private power, the same kind of power wielded by the trusts, and during the Gilded Age that power became enormous.

The nature of political machines was revolutionized by foreign immigration. Thousands of immigrants, newly arrived and alone in a great city, clustered in Irish or Italian or Jewish neighborhoods. Usually arriving penniless and with no employment waiting, they were easily mobilized by political bosses of the same ethnicity, who pledged to take care of them, see to their special needs, and shield them from hostile forces. The new immigrant thus became a devoted follower, voted unquestioningly for a certain slate of candidates because Big Jim or

Giovanni or Moishe told them to. And the bosses were usually as good as their word. They found jobs for their constituents, helped them out of scrapes, furnished occasional legal services, and saw to their basic welfare as they became settled. At Christmas time there might be a turkey for the family dinner table or new shoes for the kids.

Political machines became as well organized as corporations—and better organized than political parties. Following the city's plan of districting, they divided into wards or other units and appointed specific individuals as "heelers," to visit constituent families regularly and see to their needs. In return, the grateful constituents could be counted on to vote for their candidates, sometimes several times, at election time. Machines had their enforcement side as well. Squads of "bully boys" could be called in at a moment's notice to disrupt an opponent's activities, to trash a polling place or to threaten, beat, or sometimes even kill anyone who dared oppose the activities of the boss.

Graft and stealing quickly became another important part of the political machines. Once in power the bosses were in a position to make fullest use of their influence in their own community. Of the many and varied forms of corruption carried out by political machines, these have always been the most prominent:

- Gambling, prostitution, dope-dealing and other forms of vice, which were allowed in specific precincts as long as the machine was kept well paid by the criminal businessmen.
- Contracts awarded to "connected" construction companies or providers of services to the city, often at vastly inflated sums, in exchange for under-the-table kickbacks to the machine.
- Corporations paying bribes to the machine for special privileges, such as the right of trolley companies to run overhead wires or the right of restaurants to serve alcohol.
- Area monopolies, such as the right to be the only business in the area to sell fish, and franchises were simply sold to the highest bidder.

Tammany Hall

New York's Tammany Hall was the grandfather of American political machines. It was started in 1789 and was soon controlled by former Vice President Aaron Burr as a kind of fraternal lodge with a political agenda. Its primary purpose was to promote the election of its own candidates to state and local office, the assumption being that Tammany men were somehow more fit than other office seekers. In a world with weak political parties, Tammany came to move like a tank among archers. Come Election Day, it could almost always deliver the vote for its hand-picked candidates.

William Marcy Tweed

1823–1978
"Boss Tweed" was a leader of the Tammany Hall political machine, which rigged elections and stole massive amounts of money from New York City.

Political machines levied a heavy toll on the American taxpayer. Under **William Marcy "Boss" Tweed**, for example, the Tammany Hall machine managed to siphon off as much as $2 *billion* (in current dollars) from the city and people of New York. While other cities did not run up tallies as high as this, neither did they have pockets as deep.

In California, one railroad trust's Political Bureau worked closely with political machines in order to tighten its hold on state and local government. A famous graft investigation conducted in San Francisco in 1906 revealed that the city's major corporations worked hand in glove with the then-notorious Abe Ruef machine. The same was true of political machines elsewhere.

The Constitution and the American Founding were meant to be a long-term answer to the problem of unacceptable exercise of power. Markets were to disperse economic power to the point that no one could control prices in the economy. Government was to be controlled by repeated exercise of consent through elections and by the structural devices in the Constitution. An ordinary free American in 1800 had very little contact with government and did not see businesses or other institutions as powerful or oppressive. But the trends reviewed in the previous chapter changed the scale of America, and this change in scale created private power. Business was so much bigger and so impersonal that individuals without political or financial influence could easily feel small and threatened by this new power. Moreover, state and local government had also grown because of the growth of cities.

Populism

1880s' political movement favoring nationalizing banks and railroads to protect farms and rural towns from the private power and corruption of big corporations.

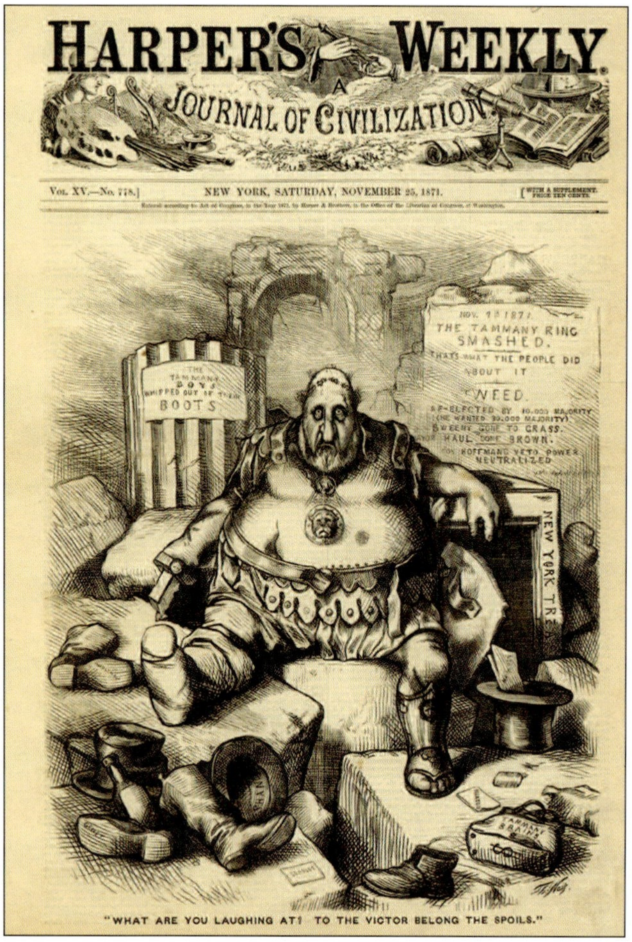

Figure 12.13 Another Thomas Nast cartoon depicting the result of the 1871 elections in which most of the Tammany Hall candidates were defeated.

Now a city government could become corrupt and tyrannical in ways not imagined earlier. Thus, Americans from 1880 to 1914 looked for ways to offset the new sources of power generated by the change in scale of America. Ultimately, they turned to the source of power that could always trump other sources—government.

The Progressive Response

Political responses to the problems of private power and government corruption took several forms. All such responses had this much in common: they assumed that the problem could only be solved by government—usually the federal government. The answer to private power, so it seemed, was public power. But if the solution was public power, then reformers had to be sure that government at all levels stayed subservient to the people.

Populism

One such early response came from an agriculture-support movement called Populists. Populism as a political party arose in the 1880s and 90s. It was centered in the South and Midwest, and was essentially a rural phenomenon. America in the Gilded Age was still a nation composed mostly of small farms. The farmers were pressed by many problems, ranging from frequent droughts (which killed their crops) to their own overproduction (which brought crop values down) to currency deflation (which made their mortgages hard to pay back). Farming was a hard life in the best of times, and these times were far from the best for those who produced the nation's grains, fruits, and vegetables.

As Populists analyzed farmers' difficulties, they concluded that the blame essentially lay with banks and railroads, two organizations that farmers couldn't control. The banks were blamed for currency deflation and falling prices, meaning that the farmers had to pay back their loans with money that was harder and harder to come by. They blamed the railroads for monopolistic freight rates that drove down the profit the farmer received for produce shipped to city grocers and millers. Eventually, the rural Populists were joined by disgruntled city factory workers, whose unresolved problems with the dangers and working conditions of industrial labor were becoming acute.

Populists were influenced by the increasing popularity of socialism, leading them to propose that banks and railroads be nationalized and operated by the federal government. They also proposed that currency be inflated by means of the free coinage of silver to augment the supply of gold that backed the nation's money. Populists won many state and local elections in the South and Midwest. However, when they attempted to form a viable third party on the national scene, their platform was essentially co-opted and absorbed by the Democratic Party, which nominated the Populist candidate, **William Jennings Bryan**, for President in 1896. The election campaign was intensely emotional and accompanied by rampant demonizing of those who held private power, people whom Bryan dramatically compared to the crucifiers of Jesus Christ.

Bryan's loss of the election to William McKinley spelled the end of Populism as a viable national political force. But the farmers' difficulties that had given it birth and attracted entire rural communities to want reform were not soon to go away.

Progressivism

A different response to the problem of private power was mounted by the Progressive movement. Progressivism was an urban phenomenon and much different in character than Populism. Where the Populists had looked backward nostalgically, Progressives (as their name implied) looked hopefully to the future, to Progress. There was much science and technology in the Progressive platform. Its political perspective was grounded in research and the views of experts in universities (notably the University of Wisconsin) in the emerging fields of sociology, psychology, political science, and social work. The movement was filled with cutting-edge intellectuals, crusading editors, and people with clipboards and stopwatches.

Figure 12.14 William Jennings Bryan at the 1896 Democratic National Convention.

Progressivism was heavily influenced by the ideas of Charles Darwin, especially as these had been refined and applied to society by **Herbert Spencer** and others. Society, like everything else, claimed the "**Social Darwinists**," is in a state of constant change and development, evolving into ever higher and more complex forms. What we perceived as social or political problems were simply the strains of evolutionary progress. The issue for the Progressives, then, was whether evolution would be random and haphazard or intelligently guided.

There were Progressives in every part of the country and in every level of government, although many who supported the

William Jennings Bryan

1860–1925
A lawyer, statesman, and popular speaker, Bryan ran for President on the Democratic ticket three different times. He was a prominent leader in the Progressive movement and served as Secretary of State to Woodrow Wilson. He may be most well known as one of the lawyers in the famous Scopes Trial about teaching evolution in schools.

Progressivism

Post-populist, urban-based political movement against private power and corporate corruption that looked hopefully towards the future, emphasizing the benefits of science and technology.

Herbert Spencer

1820–1903
Considered the father of Social Darwinism, Spencer coined the phrase "survival of the fittest" in his 1864 book *Principles of Biology*.

Social Darwinism

Belief that society, like everything else, is in a state of constant change and development, evolving into ever higher and more complex forms.

Progressive Party's ideals were elected as Republicans and Democrats. Republican President Theodore Roosevelt was widely respected for his trust-busting work and environmental programs that identified him as a progressive under his GOP exterior, and after leaving the White House he later campaigned unsuccessfully as a third-party candidate for the "Bull Moose Party," which was the Progressive Party's platform under a new name. President Woodrow Wilson, a Democrat and former president of Princeton University, was also a strong proponent of Progressive Party ideals.

Progressives addressed numerous public problems, from pressure on cities to clean up their slums to offering plans to increase the efficiency of schools. Progressives wanted industry to reduce hazards in the workplace. They wanted government to provide better care for expectant mothers, conserve natural resources, and design and implement an improved banking system. But their main struggle was with private power, with the trusts on one hand and the political machines on the other. Progressives built their new construction of America on both economic and political reform.

Economic Reform

The economic reform called for by the Progressives touched on the commonly accepted market weaknesses of imperfect information, monopoly, externalities, and economic instability. There was substantial concern about the justice of a market economy during the years dominated by progressive efforts to reform the entire economy. (See Appendix B for a discussion of market weaknesses.)

Elected progressive reformers first tackled the problem of business monopolies, the trusts. Two strategies were used. The first strategy was to break up the monopolies or trusts and create fair economic competition. In 1890, Congress passed the Sherman Anti-

Figure 12.15 A cartoon showing Roosevelt as a moose, the republican elephant and the democratic donkey looking on in the background.

Trust Act which made conspiracies to restrain trade illegal. The Sherman Act gave the government the powers to force Rockefeller's Standard Oil to divide into five separate companies, with the hope that these "baby Standard Oils" would compete with one another for a share of the consumer market by keeping oil prices attractive to the consumer. Other laws fostering competition were also passed, most notably the Clayton Anti-Trust Act in 1914.

The second strategy was to regulate business power rather than attempt to destroy it. Regulation of business started early in the Progressive Era, when in 1887 the Interstate Commerce Commission (ICC) was organized to regulate railroads and the rates they charged. States also initiated regulatory agencies to control public utilities such as power and subway systems. The regulation of railroads and utilities was like antitrust laws aimed at controlling monopoly power.

Other government regulations focused on other market weaknesses. Political pressure to clean up food supplies led to creation of the Food and Drug Administration (FDA). Consumers were no longer buying their meat from a neighborhood butcher who oversaw the processing of his meat products from a live animal to the beef roast or chicken legs or bacon displayed on ice in his glass showcase. The growth of cities made it more practical for the butcher to simply order his meats from big suppliers who may or may not reject diseased animals and use clean processing methods. This put the consumers at serious health risk, because they no longer had the information needed to guide their purchasing.

Regulation also began to control market externalities. Local governments took responsibility for providing clean water, public sanitation services, and clean milk supplies. State and local health departments began to monitor infectious diseases, leading to the organization of the U.S. Public Health Service in 1912. In 1891, the Forest Reserve Act was passed, making federal government responsible for conserving and protecting national forests and the plants and animals within them. In 1905 a chief forester, Gifford Pinchot, was appointed to oversee the national forests.

Progressives were also concerned about economic injustice. They saw the roots of social problems—crime, alcoholism, and illegitimacy—in poverty. They were confident a better distribution of income would produce countless social benefits. The first income tax implemented after passage of the 16th Amendment was a progressive tax with rates rising from 1 percent for low-income workers to 7 percent for high-income workers. To progressives, the income tax held promise as a way of redistributing economic rewards within the economy. With an income tax, rates could be increased with income, shifting the burden of supporting

Figure 12.16 1912 political cartoon depicting Teddy Roosevelt adding Progressivism as an ingredient to his political speech.

government programs to upper-income groups. It was hoped that a progressive tax would redistribute income, achieving more economic justice. Other laws, such as child labor and compulsory schooling laws, unemployment insurance, workmen's compensation, and laws controlling the hours and conditions of women's labor, were also directed toward the goal of economic justice.

Economic reform during the progressive period also focused on another complaint often heard in the late 19th and early 20th centuries regarding economic instability caused by the banking system. Ordinary Americans traced the problems of declining prices or farm foreclosures to banks. At times banks were plunged into financial crisis and unable to service customers who wished to withdraw money from the bank. Banks ordinarily hold only a small fraction of their deposits on hand and lend out the rest in order to earn interest on those deposits. In normal times, not all depositors will withdraw their money at the same time. However, an economic crisis would some-

Federal Reserve System

A quasi-governmental organization formed to regulate the money supply and help keep the economy stable.

Initiative

Progressive reform in which citizens could put propositions directly on the ballot through petition and have them become laws by garnering a majority vote.

Referendum

Progressive reform in which laws passed by legislatures can be directly submitted to the people for a vote; a majority vote against the law removes it from the books.

times cause people to loose confidence in their bank, causing a "run" on the bank with most depositors wanting to withdraw their money. Banks, unable to pay all of their depositors, would sometimes close their doors or simply refuse to give customers access to their deposits for a period of time. A serious financial depression in the mid-1890s, followed by a particularly bad banking panic in 1907 provided bankers and politicians with the example they needed to push for creation of a central bank.

In 1913, the **Federal Reserve System**, a group of regional reserve banks headed by a Board of Governors in Washington D.C., was created to regulate the banking system. Eventually, the "Fed" as it is called by the financial community would become a dominant institution in control of the economy. It is a quasi-government organization composed of an independent government agency with responsibility for control of the nation's money supply. It also has considerable influence over interest rates in the economy. Creation of the Federal Reserve was one of the most important outcomes of the far-reaching dynamics of the Progressive Era. Unfortunately, the Federal Reserve, which was created specifically to prevent money market crashes, failed to prevent the Great Depression of the 1930s or several shorter recessions of more recent decades.

Figure 12.17 The Federal Reserve System is headquartered in the Eccles Building on Constitution Avenue in Washington, DC.

Much of today's government regulation of the economy originated in the Progressive Era. Control of monopolies, regulation of business across a broad front to reduce the effects of poor information and externalities, government control of the monetary system, and ongoing concern about equable distribution of income all began with the progressive movement. Progressives chose this path of government regulation and intervention of what was still a market economy rather than adopting the socialism that was being experimented with in Europe.

Governmental Reform

Reformers of the Progressive Era were also intent on making government more responsive to the people. States amended their constitutions to inject more participation into government. The typical approaches usually included:

Initiative Citizens could put propositions directly on the ballot through petition. If the proposition received a majority of the votes it became law.

Referendum Laws passed by legislatures could be directly submitted to the people for a vote at an election. If a majority of the people voted against the law, it was taken off the books.

Recall Citizens could petition election officials to call a special election to consider the recall of an elected official. If a majority voted for the recall, the person was removed from office.

All of these measures were designed to give ordinary citizens the power to bypass a legislature or elected official who might have been corrupted by private power.

At the federal level, the most important change directed at increasing democracy was the 17th Amendment. This shifted the power and responsibility for selecting a state's senators away from the state legislature (as called for in the Constitution) to popular, direct election of senators by the people. State legislators were seen as too close to large economic interests of a state.

The most far-reaching change may have been the 16th Amendment, which finally made a direct tax on the incomes of individuals and businesses a part of the Constitution. An income tax law had been passed during the Civil War to fund the Union's need for capital to expand its army. In the 1890s, the Supreme Court had held that an income tax was unconstitutional because the Constitution prevented direct taxation. The 16th Amendment allowed direct taxation. Income tax would soon become Congress's favorite method of generating government revenue. With the income tax, economic growth itself pushed more and more money into government coffers.

During the 1920s, the Progressive spirit of reform visibly faltered, in part because the economy was booming in most sectors. Then came the event that would shake the faith most Americans had in the free market system—the Great Depression.

The Great Depression

The prosperity of the 1920s created a widespread confidence that all conceivable free market weaknesses had been overcome. Real income marched steadily upward and the nation's stock market values soared to dizzying new heights. The American Dream was becoming a reality for millions. But in 1929 there were signs that the rapid economic growth of the1920s was coming to an end. Agriculture, the provider of America's cheap, food supply, had not shared in the prosperity of the 1920s. A long drought in the Midwest left many farmers struggling for economic survival, with some losing their farms because of poor harvests and unpayable loans. In the summer of 1929, industrial production also began to decline.

The Federal Reserve, perhaps concerned about unrealistic speculation that was driving up stock prices, tightened the money supply in the economy. This decline in the money supply started pushing the economy toward a recession. Then, in October, 1929, came the stock market crash. Stock market prices had been going down gradually through September of that year, but by the last week of October, investor panic had set in. Black Thursday (October 24) was followed by black Monday (October 28) and black Tuesday (October 29). Four and five times the normal amount of shares were being traded and stock prices for even the most stable publicly traded corporations were in free fall for three years. Figure 12.17 shows the decline and collapse of stock values.

In 1930, the bad economic news shifted from the stock market to banks. Losses from falling stock prices put some bank loans in default and raised concerns about the stability of banks. Depositors, fearful about the sol-

Recall

Progressive reform in which citizens can call a special election by petition to recall an elected official; a majority vote removes the person from office.

The Great Depression

Extended recession in the 1930s that led to widespread unemployment, bank failure, and a general downturn in the economy until World War II.

Dow Jones Industrial Average January 1925 to January 1933

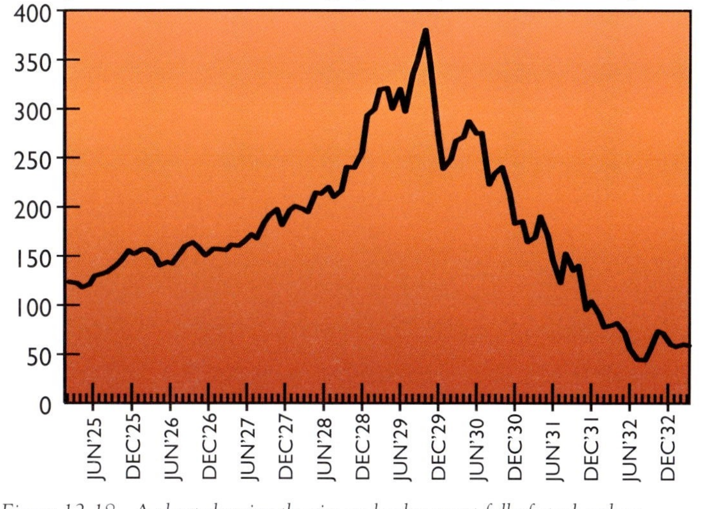

Figure 12.18 A chart showing the rise and subsequent fall of stock values marking the onset of the Great Depression.

Figure 12.19 The trading floor of the New York Stock Exchange just after the crash of 1929.

vency of banks, tried to withdraw their funds causing runs on several important banks. The banks began to fail and shut their doors. As banks failed, the economy's money supply fell and more economic disruption occurred.

Bank run

When most depositors try to withdraw their funds simultaneously from a bank.

Inexplicably, the Federal Reserve, with both the responsibility and the capacity to stop the bank failures and control the money supply, failed to do so. The economy continued to spiral downward. Unemployment was 3.2% in 1929, 8.7% in 1930, 15.9% in 1931, 23.6% in 1932, and 24.9% in 1933 before the unemployment rate began to level off. A recession in 1929 combined with the bank failures of the 1930s, resulting in total loss of consumer confidence, created the Great Depression. The American dream had changed abruptly into a decade-long American nightmare.

Hard Times

The Depression demonstrated how bad things could get in a market economy, although some later economists have laid the blame for the Depression at the feet of the Federal Reserve. Layoffs and plant closures became daily events. Big ticket purchases like appliances and automobiles were particularly hard hit. By 1932 America's booming automobile production had slumped from its 1929 level of 4.5 million units to a scant 1.1 million. Henry Ford's Ford Motor Company, once the bright star of U.S. industrial achievement, completely shut down its assembly lines for a time.

Nor was Ford alone. At the very bottom of the Depression period 185,000 businesses were bankrupt, 25 percent of the nation's work force was idled, and per capita income had been wrenched back to the earning powers of the early 1900s. Some 4,000 banks closed their doors permanently in 1933 alone, and the collapse of each had caused

Figure 12.20 An early 1930s breadline at McCauley Water Street Mission under the Brooklyn Bridge in New York City. Lines like this were common during the worst of the Depression years.

the permanent disappearance of a sizeable part of the personal savings of entire communities. American farmers, victims of chronic overproduction, had not known real prosperity for two generations, and now their plight was truly a national tragedy. Not only was the economic depression dropping prices, but the southern plains were caught in a severe drought. Massive dust storms plagued parts of Colorado, Kansas, Oklahoma, Texas, and New Mexico in the early 1930s, forcing many farmers to abandon their indebted and parched land and migrate to the Pacific coast. No group felt the pain of the Depression's falling prices more than those farmers who tried to stay on their land. Eggs were 53 cents a dozen in 1929 but only 29 cents in 1933—if the consumer had the money to buy them. Milk fell from 58 cents a gallon to 40 cents, well below the cost of producing it. Orange prices decreased from 45 cents a dozen to 22 cents. Farmers, seeing their incomes fall faster than their expenses, became desperate for some solution. California growers left oranges on the trees to putrefy in the sun, because they couldn't afford to hire anyone to pick them, and they couldn't afford to ship the fruit even if they had picked it. Minnesota dairymen emptied milk cans by the truckload onto the dusty roads. "You don't pick three-cent cotton," explained a gaunt Mississippi sharecropper. "You just leave it sit."

Almost as hurtful as poverty and hunger was the damaged pride of Americans who had been climbing up the economic ladder for generations and now had to climb down or risk being pushed off. Executives were demoted to shipping clerks, and shipping clerks to custodians. Unemployed white collar workers stood on the street corners and sold apples, still dressed in the suits and ties they had worn in their offices just a short time before.

Some hid their poverty and pain, staying indoors so that neighbors wouldn't know their furniture had been repossessed. They told—and lived—complex lies. Finally, when the house went up for sale and dissembling became impossible, they did desperate things. Suicide rates increased.

Often, though, the down-and-outer simply took flight. He left his family to look for work, and just never returned. Sometimes he wound up in a skid row hotel, paying a dollar a week and eating at streetside free soup kitchens. Some learned to ride the rails, sleeping in boxcars or haystacks as they roamed, following vague rumors of distant jobs picking apples or digging ditches. Sometimes he found work, saved up enough to send home, and later returned to family and better times. Sometimes he ended up permanently living in hobo camps or on the pavement.

The wife and family he left behind were no more fortunate. Usually she was stranded with the children, often young, and thrown back upon the charity of neighbors. Hundreds of surviving photographs of these trun-

Figure 12.21 Part of an impoverished family of nine on a New Mexico highway who were Depression refugees from Iowa. They left Iowa in 1932 because of the father's ill health. Here they are about to sell their belongings and trailer for money to buy food.

Figure 12.22 During the Depression, shanty towns (sometimes called "Hoovervilles") sprang up around the country, populated by the unemployed and homeless. Dwellings were made from whatever materials were available, including scrap metal and cardboard.

cated families frequently show thin faces reflecting hopelessness and despair. An entire generation was damaged, some members beyond repair, by the ordeal that we now call the Great Depression.

Causes of the Great Depression

Even though the Great Depression was the most important domestic event of the 20th century, economists are still debating its exact causes. A depression, which is a severe recession, happens when prices fail to adjust quickly and flexibly to a new condition in the economy. In other words, some shock or disturbance requires the economy to adjust but the adjustments, which should come gradually in the form of wage and price changes, arrive as sudden unemployment and reduced industrial production. Most economists agree that the stock market crash of 1929 would have caused an economic recession. The disagreement centers on the question of what turned a routine economic contraction into the Great Depression? Scholars now point to three causes: a severe decline in the money supply, a steep fall-off in consumption, and a disruption in international trade because of tariffs and ill-conceived international monetary arrangements.

Whatever the cause, the Great Depression rattled confidence in a self-regulating market economy by ordinary Americans and the governing elite. Market instability was to them the primary weakness of the economy. Some people were ready to jettison a market economy altogether and turn to socialism, which seemed to have all the answers. If a socialist government planned the economy and gave everyone a job, unemployment would not devastate peoples' lives as the collapse of the market economy had in the 1930s. But the majority felt that allowing America to embrace socialism (and quite possibly, letting in its violent cousin, communism) was going too far a field from the Founders' original ideal. Good, honest government could still manage the market economy—give it guidance and confidence or if need be step in and oversee it directly. A few economists counseled patience, claiming that in the long run, markets would stabilize and the nation would return to full employment and recovered prosperity. But the majority opinion was reflected in the quip of John Maynard Keynes, an influential British socialist economist, who said, "In the long run, we are all dead." Politicians quickly understood that they would be politically dead if they did not give the government the power to intervene in the sick U.S. economy of the 1930s.

Election of 1932

The Presidential election of 1932 represented a dramatic change in American politics. The candidate of the Democratic Party, **Franklin Delano Roosevelt**, put together a coalition of Southerners, farmers, labor unions, and urban centers that would dominate American politics for the next seventy years. Voters in 1932 were ready to punish Republicans, the party that was in power when the Depression began. After the 1928 election, Republicans held majorities in both the House (267 to 167) and in the Senate (56 to 39). After the 1932 election, there were only 117 Republicans in the House and 35 in the Senate. Democrats held the Presidency and larger majorities in both houses of Congress. The election of 1936 reduced the Republican presence in government leadership to a shadow, with only 89 House members and a measly 16 in the Senate. Clearly, voters blamed the Republican Party and wanted a different approach.

The New Deal

President Roosevelt, a dynamic former governor of New York State who had been crippled by polio as a young man, pledged a "New Deal" to America's impoverished citizens. The term had interesting connotations. If life was like a poker game which had somehow become corrupt, then calling back the cards, shuffling them, and dealing them out again amounted to a kind of refounding of the game, something like Lincoln's "new birth of freedom." Roosevelt, who was committed to both capitalism and democracy, probably had nothing so revolutionary in mind. But owing to the desperation of the situation, he found himself driven forward by events. His New Deal would impact the Founding as heavily as the Civil War had done generations earlier.

Figure 12.23 Franklin D. Roosevelt delivers a speech in New Albany, IN, during the 1932 Presidential campaign.

Experimentation

Roosevelt and his "Brain Trust," which included advisors of considerable political and academic sophistication, concluded that the market system had broken down in some fundamental way and either could not or would not repair itself. Therefore the system must be retooled so it would work differently. The Roosevelt advice team was open to any and all proposed solutions. Instead of allowing producers and consumers to act in their perceived self-interest and letting Adam Smith's "invisible hand" guide the outcome, New Dealers decided to use cartels with built-in negotiated agreements on prices as their principal weapon against the Depression. Convinced that higher prices would stimulate the economy, they brought competing businesses together to set prices, wages, and conditions of labor, not in a spirit of self-interest but for public benefit. It was a bold experiment in pure virtue.

The National Recovery Administration (NRA) and its country cousin, the Agricultural Adjustment Administration (AAA), put the cartel theory into action. As with all cartel monopoly, the plan was to force up prices by artificially limiting production. The resulting higher prices, with accompanying higher wages, would then stimulate demand. However, with a few exceptions, this approach to solving the Depression failed as completely as earlier efforts by Roosevelt's

Franklin Delano Roosevelt

1882–1945
The 32nd President of the United States, Roosevelt served four terms, the only U.S. President to serve more than two terms. His exuberant public personality helped bolster the nation's confidence as it struggled through the Depression and then entered World War II.

New Deal

Plan by Franklin D. Roosevelt involving the creation of various government agencies and programs designed to stimulate the economy and help the U.S. overcome the Great Depression.

Herbert Hoover

1874–1964
The 31st President of the United States, Hoover lost the 1932 election for a second term (to Franklin D. Roosevelt) when his responses to the Stock Market Crash of 1929 failed to end the economic recession and the nation slid into the Great Depression.

John Maynard Keynes

1883–1946
A British economist whose ideas would influence Roosevelt's New Deal intervention for the U.S. economy.

Keynesian economics

Economic theory in which the economy would regulate itself, but in the case of extreme depression the government would be needed to artificially stimulate demand by increasing spending or cutting taxes.

unfortunate predecessor, **Herbert Hoover**, had. Like the market system itself, the New Deal was based on academic theories about human nature, and those theories were simply not borne out. Monopoly power, whether private trust or government programs, was not in the public interest. Fortunately, the Supreme Court declared the NRA unconstitutional. By 1935 the NRA—and most attempts to fix prices—was a dead letter.

The Agricultural Adjustment Act (AAA) of 1933 tried to raise farm prices by restricting the amount of acreage that could be planted in major crops and making payments to farmers for not raising those crops or livestock. This act was also declared unconstitutional, but Roosevelt's Brain Trust found a way around that decision by tying acreage allotments to soil conservation. Because of the Supreme Court and basic human nature, the attempts to raise prices by restricting production also failed. Roosevelt's collection of professors, intellectuals, maverick businessmen, and labor leaders then turned to other plans. But they needed an intellectual home to give credence to their efforts. They were neither socialists nor followers of Adam Smith's free market economics, but they still wanted to regulate the economy in a way that would end the Depression and get the economy humming once again. Eventually they decided to use the theories of the British civil servant and economics professor at Cambridge University, **John Maynard Keynes**.

Keynesian Economics

Undoubtedly, it was trial and error as much as any formal economic theory that convinced the New Dealers to try Keynes's socialist strategies. But Keynes, who was England's most influential economist during the Great Depression, would have a lasting influence on all aspects of the United States' economy. In 1936, he published one of the most influential books of the 20th century, *The General Theory of Employment, Interest and Money*. This book and Keynes's enthusiastic followers would prove to exert major influence on governments across the world over the next half century.

Most economists before Keynes believed that a market economy would be self-regulating. They were guided by a maxim from a French economist, Jean Baptiste Say, who proposed that "supply creates its own demand." In other words, payments to those who produce goods and services would be sufficient to buy the goods produced. Hence, the prescription of most economists when confronted with recession was simply to wait, figuring that in the long run the economy would right itself.

In a technical treatise directed to economists, Keynes argued that a supposedly self-regulating economy could fall into a trap that would not return the economy to full employment. Keynes argued that government had to be activist in the face of recession. Keynes's main policy idea was fairly simple. Recession is characterized by a fall-off in demand, and it is this fall-off that causes the economy to spiral downward. Some businesses lay off workers. Laid-off workers have less money and so spend less money. Just as a small investment in a community expands as it passes through many hands, a small decrease in demand causes many more decreases as it ripples through the economy. Other businesses sell fewer goods and as a result more workers must be let go. And so on. Keynes pointed out that it is within the power of government to turn this vicious cycle around through its powers of taxing and spending.

Since government, by its very size, is a key player in the economic game, it can artificially stimulate demand by increasing government spending or cutting taxes. Government spending can stimulate demand directly as the government builds roads and buildings or embarks on a new program.

Alternatively, the government can induce businesses or consumers to spend more by reducing taxes, leaving them with more of their income. Keynes's prescription to combat recession implied that government would necessarily run deficits in its budget during a recession, adding to its national debt. But Keynes also expected the government to run surpluses in its budget during periods of economic boom. As it turned out, politicians were much more attracted to budget deficits created by increasing government spending and cutting taxes than to budget surpluses which could happen only with restraint on government spending and tax increases.

In practical terms, Keynes pushed public policy toward government taking on the responsibility of managing the economy to combat recession and toward government accepting budget deficits to "stimulate the economy," regardless of the current state of the economy. Since 1936, when Keynes published his ideas, the U.S. government has run a deficit in its budget over 90 percent of the time, a tribute to human nature in politics and to the influence of Keynes's ideas. The most important legacy of Keynes and New Deal efforts in the 1930s was to shift the burden of proof from those advocating government intervention in the economy to those arguing for government restraint in the face of economic problems. Government became the option of first resort when an economic issue surfaced.

New Deal Relief

After the initial confusion and failed experimentation with the NRA, the New Deal settled into a two-pronged effort, short-term relief and long-term economic regulation. First, there were a series of short term relief programs to soften the effects of the Depression. Unemployed young men were encouraged to join the Civilian Conservation Corp (CCC). This program was organized to complete conservation projects on federal land and to build campgrounds in the national forests. To provide work for unemployed of all ages, a succession of work relief agencies were formed: the Federal Emergency Relief Administration (FERA), Public Works Administration (PWA), and Works Progress Administration (WPA). These programs provided work for the unemployed through construction of public buildings and roads, sanitation, arts projects, writing projects, etc.

Relief of poverty and unemployment assistance were considered a responsibility of local government prior to the New Deal. Roosevelt federalized relief, making poverty and unemployment primarily national responsibilities from then on.

New Deal Economic Regulation

Roosevelt and his Brain Trust were not satisfied providing short-term relief from the effects of the Depression. They were convinced that a market system could not regulate itself. Yet they were, by no means, persuaded that socialism—complete government control of the economy—was better than capitalism. They simply wanted the government to exert guidance over a market economy. They wanted to replace Adam Smith's invisible market hand with a visible

Figure 12.24 Roosevelt's New Deal included many public works agencies that offered employment on various kinds of projects.

government hand operating with a light touch. With the support of an overwhelming Democratic majority in Congress, they rapidly passed law after law to regulate the economy. Some of the more important regulatory changes included:

- Securities and Exchange Commission (SEC) to oversee the stock market.
- National Labor Relations Board (NLRB) to regulate negotiations between unions and corporations, giving unions better opportunities to unionize new industries.
- Federal Deposit Insurance Corporation (FDIC) to oversee bank safety and guarantee bank deposits.
- Fair Labor Standards Act, which established a federal minimum wage of $.25 (about $3 in 2000 value).
- Social Security (called Old Age Survivors Insurance), a tax on wages and salaries to provide insurance against unemployment, disability and—most important in the long run—to provide some retirement funds for most Americans.
- Federal Communications Commission (FCC) to oversee radio, telephone, telegraph and, later, all electronic media.
- Civil Aeronautics Board (CAB) to regulate airfares (dissolved in the 1980s).
- Tennessee Valley Authority (TVA) to provide power, flood control, and industrial development to the upper South. The TVA was meant to be a model for other massive federally organized and funded reclamation and modernization projects.
- Rural Electrification Administration (REA) to subsidize electrical power lines and service being brought into rural areas.

Figure 12.24 Some of the many government agencies created during the New Deal era.

Most of these regulatory efforts are still a part of the federal government, although some have undergone name changes.

The New Deal was followed by World War II, with the economy still controlled by the same president and a Democrat majority in Congress. World War II required the full mobilization of the economy for the war effort. Government once again used strategic, centralized planning instead of the free-market idea to direct the economy. Government bureaucrats directed corporations in their efforts to produce ships, planes, tanks, and other armaments. Prices were set by government officials and strict rationing was instituted for automobile gasoline, sugar, and many other goods. Markets had to take a backseat to the planning of war production. At the close of the war, many of the wartime regulations were eliminated, but Americans had now become conditioned to extensive government involvement in the economy.

The New Deal and World War II changed the relationship between the federal government and the economy in profound ways. Before the New Deal, the government did not take responsibility for the functioning of the market economy. After the New Deal, almost any serious glitch in the economy was open to debate and possible action by the federal government. A serious increase

Figure 12.25 Franklin Roosevelt asks Congress to declare war on Japan, December 8, 1941.

in the price of a staple or necessity would elicit calls for government action. Sometimes the government responded; sometimes it did not. Every perceived problem within the economy was seen as a possible opportunity for government management to replace the free market. Politicians and policy makers could imagine an ideal course of action where government could eliminate or at least mitigate some economic weakness or problem. They didn't occupy themselves with much follow-up as to whether or not government actually made the situation better or worse. The forces growing big government were manifestly stronger than the forces restraining its growth.

The Constitution and the Growth of Government

Opponents of the expansion of government tried to use the Constitution as a barrier to government growth. Advocates of increased government sometimes saw the Constitution as an impediment to progress and justice. For example, Progressives viewed the Constitution, at least the interpretation of it at the time, as a significant barrier to progress. Courts ruled that state attempts to curb the monopoly power of the trusts violated the due process clause of the 14th Amendment. And when the federal government tried to step in, it was told that the **commerce clause** gave it no authority to do so. In one famous case in 1895 (*U.S. v. E.C. Knight*), the Supreme Court had held that the American Sugar Company, which sold its product in every state of the Union, did not engage in "commerce" insofar as the Constitution was concerned.

Progressive historians turned to a direct assault on the Founding. In 1911, at the height of Progressivism, **Charles A. Beard** wrote a book titled *An Economic Interpretation of the Constitution*, arguing that the Framers in Philadelphia had had personal motivations for writing the Constitution, mostly having to do with investments. Beard's premise was that the use of the Constitution to confer private advantage was nothing new; the Robber Barons and the Founders had much in common. If the Founding was going to be used to block progress, Beard and others seemed to conclude, then the Founding must be shown for what it was, an artifact of the 18th century, and a shabby one at that.

Woodrow Wilson adopted a similar stance. In a famous speech before the Commerce Club in Washington, Wilson, who had been a professor of political science early in his academic career, argued that the structure set up by the Constitution was mechanical in nature, a thing of springs and wires and balances, and that in practice it tended to create gridlock rather than meaningful action. A Darwinian world, Wilson asserted, called for a Constitution of growth and development and change. How could Americans guide their own evolution if they were hampered by outdated rules?

In that same spirit, Franklin Roosevelt, himself an old-line Progressive, argued that the Bill of Rights needed to be broadened and deepened in order to address Depression Era difficulties. Americans, he said, had a right to economic security along with the personal rights on Madison's list. He became so angry with the Supreme Court's interpretations of the Constitution that he proposed to redesign the Supreme Court. But as humorist Peter Finley Dunne noted, "The Supreme Court reads the election returns." Supreme Court interpretations of the Commerce Clause and other important aspects of the Constitution changed to accommodate the progressive reforms. Furthermore, the Progressive Era led to structural reforms, direct election of senators, the Federal Reserve, and the graduated income

Commerce Clause

Constitutional clause that gives Congress the power to regulate certain types of trade, also a justification for the Civil Rights Act of 1964.

Charles A. Beard

1874–1948
A leader of the "Progressive School" of historiography, Beard attacked the Founders as being motivated by economic self-interest. Although initially supporting New Deal policies, Beard did not agree with Roosevelt's foreign policy and felt that the U.S. should be more isolationist.

tax. All those reforms would lead to profound changes in the theory and practice of national government in successive decades.

In practice, the Constitution proved to be no serious impediment to expansion of government power over the economy. The Commerce Clause (regulation of interstate commerce) was interpreted broadly, providing plenty of stretch for government involvement to be declared constitutional. By the end of the New Deal, almost any government intervention in the economy could be justified by the Supreme Court's interpretation of the Commerce Clause. The "takings clause" in the 5th Amendment of the Bill of Rights was interpreted narrowly, leaving private property subservient to government desires or intent. Rights considered economic were seen by the courts as less substantive than the rights of free speech, privacy, or personal exercise of religious expression. Criminal due process worried the courts much more than due process involving economic values.

Thomas Woodrow Wilson

1856–1924
Wilson was the 28th President of the United States. He felt the Constitution was too rigid and outdated.

As government grew after the Great Depression and World War II, Americans gradually became more cynical and questioning about government involvement in the economy. Leading intellectuals such as the University of Chicago's Milton Friedman argued that government intervention in the economy was nearly always a mistake, that the economic forces of the market were superior to mistake-prone government planners. In the late 20th century, starting in Jimmy Carter's Democratic administration, but accelerating in the Ronald Reagan Republican administration, deregulation was the watchword. Government pulled back in many areas, such as airline ticket pricing and regulation of media ownership. The Post Office, a government monopoly for 200 years, faced heavy competition from FedEx and UPS. Some people even began to call

Monetarists

Supporters of an economic theory emphasizing the role of money supply in an economy.

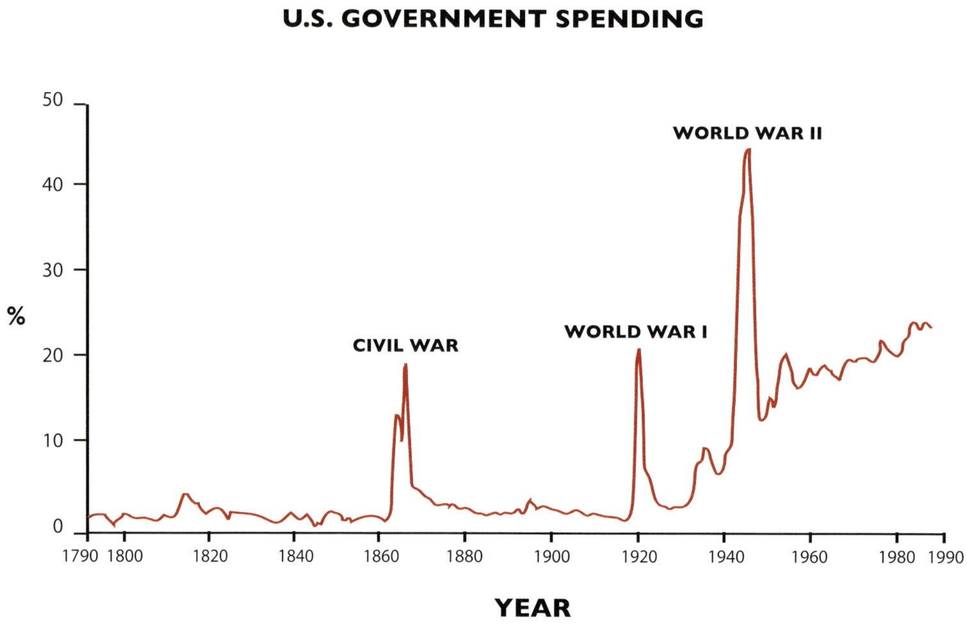

Figure 12.26 A chart showing U.S. government spending as a percentage of the gross domestic product. Ignoring the spikes from the three major wars, government spending has shown a dramatic upturn since the New Deal era.

Figure 12.27 Jimmy Carter signs the Airline Deregulation Act of 1978.

for economic competition in education in place of public schools.

But neither side has thus far gained complete victory or even complete vindication of its theories in any area where those theories have been tried. Every economic issue has provoked the old debate of free markets vs. government planning. Many advocate government planning as a way to produce security and the right distribution of income or consumption. Others point to the efficiency of markets and the capacity of entrepreneurial energy to innovate, solve problems and create new products and services to stimulate the economy. Health care and education continue to be common battlegrounds, as is social security with its fuzzy Congressional accounting for the way they have used SSA funds.

The promise of the market economy was closely akin to the promise of political democracy. Neither guarantees the right answer in a given situation. The best either can provide is what the people themselves seem to value. If there are faults in the public character—blind spots, distortions, moral lapses—they can't be papered over in the marketplace any more than in the voting booth. That is because both institutions are a reflection of the human soul and the desire for both liberty and economic security. No crystal-clear policy to manage the nation's ongoing economic hopes and problems is yet in view. Nonetheless, government think tanks, individual economists and political theorists keep generating new ideas and new prescriptions for economic bliss. Chances are, however, that the American economy will continue to muddle through choosing the market in one area and government planning in another. The Founding Fathers would probably understand because they were also suspicious of utopian solutions to real-world problems.

Timeline

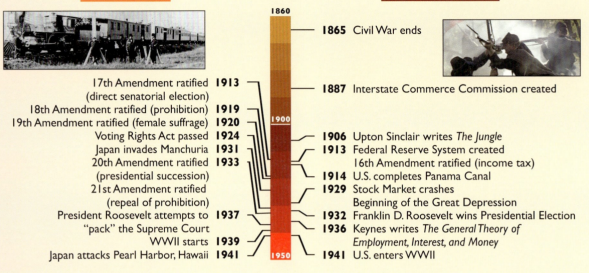

Contextual Events

17th Amendment ratified (direct senatorial election) **1913**
18th Amendment ratified (prohibition) **1919**
19th Amendment ratified (female suffrage) **1920**
Voting Rights Act passed **1924**
Japan invades Manchuria **1931**
20th Amendment ratified (presidential succession) **1933**
21st Amendment ratified (repeal of prohibition)
President Roosevelt attempts to "pack" the Supreme Court **1937**
WWII starts **1939**
Japan attacks Pearl Harbor, Hawaii **1941**

1860 — 1900 — 1950

Chapter Specific Events

1865 Civil War ends
1887 Interstate Commerce Commission created
1906 Upton Sinclair writes *The Jungle*
1913 Federal Reserve System created
16th Amendment ratified (income tax)
1914 U.S. completes Panama Canal
1929 Stock Market crashes
Beginning of the Great Depression
1932 Franklin D. Roosevelt wins Presidential Election
1936 Keynes writes *The General Theory of Employment, Interest, and Money*
1941 U.S. enters WWII

Key Terms

political machine
muckraker
Populism
Federal Reserve System
recall
monetarists
trust
robber baron
Progressivism
initiative
bank run
The New Deal
monopoly
markets
Social Darwinism
referendum
Keynesians

Key People

Andrew Carnegie
Herbert Hoover
Eugene Debs
Franklin Delano Roosevelt
Boss William Marcy Tweed

Questions

1. How was business enterprise different before and after the Civil War?
2. How did foreign immigration contribute to the power of political bosses?
3. Describe the two ways in which the U.S. government tried to control the power of monopolies.
4. Why did government grow during the Depression and World War II?
5. In what ways is the Constitution interpreted to justify or inhibit the growth of government?
6. Pick one of the market weaknesses and show how it led to the growth of government.

Chapter 13

America and The World

Abstract

America's policy regarding its interaction with the rest of the world has gone through several significant changes. In George Washington's Farewell Address, the principle of isolationism—a reluctance of becoming "entangled" in foreign affairs—is established. During its early years, America was more concerned with domestic issues and westward expansion. This chapter examines the course America has taken in its foreign policy since those early isolationist days and examines America's struggle between the idealistic responsibilities of spreading democracy through foreign involvement often accompanied by costly wars and its isolationist tradition.

The American Founding had to adapt to deep change in political accommodation, constitutional clarification, and the rise of the market system. The stresses arising from these internal developments would be more than matched by the stresses of confronting the outside world.

Some find American anxieties about the rest of the world ironic. After all, they point out, from the close of World War II to today, the United States has been the only consistent superpower, with enough armed might to destroy legions of enemies. Its institutions have triumphed. It is looked to for guidance and inspiration. It grows stronger while erstwhile enemies flounder in their failed social and political experiments. What is there to be afraid of?

The answer to this question is complicated, and once again it grows out of the Founding. For in the course of the Founding, Americans came to embrace two contradictory beliefs:

Isolationism

Political ideology that favored not becoming entangled with European powers, either as friend or foe, but did favor robust trade relations with a variety of partners.

- That they should withdraw from the world to whatever extent possible and chart their own course.
- That they should lead the world by example and become a beacon of virtuous conduct.

Attempting to reconcile these alternatives—which are not *complete* opposites—has become the central problem in America's foreign relations.

Born in a Machiavellian World

The world into which the United States of America was born was one torn with strife. Eighteenth-century Europe, like ancient Greece, was in a perpetual state of conflict, principally between England and France, but often including other powers as well. Diplomatic relations were fraught with intrigue, with alliances being made and refuted, with betrayals, land-grabs, power plays, and ultimately with war. Between 1702 and 1763, France and England fought each other four times.

This complex conflict played into the hands of the American rebels and helped account for their success. Colonial diplomats, notably Benjamin Franklin, knew how to play England against France and Spain and mobilize all of their mutual hostilities. Yet most of the Founders realized that European contention, however useful at the time, might later prove deadly to America's cause. To forge a permanent alliance with either England or France, for example, would risk embroiling the newly hatched republic in that endless and eviscerating Euro-conflict.

An important document to emerge from this realization was Washington's Farewell Address. (See Appendix C.) It was written for the President in 1796 by Alexander Hamilton and reflected much of Hamilton's thinking, but its main principles were laid out by the retiring first President himself and he carefully worked over the text. America's relationship to the rest of mankind, the Farewell Address proposed, should be a course that we now call **isolationism**.

Isolationism did not mean that the United States ought to sever all ties abroad. It favored robust trade relations with a variety of partners, for trade was then and still is the very lifeblood of a commercial republic. Nor did isolationism imply diplomatic haughtiness. "Observe good faith and justice towards all nations," Washington counseled, "cultivate peace and harmony with all." But Americans should stop right there. They should not become, to use Jefferson's later term, "entangled" with the European powers, either as friend or foe. "Nothing is more

Figure 13.1 George Washington *painted by* G. *Stuart; engraved by H.S. Sadd, N.Y. Washington proposed a policy of isolationism in his farewell address.*

essential," wrote Washington, "than that permanent, inveterate antipathies against particular nations, and passionate attachments for others, should be excluded. . . . The nation which indulges towards another an habitual hatred, or an habitual fondness, is in some degree a slave."

We have already seen how such entanglements had beset the new nation in the 1790s. The Federalists in power during the presidencies of George Washington and John Adams were sympathetic to their late masters, the British, while being suspicious of their late military saviors, the French. Fifteen years later, the U.S. veered in the opposite direction, with sympathy toward an imperial Napoleon and engaging in a shooting war with Great Britain—the War of 1812. Washington's "good faith and justice towards all nations" proved to be a difficult path to follow.

In 1821 a second building block of isolationism was set in place. The United States was stronger now and had nearly doubled its size with the Louisiana Purchase. In the opening stages of the War of 1812 it had suffered the humiliation of watching its Capitol being burned, but the nation later acquitted itself well, especially in the Battle of New Orleans, and there was strong national feeling in the air. So when Mexico declared its own independence from Spain, and fears of a possible Spanish reconquest surfaced, the United States proclaimed its soon-to-be-famous **Monroe Doctrine**, named after the fifth President, James Monroe. To the idea of any European "recolonization," at least in the Western Hemisphere, the Monroe Doctrine emphatically said *no*. Any aggressive move by a European power in the Western Hemisphere would be regarded as a challenge and affront to the United States.

Monroe Doctrine

Ideology of James Monroe in which any aggressive move by a European power in the Western Hemisphere would be regarded as a challenge and affront to the United States.

Over time, the Monroe Doctrine evolved a broader and more general meaning, that the Old World and the New World were to be regarded as "separate spheres," not commercially but politically, as called for in

Figure 13.2 James Monroe by William James Hubbard, ca. 1832.

the Farewell Address. Europe was for Europeans. America was for Americans. European powers were not to lure the emerging republics of the West into the Old World's tribal feuds and disposable alliances. It was yet another powerful idea, and it was soon backed by vigorous popular approval.

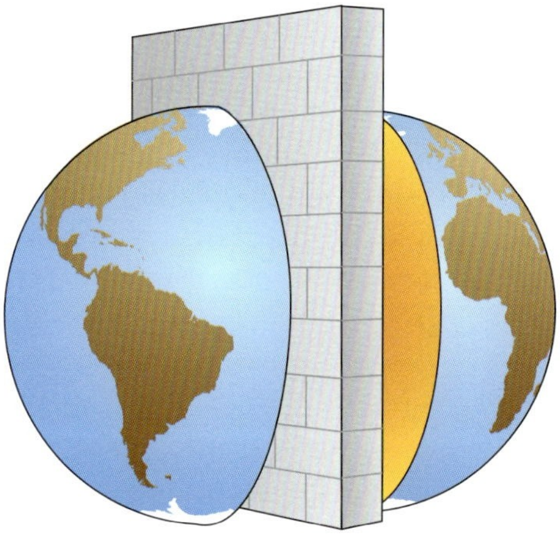

Figure 13.3 The Monroe Doctrine eventually came to mean that Europe and the Americas were to be politically separate from each other.

The Theory and Practice of Isolationism

The United States was destined to cling to isolationism for more than a century after the Monroe Doctrine. It may be argued, in fact, that the isolationist impulse still exists today, even though it has long since been abandoned as a formal policy. To understand the appeal of isolationism, we must review certain aspects of the Founding.

The founders, like all statesmen everywhere, believed that self-interest was the primary consideration in foreign affairs. Regardless of religious ideals, professions of justice, or the supposed love of mankind, conduct toward those outside of national borders had almost universally been shaped by self-regard. It has always been a case of my family, my tribe, my city, my country—as opposed to yours.

Individual self-interest generally leads to cooperative behavior, while **national self-interest** is often aggressive and destructive. Thus, the Greek city-states, despite their republican institutions and democratic values, conspired against one another ceaselessly, entering into alliances and counter-alliances, sending out spies, hatching intrigues, working up schemes of imperial dominion, and often going to war. The Athenian democracy was destroyed in a conflict that was fomented by Athens itself in an effort to dominate its neighbors.

National self-interest

An approach to foreign policy that places the value and interests of one's own country above that of others. Unlike individual self-interest that promotes cooperation and exchange in a market, national self-interest tends to be aggressive and destructive.

Alas, American self-interest was no different. From the very beginning of settlement, the colonies regarded one another not so much as fellow Britons but as rivals and competitors. Of course, they closed ranks when opposing the Indians, whom they all despised. Even here, however, they were willing to try the occasional intrigue, coaxing one tribe to join with the whites against another tribe, or later on, bribing or making promises to entice whole Indian nations into joining their war against the French or the Spanish. The Americans cherished their freedom from the British but showed little compunction when dealing with those they considered Others, i.e., Indian tribes that lived on desirable terrain further inland.

The new Republic's aggressive spirit was to become legendary after the Revolution. Americans knew that by pushing into the West they were trespassing upon Indian lands, but they rarely hesitated. They made war upon some tribes, negotiated treaties with others, converted still others to an agrarian way of life, but all the while kept the pressure on. During the Presidency of Andrew Jackson, they drove entire aboriginal nations from their ancestral homes, bundling men, women, and children into steamboats and railroad cars or herding them like cattle away to distant, and therefore not yet desirable lands.

Settlers also pushed into territories claimed by France, Spain, Mexico, even Britain, always with the thought of taking them over, politically speaking. Louisiana and "the Floridas" (east and west) were acquired by peaceful purchase from Europeans, not from the resident Indians. Texas was wrested from Mexico by insurrection. California and the rest of the Southwest were taken as the spoils of war, paid for but none too politely. And the Oregon Territory (Washington and Oregon) became American soil after a belligerent showdown—"Fifty-four forty or fight!"—with Great Britain.

American eyes were also cast upon the entirety of Canada. They were cast upon Cuba and the Caribbean. They were cast upon Central America, upon Mexico south of the Rio Grande, upon Russian holdings on the Bering Sea. Toward the end of the 19th century, Americans began coveting islands in the far-flung Pacific—stepping stones toward Asia.

Expressed in sometimes tortured logic, expansionism reflected the popular will and thus grew out of consent. Americans saw themselves as a people on the march, extending their "Empire of Liberty" throughout the western world. New York editor John L. O'Sullivan called it "**Manifest Destiny**" and the term stuck. It suggested that the real purpose of isolationism was to shoulder the Europeans aside and gain the Western Hemisphere for America's own, not always honorable, purposes. Somehow it was America's obvious destiny to rule from the Atlantic to the Pacific, if not all of the Americas. Perhaps this view was harsh, but by the end of the 19th century the United States had declared war on a tottering Spanish Empire, had taken Spain's colonial possessions in the Caribbean and Pacific, had encouraged a revolution in Columbia in order to gain canal-building rights in Panama, and was now routinely dispatching

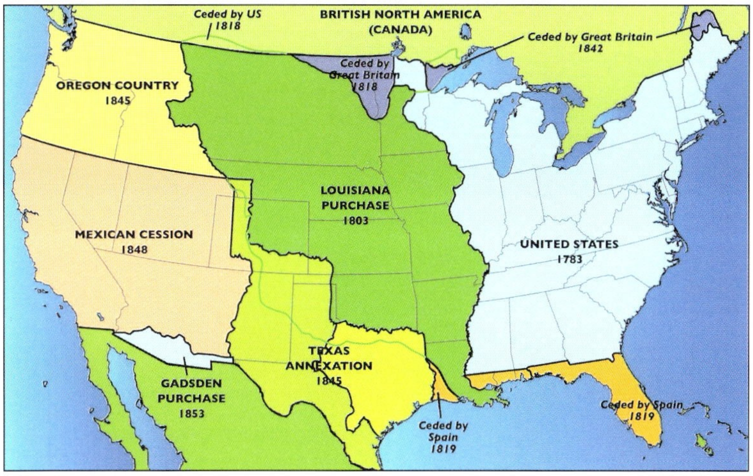

Figure 13.4 Manifest Destiny in the United States.

Manifest Destiny

The belief that American expansion of an "Empire of Liberty" through the Western Hemisphere was justified by the benefits of bringing the American way of life to acquired territories.

Figure 13.5 American Progress *by John Gast, 1872. The nation's westward expansion is symbolized by the woman stringing telegraph wires behind her while Indians, buffalo, and various animals flee before her.*

Stars and Stripes gunboats to police the Caribbean. Yes, Manifest Destiny was a view that many were beginning to share. American self-interest seemed to be playing a major role in our relationships with the rest of the world.

There was, however, a second strain in U.S. foreign policy, one which also grew out of the Founding. This had far less to do with national self-interest and far more to do with the ancient ideal of virtue. A close reading of the Farewell Address sets it forth eloquently. Good faith toward all nations, Washington argued, is "worthy of a free, enlightened, and . . . great nation, to give to mankind the magnanimous and too novel example of a people always guided by an exalted justice and benevolence." It was true that "temporary advantages" might accrue to the bold and aggressive, he allowed, but never long-term happiness. "Can it be," Washington asked, "that Providence has not connected the permanent felicity of a nation with its virtue?"

Figure 13.6 Washington felt that the U.S. should be a virtuous beacon to the world.

Figure 13.7 Blending Manifest Destiny with isolationism produced a very conflicted foreign policy.

The United States, in other words, was not justified in mimicking the style of Europe. Virtuous republics had no right to push smaller countries around, or to colonize backward peoples, or to wrest possessions from rival empires. The new Republic should mind its own business and conduct its foreign affairs honorably. "Religion and morality enjoin this conduct."

Isolationism was equally relevant to this attitude. Isolationism implied that the United States should remain aloof from the other powers of the world and disdain their shabby practices. "Why quit our own to stand upon foreign ground?" the retiring first President asked. "Why . . . entangle our peace and prosperity in the toils of European ambition, rivalship, interest, humor, or caprice?"

These words would echo through American history. They would be invoked time and again, not only by those who stood for isolationism, but also by those who stood against American aggression. For instance, the General Election of 1900 was in large measure a referendum on the question of American imperialism. William Jennings Bryan and a host of prominent voices spoke passionately against the retention of Hawaii, Samoa, Guam, Puerto Rico, and the Philippine Islands, most of them spoils of war taken by force from Spain. America should not behave like this, the "anti-imperialists" implored. It should be a light unto the nations, a beacon of human freedom, Washington's "good faith and justice" toward all.

It goes without saying that Manifest Destiny and anti-imperialism—virtuous isolationism—were essentially incompatible. Nevertheless, each had a large and vocal political constituency and accordingly,

America's guiding foreign affairs policy was doomed to be "schizoid." On one hand, Americans would continue to pursue their national self-interest, against whoever stood in their way, and on the other hand, they would periodically pull themselves up and reflect on the meaning of their example.

Nowhere would this confusing, schizoid two-sidedness be more apparent than in the roles America was to play in the bloody wars of the 20th century.

Figure 13.8 President Wilson before Congress announcing the break in the official relations with Germany. February 3, 1917.

World War I

The imperial ambition and peculiar alliances that led to the "Great War," as World War I was first called, were long in building. And those brutally simple causes emphasized—at least in the United States—the wisdom of the Farewell Address. Jostling for the high ground had become a way of life in Old World chanceries, along with territorial ambitions, naval arms races, festering animosities, imperialist swordplay, secret alliances, thirst for revenge, and an aggressive style of bragging known as **jingoism**. ("We don't want to go to war—but by jingo if we do. . . .") By the summer of 1914, Europe was a ticking bomb.

Most Americans were sold anew on their isolationism, especially when the war's *big* battles began, with casualties no longer in the hundreds or thousands but in the hundreds *of* thousands. No one could make sense of death on such a scale, nor could anyone adequately explain why the war had started in the first place or what it was all about.

The isolationist approach was to remain neutral in thought and deed, as President Wilson urged, and to insist on the right of neutrals to continue with their commerce as if nothing had happened. Modern technology, together with the desperation of the belligerents, made such an approach impossible. Neither the Allies nor the Central Powers were about to allow trade with the other side, and German U-boats made neutrals' rights obsolete. (Through the mist-shrouded lens of a periscope, one ship looked pretty much like another.) The Germans decided to abandon the old sailing era rules of stop and search and to engage in unrestricted submarine warfare, simply torpedoing every vessel in sight. This practice resulted in a supposedly neutral U.S. liner being torpedoed off the Irish coast, pushing the U.S. into the war in spite of itself.

Jingoism

An aggressive style of bragging by European governments that helped bring about World War I.

Like most Americans, Woodrow Wilson wanted no part of the various land-grabs, power-plays, and colonial entanglements that had brought about hostilities in the first place. He accordingly cast U.S. participation in highly idealistic terms. Americans would

Figure 13.9 Uncle Sam needs YOU! *painted by James Montgomery Flagg in 1916–17. This well-known image was part of the propaganda for the recruiting of American soldiers.*

fight, he said, "to make the world safe for democracy." National self-interest also influenced the American decision to enter the war. A victory for Germany and the Central Powers would have had disastrous consequences for America's trading partners and ultimately for U.S. trade itself. If it wasn't truly a war to make the world safe for democracy, it was certainly a war to make the world safe for Americans. But the confusion of America's war aims would cast a long shadow.

Battles of the Great War far exceeded anything in history for blood shed in short periods of time. In the Second Battle of the Somme, for instance, some 30 British and 15 French divisions were engaged against more than 100 German divisions breaking through the western front in a last desperate gamble for victory. Before the German thrust was countered, the British alone had suffered 200,000 casualties and lost another 190,000 as prisoners. Compare these numbers to the 20,000 Confederates lost at the Battle of Gettysburg.

American forces arrived too late to participate in such horrors as the battles of the Marne (1914), Verdun (1916), or Passchendaele (1917), but they fought in most of the battles of 1918, including the Somme. Nor was their role incidental. When the first "doughboys" arrived, the Allies were facing total calamity. The Russians had quit the war on the eastern front, and all the German troops recently facing them were being shuttled to the west for that final go-for-broke drive. It became a race between the Somme offensive—which would cost the Germans a *million* casualties—and the full deployment of the American Expeditionary Force.

The first U.S. actions were at Chateau Thierry and Belleau Wood in May and June, where Marine casualties were so heavy that the French would rename the wood "Bois de la Brigade Marine." After the sagging front was stabilized, the Allies counterattacked in mid-July, once again with fresh Yankee troops in the lead. In one of the notable successes of the war, a half dozen American divi-

Figure 13.10 A montage of World War I photos. Clockwise from top right: Navy warships; Battlefield trenches; A French Nieuport 17 C1 fighter; Royal Navy battleship HMS Irresistible sinking after striking a mine at the Battle of the Dardanelles; a Vickers machine gun crew with gas masks; a British Mark IV Tank crossing a trench; a Sopwith Camel biplane; and trenches on the Western Front.

"All the News That's Fit to Print."

The New York Times.

THE WEATHER

ARMISTICE SIGNED, END OF THE WAR! BERLIN SEIZED BY REVOLUTIONISTS; NEW CHANCELLOR BEGS FOR ORDER; OUSTED KAISER FLEES TO HOLLAND

SON FLEES WITH EX-KAISER

Hindenburg Also Believed to be Among Those in His Party.

Kaiser Fought Hindenburg's Call for Abdication; Failed to Get Army's Support in Keeping Throne

BERLIN TROOPS JOIN REVOLT

Reds Shell Building in Which Officers Vainly Resist.

Socialist Chancellor Appeals to All Germans To Help Him Save Fatherland from Anarchy

WAR ENDS AT 6 O'CLOCK THIS MORNING

The State Department in Washington Made the Announcement at 2:45 o'Clock.

ARMISTICE WAS SIGNED IN FRANCE AT MIDNIGHT

Terms Include Withdrawal from Alsace-Lorraine, Disarming and Demobilization of Army and Navy, and Occupation of Strategic Naval and Military Points.

WASHINGTON, Monday, Nov. 11, 2:48 A. M.—The armistice between Germany, on the one hand, and the allied Governments and the United States, on the other, has been signed.

The State Department announced at 2:45 o'clock this morning that Germany had signed.

Figure 13.11 Front page of the New York Times *on Armistice Day, November 11, 1918.*

sions broke through the German line near Soissons. U.S. forces were hurled into the St. Mihiel salient in September, and then into the larger Meuse-Argonne offensive in October, their numbers increasing with the arrival of each troopship. By the time German-Austrian leaders gave up and signed the Armistice on November 11, 1918, more than a one and one-quarter million American soldiers were engaged. The war couldn't have been won without them.

While American forces did not suffer as the British or French had (thanks largely to better leadership), they suffered considerably in those final months, sustaining over 120,000 casualties in the Meuse-Argonne alone—and that was enough to trigger a violent recoil back in the states. On the home front, lengthening casualty lists, wartime privations and dislocations, labor unrest, the "Red Scare," and to top it all off, the ravaging world-wide flu epidemic suddenly made Americans regretful of their odyssey "over there." All the old qualms of isolationism began to resurface.

In the spirit of "never again," President Wilson proposed a highly idealistic peace plan based on his famous "Fourteen Points"—a recipe for avoiding such tragic political conflagrations in the future. The last point, suggesting formation of a League of Nations, would have spelled the end of isolationism once and for all, for the U.S. would have had to commit its military power, in advance, to all security actions taken by the League. More Belleau Woods? Chateau Thierries? Meuse-Argonnes? With the Constitution-specified war-making powers of Congress overridden by the League Covenant's notorious Article VI, Americans began to seriously wonder.

As the peace talks unfolded in Paris, it became clear that Allied leaders shared little of Wilson's idealism. After all, their respective peoples had spilled prodigious quantities of blood and treasure, and they had better get something to show for it from the conquered Germans. Some wanted territory. Others sought movable assets. Still others wanted trading privileges, or colonial possessions, or political advantages. And everyone

Figure 13.12 Lloyd George (England), Vittorio Orlando (Italy), Georges Clemenceau (France), and Woodrow Wilson (U.S.) at the peace conference in Paris, France, May 27, 1919 following the defeat of Germany in World War I. They spent much of their time at the conference arbitrarily resolving boundary disputes and creating new countries.

wanted massive reparations—with Germany and Austria accepting all the blame.

The American Senate looked at all the vengeful diplomatic finger-pointing and backed away from the peace settlement—despite Wilson's avid support for it. This was a stunning instance of constitutional mechanisms in the braking mode. President Wilson appealed to the people themselves to change their minds about the Treaty of Versailles, but he suffered a stroke during his whistle-stop tour and ended his term of office a reclusive, broken man.

If it had been a contest between virtue, in the form of Wilson's idealism, and self-interest, in the form of neo-isolationism, Americans appeared to have opted for the latter—with a collective snort of disgust. Their memories of the war grew still more bitter with the passage of time and had much to do with the cynicism of the 1920s. "There were words like 'honor' and 'duty' you couldn't stand to hear anymore," wrote a disillusioned Ernest Hemingway. Companies that had supported the war were investigated by Congress and found to have wielded a corrupt influence. One of them, munitions-maker I. E. DuPont, became sneeringly known as "merchants of death."

Americans failed to realize how much the world had changed already, and how much their own power had come to figure in the new realities. As of 1920, the U.S. Navy was the largest in the world. The American industrial establishment now exceeded that of all the European powers combined. American technology was becoming militarized, too, and would soon lead the way in developing heavy bombers, high-speed fighters, submarines, aircraft carriers, tanks, radar, sonar, and other high-tech innovations. The rising dictators of Europe and East Asia would keep a wary eye on American power, but they would gamble that the isolationist brakes would prove to be stronger than the internationalist engine.

Figure 13.13 The U.S.S. Shaw explodes during the Japanese attack on Pearl Harbor, December 7, 1941. The attack drew the United States into World War II.

In sum, our two conflicting themes of U.S. foreign policy did not combine well in the experience of World War I. National self-interest did not illuminate a clear path for Americans, at least not one they continued to follow. And the republican virtue of Washington's Farewell Address, suggesting that the United States should model good faith and justice in a troubled world, also failed to guide them through.

America continued to pursue a high-minded foreign policy in the years following the Great War, but it was a policy of persuading by example, not of exercising moral leadership. When the Japanese invaded Manchuria in 1931, the U.S. merely scolded them for aggression and refused to recognize their conquest. Japan's response was to invade China. The story was repeated when Italy's Mussolini attacked Ethiopia, when Hitler marched into the Rhineland, when Germany annexed Austria and began clamoring for Czechoslovakia. Franklin D. Roosevelt, the most powerful figure on the planet, merely sent the German Fuhrer a letter and asked him to behave.

Isolationism had its costs.

The Good War

At the time of this writing, the veterans who fought in World War II have become elderly and are disappearing at an accelerated rate. While little has been done to honor them formally (the World War II Memorial in Washington, D.C., for example, was constructed only in 2004), they have been celebrated in any number of popular histories, novels, and movies, of which *Saving Private Ryan* is perhaps the most memorable. Such artifacts powerfully remind us of the Good War.

What made World War II the "good war" furnishes an important key to understanding America's relationship to the world. It was not until that particular experience that Americans achieved a satisfying sense of their role on the world stage. We can identify plenty of self-interest in the American desire to secure a world that was sane, orderly, law-abiding, and open to American trade. But we can also identify works of exalted virtue, especially on the part of American men and women at arms.

Despite the rise of the dictators and the horrors they unleashed, American isolationism survived right up to the eve of Pearl Harbor. (As late as 1940, a bill for military conscription passed in Congress by a single vote!) In a left-handed way, Japan performed a valuable service in surprising the Pacific Fleet at rest in Pearl Harbor, Hawaii, for in one stroke all doubt and uncertainty were erased. Americans were in the war at last and in a role of committed leadership. Even though the price they would pay on the battlefield would be dwarfed by that of the British (to say nothing of the Russians, who fought Germany in spite of signing a peace accord with the Nazis), it seemed clear in all minds that U.S. participation would prove decisive. When he heard of the attack at Pearl Harbor, a beleaguered Winston Churchill joyfully threw his hat in the air.

Figure 13.14 The sheer scope and size of World War II meant that it impacted daily life for virtually all of the American citizenry. From drafting family men with young children at home, to the rationing of staples such as gas and sugar, to planting victory gardens, the American people rose to the challenge.

In the "total war" of which this would be the world's harshest example, America held many of the winning cards. Only incidentally did these consist of skillful generals and admirals. Much more important would be:

- A colossal industrial establishment (beyond the reach of enemy bombs) and the capacity and capability to quickly adapt it to military needs. The war would be won, in large measure, by sheer output.
- Technological sophistication and the ability to mobilize it. The war would be ended by a "secret weapon" almost beyond human imagining.
- Quiet resolve on the part of millions,

Figure 13.15 LST262 and other ships during the D-Day invasion in 1944. The Allied victory was due in part to the massive production capabilities of the United States.

both military and civilian, who would fight not for conquest or plunder but for the defense of civilization.

While there were few battles in which U.S. forces covered themselves in glory, there were many in which casualties ran very high. At Monte Casino in Northern Italy; at Guadalcanal in the Solomon Islands; at Tarawa, Iwo Jima, and Okinawa along the approaches to Japan; and in the Belgian woods surrounding the town of Bastogne, Americans would prove they could be tough and tenacious in bloody, see-saw engagements. And in the D-Day landings at Normandy, they would prove their valor beyond question.

In some ways, however, the real war was the one described by Bill Mauldin's rumpled G.I. cartoon characters, Willie and Joe, who had no taste for glory and few illusions about life. Willie and Joe spent their days slogging through the mud, sleeping in the rain, trying to find comfort in a foxhole. When in one cartoon a medic offers Joe a medal, he replies: "Just gimme a couple o' aspirin. I already got a purple heart."

Equally important to the war's outcome were developments on the home front. Industrial production was the war's crucial factor, and the Allied victory was in large part a matter of mobilizing the country's vast manufacturing establishment. Henry Ford converted his automobile plant at Willow Run, Michigan, to the production of B-24 Liberators, proving that heavy bombers could be turned out on an assembly line much like Model Ts. In Richmond, California, Henry J. Kaiser, an innovative engineer-turned-shipbuilder, put together a monstrous operation for the rapid construction of armed freighters called "liberty ships." The first of these vessels required 196 days to build. The last took 4 days, 9 hours, 36 minutes.

The production miracles were made possible by high wages, grueling hours, and a labor force augmented by blacks, Hispanics, Filipinos, and women. They were also made possible by steely determination. Home-front heroism, less conspicuous than the battlefield variety, was no less relevant to the war's outcome.

As with production, so too with technology. Americans pioneered the development of jet aircraft, radar-guided gunnery, and nuclear weapons. The building of the atomic bomb was a story unto itself. To make uranium fissionable, it was necessary to separate U-235, which was radioactive, from U-

Figure 13.16 American troops disembark in the surf at Omaha Beach on D-Day, June 6, 1944.

Figure 13.17 Picture taken of the atomic bombing of Nagasaki on August 9, 1945.

238, which was not. Since the two isotopes were chemically indistinguishable, the job required the combined efforts of a thousand scientists, engineers, and production specialists, all working in utmost secrecy.

At war's end, the Axis powers were simply overwhelmed. Hitler had supposed that his "Fortress Europe" could withstand any Allied assault, but he would live to see it in the shadow of U.S. bombers advancing in endless waves over Germany itself. In 1942 his U-boats were sinking 600,000 tons of shipping a month in the North Atlantic, but American production out-did the losses and *still* kept the Allies well supplied.

For these and other reasons, World War II became "the good war" in American memory. Its character resolved many of the dilemmas Americans had faced in the past:

- With enemies as dastardly as the Nazis, Japanese imperialists, and Mussolini's reluctant legions, Americans had no qualms about "entangling alliances" or engaging in European turmoil.
- By assuming a leadership role, Americans no longer had to worry about a cynical peace settlement.
- Americans labored to purify their own motives as well as those of their allies, specifying the total surrender and democratic reconstruction of defeated enemies.

All things considered, there was little conflict between virtue and interest in World War II. Americans could fight for a better world for themselves while at the same time seeking a better world for everyone. The U.S. strengthened its own defenses while simultaneously freeing captive peoples, combating oppression, and spreading democratic principles. America's decision, unprecedented in history, to help lift defeated enemies back to their feet, resulted in the nation's gaining three valuable postwar trading partners: Germany, Italy, and Japan.

Just as Woodrow Wilson had spelled out his Fourteen Points at the end of World War I, Franklin Roosevelt enunciated his Four Freedoms in a speech that imparted a higher meaning to the latter conflict. All of mankind, he said, ought to be free to speak out, free to worship as they choose, free from want, and free from fear. As a crusade to secure such freedoms, World War II made sense.

Figure 13.18 Churchill, Roosevelt, and Stalin at the Yalta Conference, February, 1945.

The Bad War

What came to be called the **Cold War** began virtually the moment the Axis surrendered, and it stretched through the latter half of the century. It took the form of an armed stalemate between the United States and the Soviet Union, with most other nations impelled to take one side or the other. It was portrayed as a war of freedom versus tyranny, democracy versus totalitarianism, capitalism versus communism. It was complex and many-sided, and it kept changing like a chameleon as power relations shifted, or as one side or the other developed some new weapon, or as chance events favored East or West.

Cold War

The armed stalemate of the United States and the Soviet Union during the latter half of the 20th century; it was portrayed as a war of freedom versus tyranny, of democracy versus totalitarianism, of capitalism versus communism.

Where World War II had brought together the people of America, the Cold War seemed to pull them apart:

- Now there were "entangling alliances" all over the world, and anyone could see that they truly *were* entangling. The smallest spark in some remote corner of the earth might set off a general conflagration.
- America was unquestionably the leader of the free world—but leadership at what cost! Some of its allies were dictatorships themselves, angling for their own advantage. Others were simply playing one side against the other. And still others (France, for instance) openly despised the American role.
- Technological advantage was only momentary, if not meaningless, for in a world of nuclear warheads, who cared whether one side could destroy the other a greater number of times.
- America's professed ideals were often challenged in the Cold War environment, with flag-burnings, sit-ins and violent protests. Radicals' placards screamed that "Amerika" (as Hitler would have spelled it) wasn't worth dying for. It was the U.S., charged others, who was guilty of imperialism now.

Figure 13.20 President John F. Kennedy mounts a platform overlooking the Berlin Wall in West Berlin, Germany, June 26, 1963.

Figure 13.19 The nuclear-powered attack submarine Jimmy Carter (SSN 23). The Cold War included an arms race to see who could develop the most and best weapons. Nuclear energy was a focus of the arms race, not only for use as a weapon, but also as a source of power for military vehicles.

The Cold War provided a setting for several hot wars, and the U.S. was in the thick of each. The first was Korea, a long, debili-

Figure 13.21 August 3, 1965, Da Nang, Vietnam. A young Marine private waits on the beach during the Marine landing.

tating conflict that achieved little beyond a grudging stalemate. The second was Vietnam, much longer and even more debilitating, ending in abject defeat of high-tech American arms to a weaker but more determined foe. Beyond these clashes there were air strikes, punitive actions, and brushfire flare-ups from Africa to the Philippines. There were revolts to suppress, regimes to prop up, bullies to thwart, and hatreds to tamp down all over the world. There was the danger of China invading Taiwan and of Taiwan provoking China. There was the Berlin saga, a serial drama running for decades, one episode of which required a seven-month airlift to supply the beleaguered city, another resulting in the infamous Berlin Wall. There was Cuba and its self-styled Red caudillo, Fidel Castro, inciting a showdown that teetered on nuclear war. There was a revolution in Chile, a palace coup in Brazil, a Maoist uprising in Peru, a ghastly "secret war" in Argentina, and insurgencies throughout Central America. And in the midst of everything, there arose in Cambodia a tyranny more horrifying, per capita, than the Hitler era.

But for most Americans the Cold War came to center upon Vietnam. It would have been a hard war to win in the best of circumstances; in the context of an East-West standoff it was impossible. The United States, battling a communist insurgency, fired more bullets and dropped more bombs than in all of World War II, but in the end had to pack up and leave. For in a world where one false move might have triggered a nuclear holocaust, the best the U.S. could hope for was a Korea-style agreement to at least stop fighting. And such a truce couldn't be reached.

In the soul-searching that followed Vietnam, Americans reopened all the old questions. Were they really obliged to police the planet? Were their own motives truly pure? Was leadership of the free world desirable or even possible? Americans had entered Vietnam in that spirit of determined hopefulness that had seen them through World War II, and they retreated from it in futility and frustration. Maybe abandoning isolationism hadn't been such a good idea.

Figure 13.22 USAF photo of F-16A, F-15C, F-15E war planes flying over burning oil wells during Operation Desert Storm, 1991.

The Troubled Peace

The Cold War was followed by a hot peace. True, the Soviet Union quietly bowed off the world stage and the West was proclaimed the winner. Yet the same sort of chaos remained. Where once the main threat had been international communism, now it was perceived as Islamic fundamentalism. Or ethnicity on the rampage. Or the Arab-Israeli conflict, with unfathomable beginnings and scant hope of an end. Or, worst of all, it was rule by terror—the notion that any group, *every* group, could get what it wanted merely by setting off a bomb in a crowd of shoppers.

Figure 13.23 New York, N.Y. (Sept. 14, 2001) A solitary fire fighter stands amidst the rubble and smoke in New York City. Days after the September 11 terrorist attack, fires still burned at the site of the World Trade Center.

War clouds hovered over the Middle East. The trouble was Lebanon. Then it was Syria. Then it was Libya. Then Syria again. Then Iran, where a revolution toppled the Shah and installed a fundamentalist regime. And finally Iraq. Each installment was linked to Israel, which the United States had pledged to support at all costs, and which Muslims were pledged to oppose. It flared into one conflict after another, 1948, 1967, 1973, and then into a guerrilla war that sputtered and popped like a burning fuse. There were any number of peace talks, overtures, and tentative settlements, all punctuated by raids, atrocities, car bombs, hijackings, and the murder of eleven Israeli athletes at the 1972 Olympics. No solution seemed in sight.

To make a bad situation worse, it was the Middle East where much of the world's oil supply was located. So whatever course the U.S. pursued, it would be accused by someone, somewhere, of acting in its own greedy interests.

In 1990, Iraqi's cold-blooded dictator, Saddam Hussein, calculated that the Middle East situation might favor a bold move on his part, so he invaded neighboring Kuwait. Iraq's presence in the oil-rich area, together with the threat it posed to Saudi Arabia, prompted the U.S. to take drastic action. The first Gulf War was the result.

The war was unpopular with many other countries, especially those in the Arab world, where the war against Saddam's brutal regime could not be divorced from U.S. support of Israel. Islamic fundamentalists had no difficulty portraying America as "the Great

Satan," and they had more in mind than foreign policy. The Great Satan was seen as a force for modernization, for liberal politics, for religious freedom, for technological progress, for the equality of races and genders, for a view of the world that was out of keeping with Islam.

The stage was now set for a different sort of war carried out in a different sort of way. It burst forth on September 11, 2001, when Islamic terrorists, organized, trained, and financed by a renegade Saudi oil sheik, hijacked four American airliners and used them as ballistic missiles. Two of the aircraft demolished the World Trade Center in Manhattan, with catastrophic loss of life, while a third slammed into the Pentagon in Washington, D.C. The fourth failed to reach its target because of intervention by the passengers and crashed in a field near Shanksville, Pennsylvania, killing everyone aboard. The 9/11 attacks were every bit as infamous as Pearl Harbor.

Like all else in the troubled peace, the meaning of the assault was unclear, and Americans were divided over an appropriate response. Some supported all-out war against "bandit nations," whether or not they had been involved in the 9/11 attacks. Others favored a go-slow approach, aimed at specific terrorists. And in a few places, U.S. actions were seen as another bout of imperialism or another Vietnam.

At the dawn of the 21st century, then, Americans had good reason to be wary of the world in which they lived. It was a world of terrifying vulnerabilities for a nation not designed for domestic war footing, with its unarmed aircraft, unguarded stadiums, unprotected food and water supplies. Few of Americas historic allies supported the wars of preemption, although a host of newcomers took their place. There seemed to be no doctrine, no idea, no catchphrase on which the U.S. could fall back—short of a vague wish to defend a free market, and peaceful civilization. And even that was called greedy self-interest.

Figure 13.24 An Army sergeant watches out for insurgent activity over the Anbar Province, Iraq, in 2005 during the Iraq War.

But was American foreign policy really guided by self-interest? The Cold War and the War on Terror highlighted a difficult problem for America's relationship with the world. Were we being conned into fighting for everyone while others stayed on the sidelines, some cheering and some jeering? The Cold War and the War on Terror were both examples of what economists call the problem of public goods. If communism as a philosophy and social movement was defeated, the benefits were shared by every society threatened by it. If the extremist groups using terror as a weapon were suppressed or defeated, the benefits went to every country that was a target of terror. What could be better than letting the U.S. do the military spending and the fighting (and all too often the dying) while many stood back and claimed the U.S. was the biggest threat to the peace? How could each nation be induced to do its part?

Troubled peace indeed.

The America Question

All of this travail notwithstanding, Americans have earned the right to take some bows on the world stage. After all, whenever and wherever they have committed their lives and treasure on the battlefield, it has almost always been in the name of some higher good, whether it is called justice, democracy, or human rights. And Americans have been uniformly magnanimous to their defeated foes, helping them to build democratic governments and thriving economies. The U.S. has not ignored national self-interest, to be sure, but it has never allowed such interest to be its only guide. Few if any other nation-states can make such a claim.

The deeper question at this point in our history is whether Americans have anything to say to the rest of mankind. For all that their *power* has counted in global affairs, what about their *example*—that point of focus in the Farewell Address? Is the world any different from what it might otherwise have become without the United States? There can be no final answer to this, the global American question. Yet it provides a handy way of weighing the ultimate meaning of the American experiment. If mankind really did turn a corner with the ringing of the bells in Philadelphia, perhaps the horror and hardship of the past century will not have been in vain.

Timeline

Contextual Events

Chapter Specific Events

1790

1796 Washington's Farewell Address

1800

1803 Jefferson makes the Louisiana Purchase

1812 War of 1812

Painting of combat between USS Constitution and HMS Guerriere by Michel Felice Corne

Civil War begins with Battle of Fort Sumter **1861**

1898 Spanish-American War begins

1900

1900 Presidential Election with William J. Bryan

1914 WWI begins
Battle of Marne

U.S. completes Panama Canal **1914**

1916 Battle of Verdun

1917 Battle of Passchendaele

WWI ends **1918**

1918 Battle of Somme

1931 Japan invades Manchuria

WWII begins **1939**

1941 Japan attacks Pearl Harbor, US enters WWII

U.S. drops atomic bombs on Hiroshima and Nagasaki
WWII ends **1945**

1944 D-Day in Europe

1947 Cold War begins

1950 U.S. enters Korean War

1953 Korean War ends

1965 U.S. enters War in Vietnam

1972 Hostage incident during the Olympics in Munich

1973 Vietnam War ends

1990 Iraq invades Kuwait

1991 Cold War ends
U.S. enters Gulf War

2000

2001 Attack on World Trade Center and Pentagon
U.S. invades Afghanistan

2003 Beginning of the Iraq War

Key Terms

isolationism
Manifest Destiny
Monroe Doctrine
jingoism
national self-interest
Cold War

Questions

1. What are some current-day examples of how alliances create entanglements on the international scene?
2. In what ways is isolationism represented today?
3. Is it America's job to "police" the world? To what extent should America care about other countries' problems (such as Iraq)?
4. What were the differences between WWII and the Cold War that made one "the good war" and the other "the bad war?"
5. What do you feel is the proper role of America as a world leader and policeman?
6. Why is "rule by terror" effective and how has it changed the traditional American way of life?
7. How is the market weakness of public goods related to the war on terror?

Chapter 14

Living with Judicial Review

Abstract

The Constitution does not explicitly give much power to the Supreme Court. Not until a decade after the Constitutional Convention did the Court begin defining its role in the nation. This chapter describes how the Supreme Court gained the power to interpret the Constitution through *Marbury v. Madison* and what the consequences have been. The Supreme Court can be either a powerful institution that promotes acceptable change and consensus, or it can be an intrusive institution that splits America apart. What would the Founders think of today's Supreme Court and how well does the Court align with Founding principles?

Ms. **Rosa Parks** was an obscure black woman living in Montgomery, Alabama who became an authentic American hero one fateful December night when she stepped onto a city bus. She sat down in a seat near the front of the bus as it filled with passengers. When the bus driver saw a white person standing in the aisle next to Rosa's seat, he asked Rosa to move to the back of the bus so the white person could sit there. (At the time, Alabama's black citizens were expected to give up their seats to any white person.) Rosa quietly, stubbornly declined to move, resulting in her being taken off the bus, arrested and placed in the Montgomery city jail. The resulting court case became a *cause célèbre* for the entire Civil Rights Movement. She was honored with the Congressional Medal of Honor, she met with several Presidents, and when she died in 2005 her body lay in state in Washington, D.C.'s Capitol rotunda and was visited by tens of thousands. Streets and parks and schools across the country, and a host of children born in the U.S., have since been named after her.

Rosa Parks

1913–2005

African American seamstress and civil rights activist who refused to give up her seat to a white passenger while riding a Montgomery, Alabama, bus on December 1, 1955. Her subsequent arrest became the basis for challenging the legality of segregation laws.

Figure 14.1 Deputy Sheriff D.H. Lackey fingerprints Parks on December 1, 1955 during her arrest for refusing to give up her seat to a white man.

Figure 14.2 After her death in October 2005, Rosa Parks became the first woman to lie in honor in the Capitol Rotunda. Top inset, President George W. Bush pays respect to Parks at the Rotunda memorial.

Her stature at the time of her death at the age of 96, reflects the degree to which Americans overwhelmingly approve of the **Civil Rights Movement**. When Rosa Parks was arrested and clapped in jail, however, Americans were much more divided about racial justice. For many in both north and south, she was just another trouble maker. Questions of racial justice involve values—fundamental notions of right and wrong—and by the middle of the twentieth century, Americans had come to disagree about several of them. Left to themselves, they may not have fomented a revolution over civil rights—they may have continued to let sleeping dogs lie. But it so happened that in 1954, four years before Ms. Parks' famous bus ride, the U.S. Supreme Court had pronounced a stunning decision in the case of *Brown v. the Board of Education of Topeka, Kansas*, which desegregated public schools and launched the Civil Rights Movement. After that, all bets were off on whether or not blacks would continue to be second-class citizens in many states. How did the Supreme Court acquire such power? How did it come to play such a role? And were these developments good or bad for the American experiment?

From Judicial Review to Judicial Interpretation

In Chapter 7 we saw how judicial review came to be a part of the U.S. Constitution. That aspect of the legal system had not been the result of formal work of the Founders, but was instead the brainchild of an exceptionally canny and determined Federalist lawyer, **John Marshall**, an early Chief Justice of the Supreme Court. Justice Marshall had laid out the case for judicial review in *Marbury v. Madison* (1803). Someone needed to determine whether acts of Congress or actions of the President were constitutional, Marshall had argued, and that someone was logically the Supreme Court.

Many of the Founders probably agreed with Marshall's logic, but they had not specifically made provision for judicial review in the constitutional text. Marshall realized that it would be tricky to make his little innovation an accepted part of the implementation of the Constitution. He craftily made certain that the winner of the lawsuit in question would be his Democratic-Republican foes, not his Federalist friends. The Democratic-Republicans were then the people in power, and they might otherwise take issue with his ruling. The ploy worked. Judicial review came to be the accepted practice, and in the mysterious way that institutions sometimes grow, it became a part of the Constitution itself.

Even so, the actual invalidation of a legislative measure was bound to be controversial, as Marshall also realized. After all, the legislature spoke for the people, passing bills that *the people* presumably wanted. Accordingly, Marshall invoked judicial review as seldom as possible, though he routinely struck down *state* laws he didn't like. He hoped his successors would be similarly self-restrained. Later Supreme Court justices, however, were not quite as politic as Marshall had been, and in any event they had the weight of tradition increasingly on their side. With ever greater frequency, they found this or that piece of legislation constitutionally wanting.

In 1857 the Supreme Court got into serious trouble with its powers of judicial review. It attempted to use those powers to resolve a major (and as it turned out, catastrophic) constitutional issue—whether or not Congress had the power to limit the spread of slavery into the nation's western territories. In *Dred Scott v. Sandford* (1857) the Court argued, quite logically, that Congress had no such power because blocking the spread of slavery would have abridged the

Figure 14.3 John Marshall, Chief Justice U.S. Supreme Court, by Asher Brown Durand, engraver (1796–1886) and Henry Inman, painter (1801–1846).

Civil Rights Movement

Movement by African-Americans citizens in the 1960s to gain equal civil rights and to end racial discrimination and segregation.

John Marshall

1755–1835
Fourth Chief Justice of the United States, Marshall ruled that writs of mandamus were unconstitutional in the case *Marbury v. Madison*, thereby establishing a precedent for judicial review.

rights of property ownership. Southerners loved the decision and roundly applauded it. Northerners hated it and condemned it out of hand. The great controversy over slavery that inspired the decision wasn't moved an inch one way or the other, but the ruling did cause much bitterness among anti-slavery groups. The lesson learned here? Where values were concerned, and where opposing values were deeply held, the power of judicial review was limited. Ultimately, it is now seen, the Court must rely on others to bring its decisions into effect.

Equally problematic was another aspect of judicial review—the power to expound the Constitution. The review process itself necessarily involved interpreting the constitutional text, as Marshall had explained in his *Marbury* opinion. The Court determined what a given word or phrase in the Constitution actually meant. There were no settled rules for doing this, of course, but there were settled procedures for keeping interpretations under control. In theory, the Supreme Court, like any other court, was bound by the rules of precedent, meaning that once a given word or phrase was explained and applied, the explanation and application became precedents for all future consideration. Did the word *house*, for example, as it was used in the Third Amendment, also apply to a barn or a shed, or was quartering only forbidden within a human habitation? Once the Court made up its mind on that question, the precedent would be set once and for all.

But this was only in theory. In fact, what power existed anywhere to bind the Supreme Court to its own precedents? Certainly no other courts could do so, nor could the executive or legislative branches of government. In other words, the power to interpret the Constitution was also the power to *re*interpret it, again and again if need be, so that the justices could more or less expound its meaning to their hearts' content. "The Constitution," proclaimed Chief Justice Charles Evans Hughes in a prideful moment, "is what the Supreme Court says it is."

To see just how reinterpretation worked, we need only examine any passage of the founding document that happened to be loosely worded, and watch what happened when that passage came up in court. Take

Figure 14.4 Certain judicial structural components in the Constitution, such as lifetime appointment, were included to keep the justices out of the political storms experienced at the White House and on Capital hill.

the commerce clause (Article I, section 8), for instance. Depending on one's understanding of such vague terms as "commerce" and "regulate," it was possible to come down many different ways on commerce-clause issues, and over the years the Court changed its mind repeatedly.

For example, in an early decision (*Gibbons v. Ogden*, 1824), John Marshall, who was a strong nationalist, gave the commerce clause a very broad interpretation, stretching federal regulatory power to the limit.

A later Court, overseen by Chief Justice Roger B. Taney, distrusted federal power and sought to trim it back. In a series of cases decided in 1847 (the License Cases), Taney and his colleagues reinterpreted the meaning of the commerce clause to allow for concurrent *state* power in commercial regulation.

A still later Court, that of Chief Justice Melville Fuller, took issue with both precedents and advanced a libertarian view of the commerce clause. In *U.S. v. E. C. Knight* (1895), Fuller interpreted the meaning of the word "commerce" so narrowly that neither federal nor state governments were left with much authority to regulate it.

Finally, in the 1960s, Chief Justice **Earl Warren** and his colleagues used the commerce clause to advance the cause of civil rights. The power to "regulate commerce," they argued in *Katzenbach v. McClung* (1964), included the power to enact antidiscrimination laws if the claimed discrimination had some conceivable commercial impact. Of course, everything under the sun has *some* conceivable commercial impact.

Figure 14.6 Earl Warren interpreted the Constitution to mean that the federal government could enact antidiscrimination laws.

Earl Warren

1891–1974
Fourteenth Chief Justice of the Supreme Court, Warren ruled in *Brown v. Board of Education* that segregation was unequal and therefore unconstitutional.

Figure 14.5 Melville Weston Fuller was the Chief Justice of the United States between 1888 and 1910.

With decisions such as these, the Supreme Court was doing more than simply expounding the Constitution. It was using constitutional language as a springboard for innovation. The Constitution was being warped and stretched to accommodate political agendas, intellectual fashions, and strongly-felt needs of the time.

Some political scientists have explained judicial interpretation in terms of natural law. In other words, when interpreting and reinterpreting constitutional language, Supreme Court justices could appeal to general notions of right and wrong that may be derived (or intuited) from nature. They could invoke a shadowy meta-constitution—a constitution behind the Constitution—filled with sublime ideals and perfections.

If such an approach seemed full of promise, it also seemed full of danger.

Moral Consensus

The American Founders had been deeply concerned with issues of justice. The Revolution itself, we recall, had grown out of just such issues. The Founders had also appealed to natural law as a way of grounding their sense of justice. By the law of nature, for example, people could not be taxed by a government in which they lacked representation.

But how, precisely, did one come to understand the law of nature, and to ascertain what it did or didn't allow? The Declaration of Independence implied that all right-thinking people would grasp natural law the same way, without discussion or debate. "We hold these truths to be self-evident," it said. Few of the Founders—including Jefferson himself—actually believed that all men were created equal. After all, many of them owned slaves. And as for the rights of life, liberty, and the pursuit of happiness, the Founders acknowledged numerous practical qualifications. When it came right down to it, there was a good deal of disagreement over fundamentals.

Moral consensus
A general agreement on standards of right and wrong that was more prevalent in early America than it is today.

In spite of such difficulties, it was possible to achieve a rough and ready concurrence on basic issues of right and wrong. When Americans spoke about justice, for example, they rarely had to quibble over terms—justice for *whom?*—as we often do today. The common background of most colonies, the heritage of British constitutionalism, the Christian Protestant ancestry of most American religious groups—all contributed to moral consensus.

Two centuries of deep change altered all that. It is not hard to see why. As America grew in physical size, localism and regionalism became ever more apparent, so that what was said or thought in one part of the country might scarcely be known in another. As immigration increased, so too did the number of immigrant cultures and subcultures, many of them non-Protestant, non-Christian and, later on, non-European. Urbanization drew disparate groups together in circumstances of comparative anonymity, so that the traditional restraints of family and community operated with less force. Higher learning exposed Americans to new ideas and new sources of moral insight—such as the eastern mysticism that swayed Ralph Waldo Emerson and Henry David Thoreau. And finally, diverse economic systems in various parts of the country tended to mold diverging values, as we saw in the case of slavery.

As moral consensus broke down, the process of achieving justice was altered. Earlier, when Americans had more or less agreed with one another about what justice was, all the courts had to do was apply justice to particular cases. Later on, as agreement broke down, the courts were increasingly called upon to determine what justice was. Was it right, for example, for an East Asian immigrant to punish his wife's faithlessness with death, as he might be allowed to do back home?

Again, the Dred Scott case reveals precisely such a dilemma. By 1857, many Northerners had come to believe that there was something inherently wrong with slavery, that it wasn't just bad economics or flawed public policy—it was morally repugnant. Accordingly, these Northerners assumed that Congress naturally had the authority to limit the spread of slavery into the territories, over which it had constitutional powers of supervision. But Southerners had a different view of the matter. Slavery was part of the human condition, they argued, and was fully condoned in the Bible—so how evil could it be? And as for the law of nature, how about the natural right to own property? Should a man who saved and sacrificed to buy a few slaves be

Figure 14.7 Dred Scott. Wood engraving in Century Magazine*, 1887.*

told that he couldn't take his property into territories that were owned by all? In a world of such fundamental discord there was no consensus for the Supreme Court to fall back upon—it had to choose between opposite points of view. No wonder the Dred Scott decision failed to bring peace.

In the years following the Civil War, American courts pronounced other moral judgments and the Supreme Court backed them up. Courts decided, for example, that labor unions were bad and thus that union activities, such as strikes and boycotts, were largely beyond legal protection. The courts decided that government intervention into the economy was bad, and accordingly that state laws seeking to limit the work week, promote safety, or compensate workers for death and injury were unconstitutional. Courts decided that big business was good, and as a result congressional attempts to curtail monopoly were routinely emasculated.

Most importantly, the courts decided that Reconstruction had been a mistake. It followed that laws—and even constitutional amendments—designed to protect the freed slaves and guarantee their newly bestowed rights could be challenged successfully. The provisions of the Fourteenth Amendment that granted the freed slaves the same "privileges and immunities" as other citizens, the "equal protection of the laws," and the "due process of law" were interpreted by the courts to have little or no practical effect. As for the right to vote, which was supposedly secured by the Fifteenth Amendment, the courts approved all sorts of ploys to nullify the authors of the amendment's plain intent and strip suffrage away. Even the Thirteenth Amendment, abolishing slavery, was seriously undermined. Southern laws made it possible for Blacks to be rounded up, charged with spurious crimes—such as not having a job—and sentenced to labor in road gangs. All of this could be achieved by constitutional interpretation. The courts would simply locate some vague clause in the Constitution and give it whatever construction they chose in order to produce the desired result.

Consider, for example, the landmark case of *Plessy v. Ferguson (1896)*. The issue at stake was the growing number of so-called "Jim Crow" laws that specified one sort of accommodation for whites and an entirely different one for African Americans. Hotels, theaters, dining establishments, beaches,

Figure 14.8 In 1939, an African American drinks out of a segregated water cooler designated for "colored" patrons at a streetcar terminal in Oklahoma City. "Jim Crow" laws of racial segregation were based on the idea of "separate but equal" facilities.

transportation systems, and public schools were all becoming subject to racial separation by this time and with calculable effects. In the Plessy case, African Americans challenged Jim Crow head-on, arguing that segregated facilities on a New Orleans railway line violated the "equal protection" clause of the Fourteenth Amendment. The Supreme Court did not agree. "Equal protection," the Court said, had to do with legal, not social, matters. "If one race be inferior to the other socially, the Constitution of the United States cannot put them upon the same plane." Accordingly, so long as the facilities in question were comparable, it didn't matter if they were separate. This decision would have serious repercussions which are still felt today.

The Civil Rights Movement

NAACP

National Association for the Advancement of Colored People; civil rights organization on behalf of African Americans to protect their rights.

While the Supreme Court did not start the Civil Rights Movement all by itself, some sixty years later it clearly did drop a lighted match in the proverbial gas can. The gas itself was prepared more or less in the way that the Founders might have imagined—through political action grounded in a shared sense of justice. The world of *Plessy v. Ferguson* (1896) had first to be transformed into the world of *Brown v. the Board of Education* (1954). That transformation is instructive in the way human nature adapts and changes.

To begin with, we have already seen how the painstakingly chosen, noble, and ringing words of the Declaration of Independence began to take on a life of their own almost before the ink was dry. Even while it was difficult for Americans to reach agreement on moral particulars, they agreed more and more about the self-evident truths of human freedom, human equality, and human rights. They agonizingly debated such matters in the decades preceding the Civil War, and in the carnage of that war they asked themselves repeatedly if the bloodshed and misery inflicted by Americans on Americans wasn't punishment for slighting their own founding principles. As Lincoln put it, how could a nation conceived in liberty hold human beings in bondage?

The Civil War's legacy of racism, discrimination, and segregation did not wear well against this growing consensus. Just as the sight of a public slave auction had horrified northern visitors before the war, the sight of Blacks sitting at the back of the theater, riding in the back of the bus, or checking into the shabby hotel for "coloreds" made many Americans' skin crawl. What sort of country, they asked, could do this sort of thing?

As early as 1918, groups such as the National Association for the Advancement of Colored People (**NAACP**) began to make their voices heard. They protested against segregation, against lynching, against harsh and unfair treatment in the courts, against

Figure 14.9 Leaders of the Civil Rights Movement march on Washington, D.C., August 28, 1963. Although Lincoln signed the Emancipation Proclamation, it would take almost a century before the Supreme Court would strike down segregation laws.

bigotry in popular culture, against intimidation by the Klan, against the denial of opportunity, and the hopelessness of dead-end lives. At first these were solitary cries in the wilderness. In time, however, Americans began to listen. The impressive accomplishments, against towering social odds, of Booker T. Washington, W. E. B. Du Bois, George Washington Carver, Ralph Bunche, Louis Armstrong, and Langston Hughes reminded the white majority of the Declaration of Independence's deeper wisdom. Much of the nation was becoming increasingly uneasy about discrimination based on skin color.

Figure 14.10 March on Washington for Jobs and Freedom, *August 28, 1963. Hundreds of thousands participated, and Martin Luther King Jr. delivered his famous "I Have a Dream" speech from the Lincoln Memorial. (See Appendix C.)*

World War II, pitting America and its democratic allies against Nazi Germany, was in large measure a war against racism. In 1945, however, when the horror of Hitler's death camps came to light, U.S. armed forces were still rigidly segregated, and African-American servicemen furloughed home were once again forced to the back of the bus. One of them, Private Colin Powell, who would one day stand at the Army's head, would later recall being denied the use of restroom facilities on a drive across North Carolina.

President Truman's decision to integrate the armed forces in 1947 was an important first step. Here was a Democrat President, politically dependent on a strong Democrat Party machine in the southern states, saying that discrimination because of race was over in the nation's military forces. That same year the owner of the Brooklyn Dodgers signed on a talented baseball player and WWII vet named Jackie Robinson, breaking the color barrier in professional sports. Civil Rights activists, including brilliant African American lawyer Thurgood Marshall, began building a case for desegregating all public schools.

The great leap was finally taken by the Supreme Court in 1954, for after *Brown v. Board of Education* there was no turning back. Integrated schools, as Southerners well knew, would threaten all other forms of segregation, and their own long-held and honored social traditions were at risk. Chief Justice Earl Warren also grasped that point. Years before, when he had been California's Attorney General, he had acquiesced in the roundup and incarceration of Japanese-American citizens after the Japanese attack on Pearl Harbor, and he still seemed to suffer from guilt for that unconstitutional and blatantly racist act. His personal story was like that of many Americans. Segregation had since then become a moral issue for him—*the* moral issue—and he could not rest until it was abolished. By the force of his dynamic and charismatic personality, Warren convinced *all* of his brethren on the high bench to support the Brown decision—a rare event indeed.

But there was the matter of precedent set by *Plessy v. Ferguson*. "Separate but equal" facilities were permissible, according to that ruling, and did not violate the Fourteenth Amendment. Warren and his fellow justices barely paused before overturning it. Instead of debating the fine points of constitutional law (as had the justices back in *Plessy*), Warren simply returned to the "equal protection" clause of the Fourteenth Amendment and gave it a natural law interpretation. Segregation, he said, was wrong—end of story. "Separate educational facilities are inherently unequal."

Because most Americans had come to agree with such thinking, it was easy for the Brown ruling to operate as a catalyst. When African Americans began sitting at segregated lunch counters or boycotting segregated bus systems, they were cheered on by an increasingly favorable public opinion that had been precipitated in some sense by the Supreme Court's decision. That is, Americans now had a specific issue to address, not just a vague sense of unease, and upon addressing it they found themselves committed. In following the Supreme Court, they saw themselves following the Constitution.

The way ahead was still long and hard. There were marches, demonstrations, assaults, arrests, and physical attacks on the demonstrators by baton-wielding police, their attack dogs, and fire hoses. Black activist Medgar Evers was gunned down in Mississippi. Three idealistic young civil rights workers were murdered by the Klan, their bodies hidden for a time in an earthen levee. Dozens of Black churches would be bombed—in one of which four little girls died while preparing their Sunday School lessons—and the chief spokesman for the entire civil rights movement, Rev. **Martin Luther King Jr.**, was killed by a sniper on the balcony of a Memphis motel. Yet the words of the Pledge would eventually prove prophetic. There would someday be liberty and justice for all.

Figure 14.11 Martin Luther King Jr. at a press conference at the Capitol, Washington, D.C., March 26, 1964.

Martin Luther King Jr

1929–1968
A Baptist minister and political activist, King was also a leader of the civil rights movement in the 1960s. His most famous speech was delivered on the steps of the Lincoln Memorial, entitled "I Have a Dream."

***Roe v. Wade* (1973)**

Case in which the Supreme Court decided that abortion was protected by the Bill of Rights.

Jane Roe

The anonymous pseudonym used for Norma Leah McCorvey in the landmark case *Roe v. Wade*, which legalized abortion. McCorvey initially claimed to have been pregnant by rape, but in the 1980s confessed that she had fabricated the rape story. She also became a pro-life activist and lobbied the Supreme Court to reverse its decision.

Roe v. Wade

The stupendous success of the *Brown* decision may have encouraged the Supreme Court's activism. By the end of the 1960s the justices were ready to take on another large and troublesome issue, the legalization of the hitherto criminal act of abortion. In the 1973` landmark case of *Roe v. Wade* the Warren Court ruled that the practice of abortion could not be prohibited by state laws. The right of a woman to choose to terminate her pregnancy, the Court held, was protected by the U. S. Constitution.

As with many cases that posed thorny questions, *Roe v. Wade* was essentially fabricated. A young woman represented in the court records as "**Jane Roe**" walked into a Texas counseling center in late 1969 with a lurid tale of having been raped by three drunken men on a lonely Georgia roadside. The reported rape was bad enough, but the resulting pregnancy was even worse. "Jane Roe," thought alert abortion-rights activists, seemed like a good candidate to challenge the Texas statute outlawing abortion. Only much later did the public learn (from the supposed victim herself, who had since changed her mind about abortion rights) that "Jane Roe" had fabricated the rape story.

But this was not known in 1969, and "Jane Roe's" case was ably argued by two young Dallas attorneys, both devoted to what would soon be called "freedom of choice" for

Figure 14.13 Marchers in Washington, DC at the March for Life *on January 22, 2002. How is the abortion issue like the slavery issue before the Civil War? How is it different?*

women. One of the attorneys, **Sarah Weddington**, became so involved in the case that she gave up the rest of her practice and stumped the state for financial support, while the case made its way up the ladder of appeals. By December of 1971, Weddington, who had never yet argued in court, found herself standing before the highest bench in the land.

Happily for her, Weddington found most of the justices on her side, willing to view abortion the way they had once viewed segregation. Here, however, the constitutional authority was even more vague. In the *Brown* case, there was language in the Fourteenth Amendment which, even to a layman, suggested that racial separation violated "the equal protection of the laws." With *Roe*, no such language existed. The right to an abortion, the attorneys argued, grew out of the right of privacy—which was only *implied*, not *stated*, in the Bill of Rights.

The doctrine of **implied rights** was relatively new. For all that, it might well have found favor with the Constitution's Framers, for as we recall, they believed that the enumeration of some rights didn't necessarily deny the existence of other rights. Privacy was like that. If the First Amendment protected the right to speak one's own thoughts and heed one's own conscience, if the Fourth Amendment protected one's property from random searches and seizures, and if the Ninth Amendment specifically mentioned "other [rights] retained by the people," then an implied right of privacy wasn't so far-fetched.

Figure 14.12 The debate over abortion rights has occasionally sparked violence, including the bombing of abortion clinics.

Notice that the movement by some to claim a right to an abortion was a response to changes in the technology of contraception and the decline in fertility. The success of contraceptive methods caused individuals to believe that they were entitled to sexual intercourse without the possible consequence of pregnancy. Earlier generations would not have accepted that entitlement. Moreover, in the second half of the twentieth century, many men and women were losing interest in having children, or at least large families. Abortion was seen by many as the ultimate safeguard against unwanted children. These changes in values and technology fueled the search for the implied right to choose.

The very vagueness of implied rights enabled the Supreme Court to fall back on natural law reasoning once again. The Court simply concluded that abortion was not wrong and therefore shouldn't be prohibited. This conclusion pointed out the difficulty of natural law. As the Founders saw it, natural

Sarah Weddington

b. 1945
Attorney with Linda Coffee in the *Roe v. Wade* case.

Implied rights

The doctrine that the Constitution protects rights that are not explicitly stated or enumerated therein.

law was supposed to be accessible to everyone regardless of political affiliation or personal interest. Anyone who was not mentally or morally impaired was supposed to be able to tell right from wrong. Yet the Founders themselves would very doubtfully have endorsed abortion. They might have disagreed about it based on a variety of considerations—just as Americans do today—but they wouldn't have viewed it as a "self-evident truth."

Just as there was a history behind the Civil Rights Movement, so too was there a history behind the *Roe* decision. It traced back through early feminism, through the suffragette movement, through the efforts of Margaret Sanger and others to win acceptance for birth control, through the political and economic changes brought about by the Great Depression and World War II. Abortion—now more loosely called "freedom of choice"—became one of the great liberal causes of the twentieth century.

The Founders undoubtedly intended such matters to fall within state sovereignty, having nothing whatever to do with the federal mandate. However, in a world of automobiles and freeways, state action on the abortion issue became irrelevant. After all, if abortions couldn't be legally performed in one state, it was easy enough to go to another. Beyond this fact (and further complicating the question) there were unreported, uncounted tragedies of back-alley abortion mills, cases of rape and incest, underage children getting pregnant, and orphanages brimming with unwanted offspring.

Judicial legislation

When courts do not feel bound by the letter of the law nor by their own precedents, and instead appropriate the legislative function of making laws in resolving issues.

Still, for approximately half of the American people, abortion was nothing less than murder. In other words, for them natural law spelled exactly the opposite conclusion: the right of an unborn child to life. And the right-to-lifers could tell their sad stories too. What about all those childless couples who would only too gladly adopt? What about those aborted fetuses who gasped air into their lungs as they were ripped from their mothers' wombs? And what about those millions of lives destroyed—those potential Thomas Jeffersons and Albert Schweitzers and Mother Teresas—who ended up in abortion clinic dumpsters?

Accordingly, the Supreme Court's action in *Roe v. Wade* did not have the catalytic effect on public opinion that it had in *Brown v. Board of Education*. Instead of bringing forth moral consensus, the decision deepened moral division. A chasm opened between left and right that a full generation of discussion, argument, and social statistics has failed to close. "Right to life" and "freedom of choice" advocates have shouted at one another across that chasm ever since. And, oddly enough, both sides have rooted their arguments in American fundamentals.

Judicial Legislation

To those concerned with the Court's power, *Roe v. Wade* also raised anew the specter of **judicial legislation**. If courts did not feel bound by the letter of the law (in this case the Constitution) nor bound by their own precedents, what was to stop them from abandoning the judicial function entirely and appropriating the legislative function instead? The urge to do so always haunts a democracy. That is, the people may be so hopelessly divided on critical issues that the legislative process simply won't work. How nice, one might suppose, if the courts just stepped in and settled matters for all of us.

Some Americans believed that *Roe v. Wade* represented just such an attempt. To begin with, the question of abortion seemed to be legislative in character, not judicial, the sort of issue that the people needed to decide for themselves. The way the Court approached it seemed safely legislative as well. Justice Harry Blackmun sat down, as a

Figure 14.14 Where is the boundary between interpreting the law and creating it?

congressional committee might do, and worked out a system for determining which abortions were constitutionally protected and which were not. This was analogous to the various public school-bussing plans that federal courts had worked out earlier, determining which children needed to be bussed to what schools in order to achieve full integration. There was nothing wrong with the general idea about determining the rules regarding abortion, most Americans conceded, but should the *courts* be in charge of it?

Unlike school segregation, there was no clear and obvious answer to the abortion question. A viable fetus was obviously a person in some fundamental sense while a fertilized egg cell was arguably something else. Moreover, a woman's freedom to choose her own fate (as men routinely did when they fled from an unwanted pregnancy) was nothing to sneer at. These would be tangled issues for the brightest of legislators and they were no less tangled for judges. But the fact that judges were not responsible to the people (and did not have to stand for reelection) made some of them think they could accomplish what legislatures could not.

Various groups struggled to command opinion on the subject, soon moving the political battle into the judicial arena and making divergent views about *Roe v. Wade* a test for supporting Supreme Court nominees. Instead of the Court standing above politics as the Founders had hoped, the Court moved to the very center of politics. For it is through the political process, however clumsy and chaotic, that a democratic society ultimately sorts out issues of right and wrong.

Judicial Review and Natural Law

The question, then, becomes this: is there a good and bad way to practice judicial review? For example, did the Supreme Court do something "right" in the case of *Brown v. Board of Education* and "wrong" in *Roe v. Wade?* And if so, what was it? No one can say for sure, of course, for these are highly personal freighted value issues. But there may be an intriguing perspective from which the question might be addressed. How would the nation's Founding Fathers have resolved both cases had they lived in *our* time?

Evidence seems to suggest that the Founders would have approved the practice of judicial review. After all, forms of the practice had been used in colonial times, and several states had included it in their constitutions. However, according to Hamilton in *Federalist* 78, the role of the Supreme Court in conducting such review would be limited to laws that violated the "manifest tenor" of the Constitution—that is, laws that clearly conflicted with the plain meaning of the constitutional text. In all other cases, the justices ought to restrain themselves. In other words, the Founders might have taken a dim view of constitutional reinterpretation.

Still, as long as there were phrases in the Constitution like "necessary and proper," "general welfare" and "due process of law," a Supreme Court with the acknowledged authority to perform judicial review was bound to reinterpret the Constitution sooner or later. Indeed, those stretch-point phrases and others like them had probably been placed in the text precisely to allow for growth and adaptation. So, to repeat the question, would the Framers have acknowledged a right way and a wrong way to do what they tacitly admitted must eventually be done?

Charles Kesler

A senior fellow of The Claremont Institute and editor of the Claremont Review of Books. He received his Ph.D. in Government from Harvard University and is currently Director of the Henry Salvatori Center at Claremont McKenna College. A Constitutional scholar, Kesler asserts there is a moral text in the Constitution.

Constitutional scholar **Charles Kesler** has argued cogently that the answer to that question is *yes*. Kesler points out that the Framers, as believers in natural law, accepted that there was a moral text embedded in the American Constitution, just as there had been a moral text embedded within the English Constitution—and that kings, presidents, parliaments, congresses, and courts of every kind were bound to adhere to that unwritten charter of justice. In other words, if there was ever any interpreting to be done, the "right" way to do it was reasonably clear. Consult the Constitution's "moral theory" as a guide.

Here is the substance of Kesler's argument. "Interpreting the Constitution," he writes, "is . . . an act of inferring the meaning of particular provisions from the purposes of the Constitution as a whole." The Constitution was more than just a plan of government, as Madison explained in *Federalist* 43, it was an embodiment of republicanism's "true principles." Its purpose was to promote "the common good of society" and "the happiness of the people." It comprehended "a due sense of national character" and contemplated "extensive and arduous enterprises for the public benefit." These phrases remind us that the Constitution was regarded by its authors as a means for securing a larger end, one having to do with natural law, and as a consequence any vague or confusing language should always be viewed with that larger end in mind.

The question might be put this way: *How can the Constitution be interpreted so as to advance the republican cause?* Does a given interpretation enhance civic virtue, strengthen the rule of law, reinforce moral self-governance, promote opportunity? Does it burnish freedom? Expand choices? Facilitate human dignity? If so, it probably reflects the Framers' "original aspirations."

By this test, *Brown v. the Board of Education* would clearly seem to pass, even though it reflects a world far removed from the Founding. *Roe v. Wade* might be more problematic. It exalts freedom, to be sure, but at the expense of civic virtue—and at the cost of human life.

What Limits the Court's Power to Decide?

There are essentially two constitutional mechanisms to limit the power of the Supreme Court. First, the President and the Senate, both closer to the political preferences of the people than the Court, respectively appoint and confirm Supreme Court justices when vacancies occur. Thus, if the Court makes decisions that do not reflect popular sovereignty, future justices with judicial philosophies more attuned to the politics of the time may reverse those justices. Judicial appointments are for life and normal life expectancy has increased significantly, so most Presidents appoint very few, if any, Supreme Court justices. Change in the judicial philosophy of the court occurs very slowly. Second, decisions of the Court may be reversed by Constitutional Amendment. However, there is a strong bias against amendments that are directed at particular court decisions. The Supreme Court in practice has few limitations on their decisions other than the power of precedent and constitutional interpretation.

Liberalism, Conservatism, and the Supreme Court

We have seen that mechanisms of the two-party system often narrow the scope of political choice, for both parties vie with one another for those voters in the middle who can assure them of victory. Such being the case, the political dialogue in America has often come down to a conversation between liberals and conservatives.

Historically, liberals and conservatives have been found in both parties. Toward the latter half of the twentieth century, there has been a tendency for conservatives to feel more at home in the Republican Party, with liberals gravitating to the Democratic Party. Liberals and conservatives have many points of difference, ranging from the scope of government action to postures in foreign affairs. They have markedly contrasting views of constitutional interpretation—views growing out of the original debate between Hamilton and Jefferson.

Conservatives, following Jefferson, generally argue that the Constitution ought to be **narrowly** and strictly interpreted, for only in that way can it remain a guardian of liberty. Liberals, following Hamilton, generally argue that the Constitution ought to be **broadly** and loosely interpreted, for only then can it accommodate new and unforeseeable cultural and political circumstances.

As with the abortion question, there is no simple right and wrong to this debate—it may go on forever. As it does, the language of the Constitution is loosened or tightened according to the politics of the time.

The long-term trend, however, has been distinctly on the side of loosening. As we have seen, certain clauses in the Constitution and certain others in the Bill of Rights were more or less drafted with stretchability in mind. As the courts have interpreted and reinterpreted these clauses, the tendency has been to make them ever broader, vaguer, and more inclusive. There have been several important results.

One has been the **growth of government**. Power and responsibility have drifted steadily from the states and toward the fed-

Narrow construction

Constitutional clauses that were written to be interpreted in a more narrow or direct manner.

Broad construction

Constitutional clauses that were written to be interpreted in a more broad or general manner.

Growth of government

The steady drift of power from the states to the federal government, with increasing involvement of the federal government in American life.

SUPREME COURT LOOSENING TREND

Growth of Government

How does today's size of government compare with Locke's vision of government's purpose?

Growth of Personal Rights

How do today's rights compare with Jefferson's natural rights?

Growth of Privacy

Did the Founders ever dream that issues such as same-sex marriage or abortion would be an issue in America?

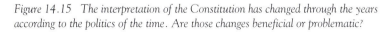

Figure 14.15 The interpretation of the Constitution has changed through the years according to the politics of the time. Are those changes beneficial or problematic?

eral government, and federal government in turn has grown ever larger and taken on responsibility for more aspects of modern life in America. Liberals have generally applauded this development, for they tend to see government more as "the answer" than "the problem." Conservatives have often vehemently disagreed—reversing those liberal perceptions. Even so, government has continued to grow steadily, through conservative administrations as well as liberal.

Growth of personal rights

The broadening judicial interpretation of personal rights that were construed fairly narrowly in the past.

A second result has been the **growth of personal rights**. Rights that were construed fairly narrowly in the days of Thomas Jefferson have continually expanded through adjudication. Freedom of speech, for example, which in Jefferson's time was exclusively limited to *political* speech, has burgeoned into "freedom of expression," encompassing everything from four-letter words to topless dancing.

Growth of privacy

A broadening of the toleration that ought to be extended to an array of lifestyles, behaviors, choices, and value systems as well as a decrease of government prescription in individuals' lives.

A third result has been the **growth of privacy**. Technically, of course, privacy is a right itself, as discussed above. But privacy is also a way of thinking about civil society. Privacy implies that society is pluralistic by nature and that, within broad limits, toleration ought to be extended to an array of lifestyles, behaviors, choices, and value systems. Privacy also implies that government should simply keep its nose out of people's business. Liberals often think of privacy, so defined, as the very essence of freedom. Conservatives often think of it as licentiousness.

By and large, judicial activism has resulted in constitutional protection for all three. That is, over the course of time, Americans have come to live with, if not always to embrace, a large and proactive federal government, comparatively weaker state governments and local communities, expanded rights for individuals and groups against society as a whole, and an ethos of anything goes.

This liberal "agenda," as it is sometimes called, had become a political issue by the end of the Vietnam War, and since then has figured in every election from 1980 to 2004. It fans out into questions of abortion, gay rights, gay marriage, drug legalization, obscenity, pornography, school prayer, religious symbols in public life, criminals' rights, victims' rights, multiculturalism, multilingualism, affirmative action, racism, sexism, ageism, political correctness, and discrimination against the handicapped.

Because all these issues have appeared at one time or another on the Supreme Court calendar, the Court now occupies a place at the very center of American political debate. While the issue is framed in various ways, it comes down to a single essential question: is this constitutional democracy—or has it become something else?

Conclusion

We should not lose sight of Rosa Parks and why she is celebrated. Without the U.S. Supreme Court her valiant action might have been in vain. Nor should we lose sight of the role the Court has played in setting boundaries, resolving ambiguities, adjudicating conflicts, and providing a stable and yet flexible body of constitutional law.

Yet, as John Marshall argued in *Marbury v. Madison*, the Court could perform these functions precisely because it was isolated from politics and shielded from baneful influences. That very isolation and shielding has created a situation in which the Court could also rule, on occasion, as though untouchable.

Where the Supreme Court has tapped into the law of nature and touched the moral intuitions of the American people, it has unquestionably been a catalyst for good. Where it has simply intruded upon the debate of the moment and shouldered legislatures aside, it has risked usurping something close to the heart of democracy—the sovereign will of the people.

Timeline

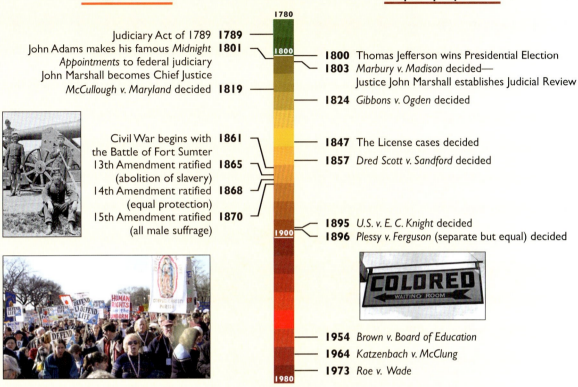

Contextual Events

Judiciary Act of 1789 **1789**
John Adams makes his famous *Midnight Appointments* to federal judiciary **1801**
John Marshall becomes Chief Justice
McCullough v. Maryland decided **1819**

Civil War begins with the Battle of Fort Sumter **1861**
13th Amendment ratified (abolition of slavery) **1865**
14th Amendment ratified (equal protection) **1868**
15th Amendment ratified (all male suffrage) **1870**

Chapter Specific Events

1780
1800

1800 Thomas Jefferson wins Presidential Election
1803 *Marbury v. Madison* decided—Justice John Marshall establishes Judicial Review
1824 *Gibbons v. Ogden* decided
1847 The License cases decided
1857 *Dred Scott v. Sandford* decided

1900

1895 *U.S. v. E. C. Knight* decided
1896 *Plessy v. Ferguson* (separate but equal) decided

1954 *Brown v. Board of Education*
1964 *Katzenbach v. McClung*
1973 *Roe v. Wade*

1980

Key Terms

Commerce Clause
natural law
judicial legislation
growth of privacy
narrow construction
moral consensus
growth of government
broad construction
implied rights
growth of personal rights

Key Supreme Court Decisions

Marbury v. Madison (1803)
Dred Scott v. Sandford (1857)
Brown v. Board of Education (1954)
Gibbons v. Ogden (1824)
U.S. v. E.C. Knight (1895)
Katzenbach v. McClung (1964)
The License Cases (1847)
Plessy v. Ferguson (1896)
Roe v. Wade (1973)

Key People

Rosa Parks
Earl Warren
Sarah Weddington
John Marshall
Martin Luther King
Charles Kesler
Dred Scott
Jane Roe

Questions

1. How would our government be different today had the Supreme Court never gained such power as judicial review? What would be missing? Which branch of government would be most powerful? Was it inevitable for the Supreme Court to gain such power?
2. *Dred Scott v. Sandford* displays how difficult it is for the Supreme Court to rule on values. Does the Supreme Court shy away from controversial issues or seek after them? What issues today are controversial or emotionally-charged?
3. After reading about the Supreme Court's ability to interpret and re-interpret the Constitution, do you feel the Supreme Court has too much power? Should the Constitution be defined as whatever the Supreme Court says it is? What ways are there to limit the Court's power now that precedents have been established?
4. Is the Supreme Court really immune from politics as justices wish to appear? Do justices do anything that makes you question this appearance?
5. Why is natural law so hard to agree upon? In a diverse society, what causes natural law to be anything but self-evident?

Chapter 15

America's Founding Heritage

Abstract

This chapter explores the heritage and the moral founding of America in the context of two questions: First, what were the underlying hopes and expectations of the Founding? Americans saw themselves as unique, bound together by a social compact, virtue, the rule of law and liberty. These ideals created a national character that facilitated the creation of a deliberately founded free nation. Second, how have those founding ideals fared over the past 230 years? Do America's founding documents still protect and uphold the notions of freedom, liberty, social compact and rule of law as they were designed to do?

We generally think of *heritage* as the sum of an entire history, of everything that has gone before and everything yet to appear. In that way, perhaps, the United States of America is different. For while its short but colorful history is certainly filled with heritage—the advent of democracy, the abolition of slavery, the mantle of world leadership—there remains a separate and distinct *founding* heritage. Unlike most great nations, the USA was clearly founded. People sat around tables and worked it out; a few remarkable men invented our nation. Accordingly, they intended it to be something specific, something particular, something that embodied their high-minded values expressed in a set of hopes and expectations. Whatever happened to it from that time forward would necessarily be judged in reference to those founding ideals.

American Exceptionalism

The idea that the American experience was different or unique from others, and therefore America had a unique or special role in the world, such as a "city upon a hill."

In this final chapter, we will address two questions regarding the American heritage. First, what exactly was the Founding about and what particular hopes and expectations lay behind it? Second, how have those founding ideas fared in the course of some 230 years of history? Has America, in other words, lived up to its founding heritage?

The Philosophy of the Founding

The American Founders were not philosophers. They worked from no carefully formulated set of premises, embraced no overarching design. On the contrary, much of what they did was simply a response to problems and situations as they arose and as they perceived proper, long-lived solutions. They muddled through as best they could. Still, the Founders shared a common background and worldview, and despite their many quarrels and reluctant compromises, they fundamentally agreed on many things. Had they not done so, the Founding would not have been possible.

Knowing what we do of the Founders and their world, we can point out at least a few elements of their common philosophy.

American Exceptionalism

The Founders saw themselves as transplanted Englishmen. Like colonials everywhere, they were eager to preserve the traditions of their homeland, and they were more than a little fearful of losing those traditions in the American wilderness. Yet they also came to recognize the American experience as being important in its own way. They saw something purifying in it, something that Columbus had glimpsed long before, an escape from long-established human corruptions and follies. Viewed from America, Europe's way of doing things looked different. The corruptions and follies of the Old Country seemed a lot clearer. To John Adams or Thomas Jefferson, Americans were not destined simply to replicate the Old World.

During the great turmoil that swelled into the American Revolution, Americans became convinced that British tyranny had specifically set about to corrupt *them*. Some of this paranoia was undoubtedly a reflection of the Puritan values that had gradually become the moral foundation of many of the colonials. But the essence of the corruption they feared was the loss of their autonomy and self-government as individuals. The patronage and good offices of the King of England came with the price tag of dependency, a price many American colonials thought was too high. Much of American exceptionalism was connected to the land, for the American soil had become sacred to them. We can still hear echoes of that sacredness in such hymns as "America the

Henry VIII

James I

Louis XVI

George III

Figure 15.1 After years of tyranny, corrupt monarchies were not a tradition the Founders intended to copy.

Beautiful." It helps us understand why Americans had such a strong affinity for the commonwealth ideology, with its emphasis on the ideals of the country party. If those "purple mountain majesties" and "fruited plains" were not yet a country in the sense of ***patria***, they were surely a country in the sense of a life lived close to the land.

America's land was indeed as beautiful as it was bountiful. More important, the colonists, perched on the Eastern seaboard and gazing west at their vast timbered wilderness, believed it to be virtually endless. Land at the plenty for ordinary settlers had a powerful meaning for colonial Americans. They had come from a world in which homelessness and joblessness had long been accepted conditions of life. In America, colonists were thus able to acquire the thing that for thousands of years had given substance and dignity to human existence. Land accorded them space, livelihood, opportunity, self-mastery, and the command of their own destiny. Small wonder that Americans saw the Old World's corruptions and follies so clearly.

Free land implied free institutions. We have seen in an earlier chapter that Thomas Jefferson, for one, subscribed to the myth of an ancient Saxon democracy. In the lost world before recorded history, Jefferson believed, men tilled their own land, managed their own lives, and *elected their own kings*. Things could never be that way again, not in Europe. Kings had become tyrants, lords had become oligarchs, and freedom had perished, basically because there wasn't enough land to go around. In America, though, free land stretched clear to the Pacific—and with free land came the promise of liberty.

As the Founding took shape, Jefferson and Madison pondered such matters. How, they asked, could they preserve the best in their English heritage and rid themselves of the worst? How could they create a modern society? Certainly not by starting from scratch, as French revolutionaries would attempt to do a few years later, but perhaps by selectively pruning away those Old World corruptions and follies. Get rid of kings. Get rid of lords. Cut the tie between church and

Patria

Latin for "fatherland," from the Greek *patris*, also the root for the word *patriot*.

Figure 15.2 The Colonials' relationship with the land was an important influence in the Founders' guiding philosophy.

state. Lift the bans on speech and opinion and throw open the windows of the soul. In America all this seemed possible.

Such was the meaning of *American exceptionalism*. The United States would become that City upon a Hill dreamed of by John Winthrop. The inhabitants of that City would not have to follow the dreary, blood-soaked road of European history. Their City would be built beyond the confines of past history.

The Social Compact

As Locke and others imagined the social compact, there was more to it than a hypothetical agreement to form government. Equally important was the idea of an active and ongoing agreement among the living, to accept the terms and conditions of their society.

Accordingly, in places where the social compact was sound, people abided by the laws and supported the governing system—both signs of social health. In places where the social compact had faltered, there was lawlessness, dissension, and political turbulence. People rejected the failed system in ways both large and small, and in the end they often rose up against it, but those uprisings, even if they were successful, always seemed to settle back into systems that might be different, but not much better. A broken social compact led to a broken society.

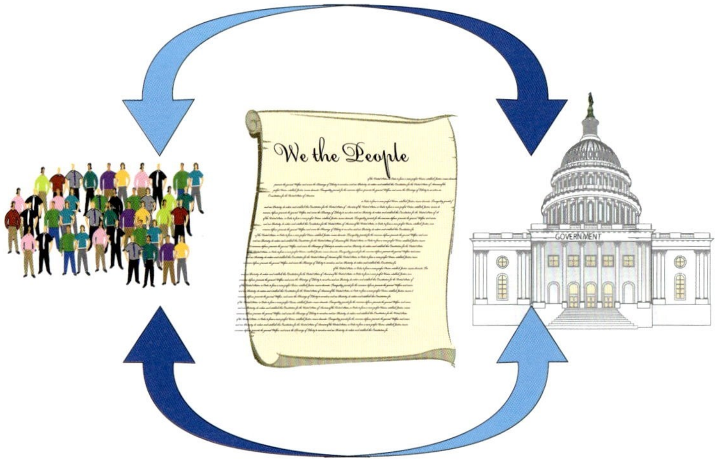

Figure 15.3 Because the Founding was created by the people, their Social Compact was respected and obeyed.

Whatever else they faced, the American Founders believed their social compact was robust. Perhaps this was due to the essentially middle-class nature of American society. In the wide-open vistas of America there were no lords peering down from great palaces, no streets full of wretched beggars. Or perhaps it was due to the relative homogeneity of colonial culture, marked by agricultural pursuits, a simple lifestyle, and a straight-laced Protestantism. But something else was at work as well.

A prominent feature of American exceptionalism was the way the social compact operated. Because Americans had more or less fashioned their own institutions, they were basically happy with them. Some of the colonies—Connecticut, for example—had essentially governed themselves from the beginning. Some of the colonies had drafted their own charters, not only ruling themselves but inventing themselves as well. And out on the frontier, it was common practice for settlers to work out their own government. In America the social compact was neither implied nor theoretical; it was an everyday reality.

The colonies weren't democratic by modern standards, but political participation was high in most of them, and much higher than in Great Britain in those years. Most adult males not only voted, they occasionally held local office, such as town selectman, and as a result were treated with dignity. The neglect the colonies received from England strengthened their sense of self-mastery. People migrated *to* them, not *from* them.

Classical political theory had often stressed a connection between the city and the soul. Autocratic ***poleis*** made for autocratic personalities among their citizenry (and vice versa), while the same held true for democratic or aristocratic *poleis*. The de facto freedom of the American colonies encouraged the personal freedom of their inhabitants, and personal freedom led to a social compact that worked. The American "city" cultivated a distinct American "soul."

This background proved crucial to the American Founding, for the Founding had to be created by the people themselves. They had to decide, citizen by citizen, whether or not to support the patriot cause, to join the militia, to acknowledge the authority of Congress. They sent their own delegates to the state conventions, and when the proposal for a federal government emerged, they sent representatives to accept or reject it. There was nothing here about government descending from on high to tell them what to think and when to think it. It was something the people themselves had worked up, tinkered with, overhauled on the work bench when it misbehaved, and eventually lived under. Never before had the social compact operated so literally.

Republic of Virtue

Like the English Whigs before them, the American Founders paid close attention to the ancient world, for its lessons seemed to address their cause. They were gentleman farmers, merchants, and speculators, not classical scholars, but part of being a gentleman was having the essentials of a classical education.

Thus equipped, the Founders couldn't fail to be impressed by the significance of virtue in the writings of the great men of history. It was everywhere they looked. Nor could they fail to see its relevance to their own situation. Republics were hardly the order of the day in 1776. No one knew exactly how they worked. However, if Plato, Aristotle and Cicero all affirmed that republics operated by the power of civic virtue, that was good enough.

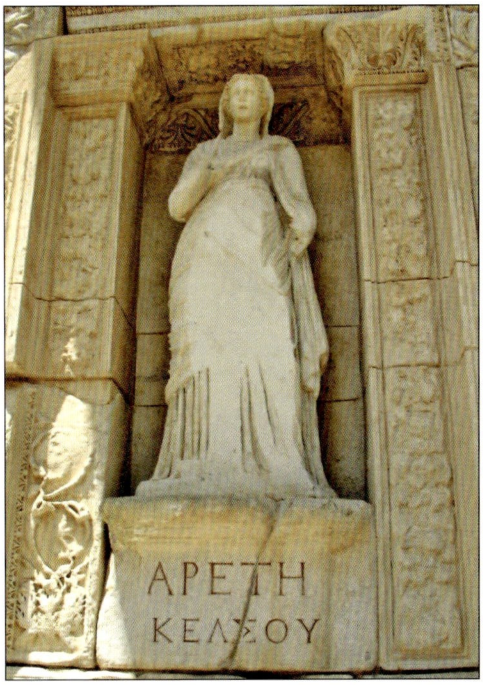

Figure 15.4 Personification of virtue (Greek aretê) in Celsus Library in Ephesos, Turkey.

There was an important shift, however, in precisely what Americans took virtue to be. For the ancients, virtue had been *areté*, "excellence," and the whole point had been to fortify the *polis* with the wisdom, courage, temperance, and justice of virtuous citizens. Examined closely, these virtues were competitive in nature, reflecting the Greco-Roman passion for contest. There were winners and losers in the virtuous society, as each tried to outdo all others in civic zeal. Christian virtues, on the other hand, were less about winners and losers and more about accommodation. Love your neighbor, go the extra mile, turn the other cheek, forgive seventy times seven—there was nothing here about outshining someone else, only about making your community a good place to live. Societies that practiced ordinary kindness, however imperfectly, turned out to be more durable than those that strove only for excellence.

Greek cardinal virtues

Elements of Greek areté: Temperance, Courage, Wisdom, Justice.

When the American Founders spoke of virtue, they may have used Greek terms, but the substance they alluded to was mainly Christian in character. The religious background of the colonies had amplified **Christian** values generally, and the Great Awakening of the 1740s had ignited an ardent Christian evangelism. There were incentives to practice virtue as never before.

This spiritual setting became a source of confidence for the Founders. Interlocking with American exceptionalism, it created a world in which republicanism could thrive, not because the average American could match a George Washington in classical excellence, but because they could care for their families, attend church, look to the health of their communities and live fulfilled lives of common decency. Virtue such as this truly could be made to work, the Founders told themselves, unlike the *areté* of ancient Greece—which had burned out.

The Revolution had proved the power of American virtue, at least to some of the Founders. Take, for example, the sacrifices Americans had been willing to make, everything from backing the boycotts to facing enemy guns. Or take the Joseph Warrens and Nathan Hales—ordinary men but also patriots who had serenely faced death. Or take the young men who suffered privation in those unspeakable British prison camps, where they died of illness and starvation and were left lying in their own filth.

Other Founders weren't so sure about the depth of virtue in the colonies, however. Some aspects of the war had not cast American virtue (Greek *or* Christian) in a favorable light. What about all the spies and turncoats? What about traitors such as Benedict Arnold? What about the self-important generals who botched battle after battle, the militiamen who fired once and took to the woods, the merchants who sold to the highest bidder? What about all the greedy, grasping politicians who swarmed like ants in the state governments and cared only

Christian virtues

Ideals of **morality** based on Christian principles (meekness, compassion, love for one's neighbor, etc.) that people should lead lives of common decency and public uprightness.

Morality

A system of conduct based on beliefs of right and wrong.

Figure 15.5 Benedict Arnold, a general in the Continental Army, plotted to surrender the fort at West Point, NY, to British forces during the Revolutionary War. Such actions probably cast doubt in the minds of some as to the stability of virtue.

about self-interest? Could this be called virtue?

Upon such reflections, some of the Founders scrapped the idea of virtue. People, they concluded, were too narrow and selfish to care about public values, much less to love their enemies. These Founders placed their faith in a structured solution to the Republican Problem, seeking out ways to set interest against interest and ambition against ambition, by establishing a constitutional government. If the structure were good enough, they thought, human nature could be made to control *itself*.

Most of the Founders, however, didn't abandon virtue completely. They came to see it as being largely situational. In other words, the human capacity for noble conduct seemed to emerge better in some situations than others, as, say, with the politician who fearlessly does the right thing but only in the glow of the limelight. This understanding of virtue provided a new approach to the use of structure. Rather than simply playing off interest against interest, structure could be used to create circumstances in which virtuous behavior becomes more likely. Here is an example:

Suppose Congress passes a bill that is blatantly discriminatory against some group out of favor. (Our history is sprinkled with such legislation.) The President abhors this bill for both political and moral reasons. Politically, he was elected by all the people, including the group out of favor, in a very tight race. Morally, he views the world from a larger perspective than most congressmen do, and thus he sees greater value in tolerance. And so the President decides to veto the bill in question. In his veto message, he pointedly appeals to the virtue of the lawmakers, urging them to do justice. Note that he doesn't have to convince all of them, only a small fraction. In fact, think how small it is. In democratic politics, a substantial number will side with the President anyway, regardless of his position, so all he has to do is add enough to that number to total one-third of the whole, enough to block an override of his veto. Chances are good that he will succeed. If he does, we have the constitutional structure to thank (along with the President's own virtue, of course), for it has enabled the President to mobilize latent virtue.

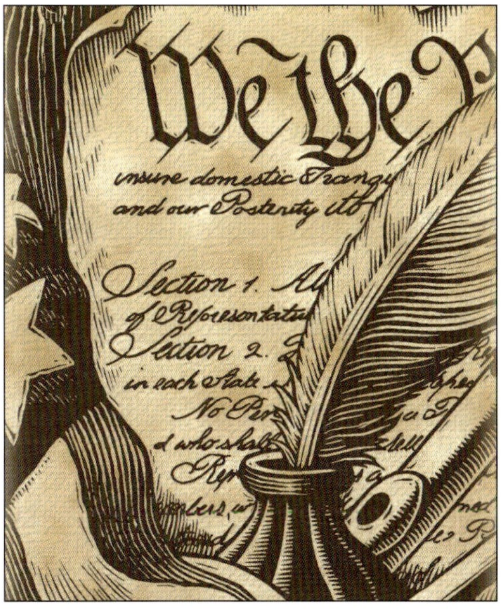

Figure 15.6 Has the balance between structure and virtue created by the Founding worked?

With such scenarios in mind, most of the Founders continued to believe in the importance, even the necessity, of republican virtue to sustain and fortify the American nation. But they wanted that virtue within a structural framework. They were too knowledgeable to rely on either structure or virtue. They knew both were needed.

Rule of Law

In the ancient world, political societies were commonly run by tyrants or oligarchs. In either case, it was the rule of will. Tyrannical will was not necessarily bad, and in a few cases it turned out to be rather fortunate, but it was generally bad enough, especially for anyone who opposed the ruler. The rule of will, so it seemed, was often capricious, nasty, and malevolent.

Among their contributions to political thought, the classical Greeks developed the idea of the rule of law. They discovered that when a society is given general, prospective laws that everyone recognizes and understands, the rule of will fades away and the laws take its place.

For the ancients, the rule of law encouraged peace and prosperity. It could also encourage personal freedom, for the rule of law tends to create a private world in which the individual is free of state interference. Once the laws become stable and predictable like the laws of nature, individuals become agents of their own destiny.

Since John Adams's time we have come to realize that maintaining the rule of law can be tricky. While the principles of generality, prospectivity, publicity, and the like seem simple enough, they are often scarce in the world of politics. Take generality, for instance. The concept is clear enough: the laws must apply to everyone. In practice, though, we come up with all sorts of reasons why "everyone" shouldn't really include *everyone*. Some groups would seem to deserve special protections. Other groups would seem to pose special risks. We feel sorry for those who have faced, say, a natural disaster and need extra help. We want the rich to pay an inordinate share of the taxes, and the poor to receive an inordinate share of the benefits. Generality gets very dicey.

We have devised clever ways of getting around generality. If we vote an aid package for all cities whose elevation exceeds 5,000 feet and whose metro population exceeds two million, we don't have to mention Denver by name. With such cleverness in mind, the British government often violated the rule of law in what they thought were the interests of the empire. The result, for England's American colonials, was tyranny.

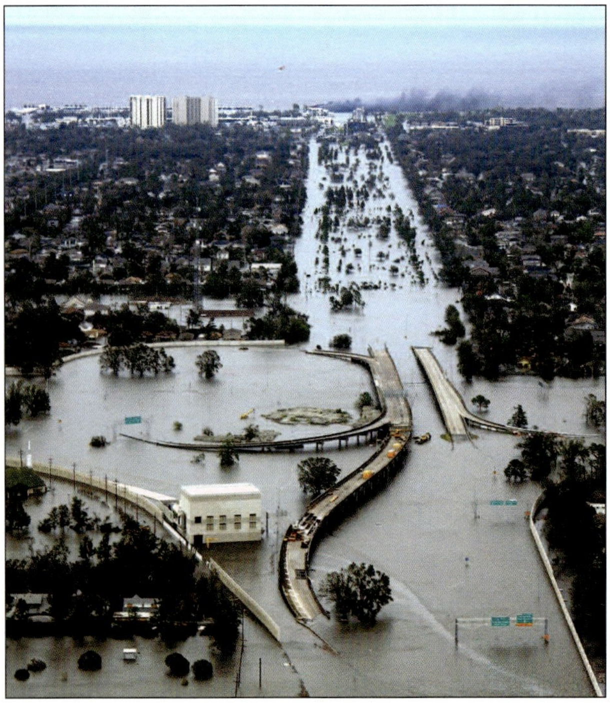

Figure 15.7 New Orleans in the aftermath of Hurricane Katrina in 2005. Katrina caused an estimated $81.2 billion in damages, making it the costliest natural disaster in U.S. history. Should the principle of generality be ignored if it helps the victims?

The principles of the rule of law are philosophical ideals, not laws themselves that could be enacted by a legislature. Their only existence is in the human heart. This explains why some societies have little awareness of the rule of law. Their people get so used to being pushed around that they barely realize it is happening. Other societies sense immediately when the rule of law is violated, even when they can't explain all the theory, for they perceive a glaring injustice. This describes the American colonies when Parliament passed the Stamp Act and revoked trial by jury.

The Founders were concerned about the rule of law and recognized it was (and still is) fragile. They included provisions in the Constitution specifically prohibiting bills of attainder (violating generality) and *ex post facto* legislation (violating prospectivity). In the interests of publicity, the Founders required that Congress publish its own proceedings, and that the President give an annual accounting of the Union. In the Bill of Rights they spelled out precisely what they meant by due process of law.

Structure may also be understood in terms of the rule of law. With faulty government structures, both the states and the Confederation had failed to establish the rule of law convincingly, violating one after another of its principles. The Founders believed that a federal government would do much better. With all of its structural mechanisms, the government still might err, but it would be unlikely to rule by simple will. For instance, a bill targeting a particular group violates the principles of generality *and* prospectivity by singling out a specific group and punishing it for past actions. Simple structures consisting of a unicameral legislature often did precisely that very thing. By adding a bicameral legislature and an executive with the veto power, the Constitution made it less likely that the rule of law would be ignored.

The Nature of Liberty

If there was a centerpiece among the ideals of the American Founding, it was the idea of **liberty**. The Founders understood the term in two different ways. There was the liberty of the colonies from an empire that no longer cherished them, and there was the liberty of individuals to live their own lives and pursue their own fortunes.

Liberty's flip side, tyranny, also came in two varieties. The first was the tyranny of Crown and Parliament, the age-old tyranny of the lash, refined into the form of unjust laws, corrupt officials, muzzled dissent, and special privileges for the few. This was essentially aimed at the colonies.

The second form of tyranny was more subtle. The Founders had to learn about this, the tyranny of the majority, on their own. It was the tyranny of democracy, of the people themselves, the tyranny of suspicion, fear, greed, resentment—the tyranny within. It was far more difficult to deal with, because it arose out of human nature itself and thus tainted everyone. Most of the work of the Founding was directed toward eliminating this darker and more elusive form of oppression.

If liberty was the absence of tyranny, it was also something bold and affirmative, encompassing many things. It was economic opportunity. It was a rough and ready social equality. It was society's permission to create, to improve, to reform, to imagine, to break away. It was hope for a better world, both for oneself and for one's community, innovating this, discarding that, trying something else, seeing what worked. It was the belief in a higher justice. It was the pursuit of happiness.

Liberty thus described had an enemy in the world, as the Founders understood it, and that enemy was government. *Any government*. For by definition all government had the capacity—and perhaps the inclination—to drift from the path of good intentions and wind up in the quicksand of tyranny. As Aristotle noted, monarchies corrupted into autocracies, aristocracies into oligarchies, democracies into ravenous mobs. This was because all of them were composed of men, not angels.

The Founders curtailed the powers of their own government in every way they could think of. Virtue was their primary reliance, as noted, but virtue worked best close at hand, and the government of the

Liberty

An ideal of freedom from oppression, tyranny, and government, allowing individuals to pursue happiness through positive action.

Figure 15.8 In order to secure liberty, the Founders had to guard against tyranny from both the leaders and the people.

Founding was necessarily government removed from the local level. It was also, perforce, a government of considerable power since it could draw upon the combined effort of all the states and it had the authority to make war.

A few of the Founders, notably Hamilton, rejoiced in this situation. They would have strengthened federal powers even more, but most Founding Fathers were far more wary. They were serious about reading the Constitution narrowly and keeping the government as trim and watchman-like as possible. Witness, for example, how a nationalist like James Madison could so quickly be afraid and join the opposition the moment the government stepped across that first line. Just a few short years after he led the Constitutional Convention's increase of national power at the expense of the states, he was drawn into implacable opposition to Hamilton's plan for a U.S. bank. Madison did not see government power as some sort of benign benevolent force.

Alexis De Tocqueville

1805–1859
Frenchman who wrote Democracy in America (1835), in which he explored the uniqueness of American character and its sources.

Finally, the Founders understood the ambiguities of liberty. They realized that just as government could corrupt into tyranny, freedom could corrupt into licentiousness, and then into chaos and anarchy. Here they placed their trust in virtue once again and specifically in the virtue of free citizens. True freedom, they believed, tended to be self-sustaining, calling forth the inherent nobility of humankind.

National Character

In 1831 a young Frenchmen named **Alexis de Tocqueville** toured the United States and took careful notes. His point of interest was the character of Americans in their new nation. Tocqueville, like the ancient Greek thinkers he had studied, assumed a connection between city and soul. To the extent that a founding establishes a people, not just a government, what would the character of Americans be, he wondered. In his two-volume report, *Democracy in America* (1835), Tocqueville pointed to Americans' ingrained optimism, to their restlessness, to their sense of equality and community, to their ready display of mechanical ingenuity, and to their distrust of Old World learning.

Tocqueville's book marked the beginning of an abiding interest in the American character. Historians continued this interest in what was American. Frederick Jackson Turner explained that character in terms of the frontier, pointing out that self-reliance, an ambitious spirit, and an egalitarian turn of mind would naturally develop among those who cleared forests and plowed land. Other historians argued that it was mobility that explained American uniqueness, still others that it was material abundance, and yet others that it was immigration and the melting pot.

There is a concept that encompasses all of the foregoing perspectives—*self-invention*. The self-invented country naturally gave rise to the self-invented citizen. After all, at the heart of the Founding lay the liberty of indi-

Figure 15.9 Alexis de Tocqueville wrote Democracy in America *in the 1830s. In it he explored the characteristics of the fledgling nation that made it unique.*

viduals to work out their own destinies. So if Americans began inventing themselves—if they studied nights to learn accounting, or went to Hollywood for a screen test, or sold the farm and moved to the city—they were simply pursuing happiness.

The pursuit of happiness made it difficult for Americans to think with a single mind or speak with a common voice. By definition, individuals see the world in their own separate ways. From colonial days, Americans have questioned and disputed every aspect of their beliefs, their values, their politics, their morals, their very identity, never settling anything to the satisfaction of all. Never was there a body politic open to such self-scrutiny.

By creating a world of rock-like stability—together with agreement on fundamentals—the Founding made it possible for Americans to divide over everything else, and for each to search out his or her own sense of meaning.

The Founding as a Moral Enterprise

One of the distinctive aspects of America's founding is the importance of the documents that marked certain steps in the founding—the Declaration of Independence that justified the Revolution, the Constitution itself that specified the processes of government, and the Bill of Rights that enumerated individual rights to be protected from the encroachment of government. These documents were in one sense official and legal. However, in a more important sense they were documents that layed down a set of moral propositions about our relationships to each other and to the government. The propositions of the Declaration of Independence and the Bill of Rights are quite straightforward. The Declaration affirmed that all men were created equal and

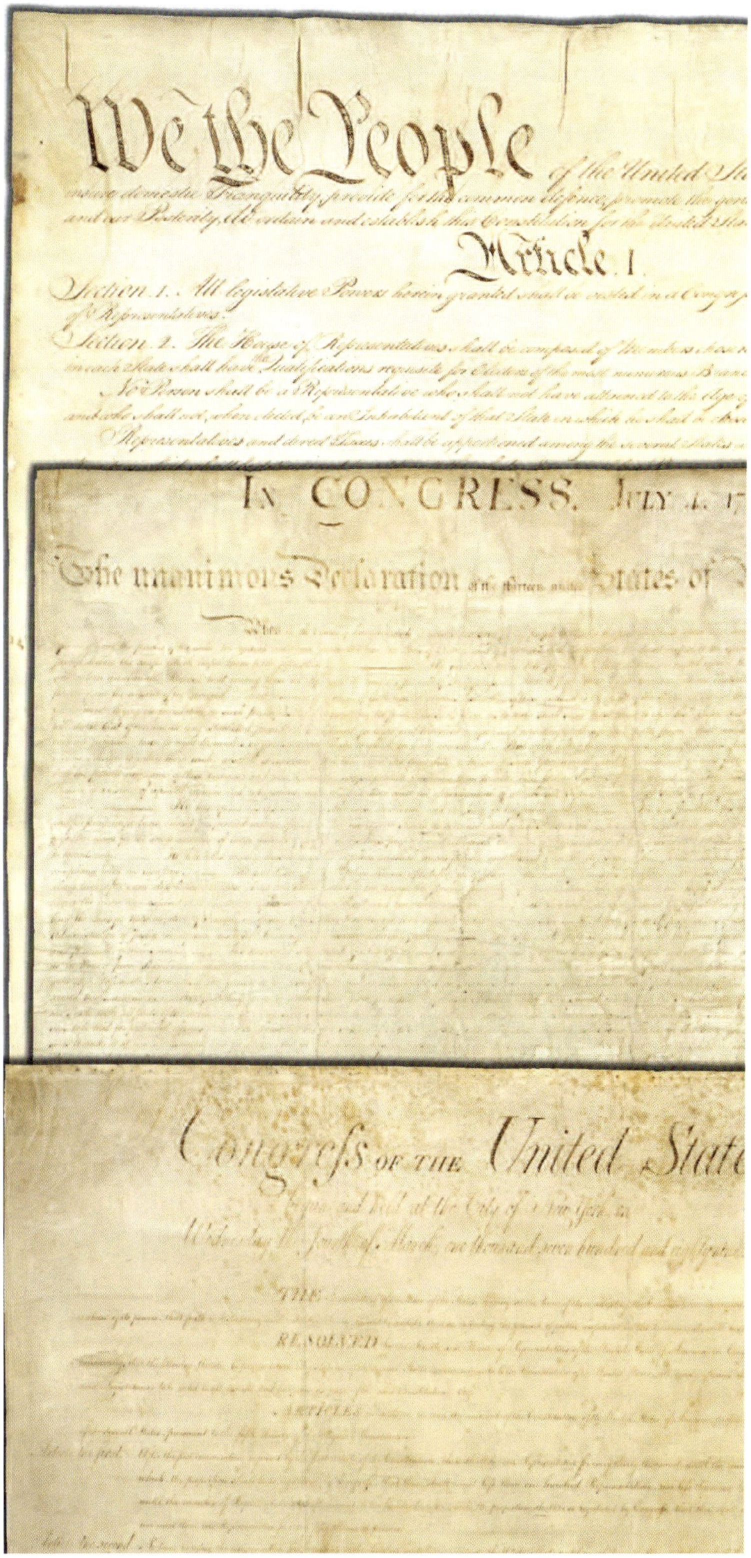

Figure 15.10 All three of our Founding documents are "moral" texts.

endowed with basic rights, that they were free to create their own government, and that they could expect that government to honor and protect the rights in question. The Bill of Rights detailed specific rights and further underscored their inviolability. Both documents became banners of freedom.

But what of the Constitution? Was it not a moral document as well? Some have noted that the Constitution, unlike the other founding texts, appears only to lay out a structure of government, proposing no ethical principles. Readers of *The Federalist* came to know better. The Constitution's moral significance was there, even if subtle. The framers had never proposed to lay out just *any* government, not even any republican government, but a government that somehow embodied Truth with a capital "T."

To grasp the ethics of the Constitution, we need to begin with the fragility of free institutions. Both in the state governments and the Confederation, Americans had seen such institutions fail, and history showed the failures to be common. Thus, for Americans, the Republican Problem wasn't just an exercise in logic; it was a pressing and daunting difficulty, and their very lives depended on finding a solution. How could they gain the benefits of republican life without incurring its dreadful costs?

Madison's insight about the extended republic comes into play here. The picture he painted in *Federalist* 10 of faction contending against faction was specifically the picture of a *free* society. Similarly, when Adam Smith described the bustle of producers and consumers in the marketplace, he too was depicting freedom. For both Smith and Madison, liberty was robust and dynamic. It was not about mystics in a state of bliss, but ordinary people in the scramble of life.

For Madison, as for Smith, the pursuit of happiness could, and sometimes did, lead to conflict. That was precisely the problem. Much of the tyranny and anarchy in the world derived from the clash of self-interests. But both Madison and Smith believed that cooperation among human beings was more likely than conflict *if* the conditions were right. Everything depended on the conditions. The point of Madison's extended republic, like the point of Smith's free market, was to achieve those conditions—to create an empire of liberty.

Smith noted that market behavior tended to harmonize individual interests so that competition did not lead to conflict. Madison might have said the same of his extended republic. Yes, there would be factions, thousands of them, but in the end their pulling and hauling against one another would render them strangely harmonious, like producers and consumers in the marketplace. There was another point, too. People accumulating wealth and goods in the marketplace had their minds on something other than politics. It was not that way in ancient Greece. The Greek politicians and thinkers had frowned on materialism, leaving the sweat and toil to craftsmen, farmers, and slaves. As a result *their* factions were composed of true-believing fanatics.

The large, extended, *commercial* republic, then, was a crucial element of that regime of liberty. Both as marketplace and political arena, the commercial republic tended to neutralize self-interest. That left less self-interest for government to deal with, and less government became possible. Yet government must still exist, if only to deal with the hard cases. One couldn't simply proclaim human rights and hope for the best. The question was, bluntly, how to make sure that government refrained from compromising the very freedom it was supposed to protect, as had happened so often in the past?

Here we glimpse the moral text of the Constitution. By creating a government that was powerful, well disposed, and securely under popular control, the Constitution made real freedom possible. The federal government was made powerful by design. It was made well disposed by its ability to mobilize virtue. Through structural counterpoise, the government was kept under the control of the people. Rights now meant something, for there was an entity capable of defending them.

We see, accordingly, that our triad of founding documents truly does fit together. The Declaration of Independence, the Bill of Rights, and the Constitution work in concert, allowing Americans to enjoy the dignity and self-mastery that freedom alone could bring. It was in this sense that the authors of *The Federalist*—and the Founders themselves—could regard the American Republic as "True."

The Inspired Founding

Several of the Founders also believed that the American Republic needed to be inspired by God. After all, divine authority was the ultimate authority, as King James knew so well. Government might be a human invention, as the social compact affirmed, but without some higher sanction it forfeited the sense of mystery which set it above mortal institutions.

The specific problem here was the Constitution's amendability. Unlike the Articles of Confederation, the Constitution was designed to be altered. This fact emphasized the social compact within. The people themselves had written and approved it; in time they could amend it as they saw fit. But if they did so too often or too casually, the mystery would be gone. That's why the framers made the amending process just cumbersome enough to discourage habitual use and tinkering.

There is a deeper question, however. Was the American Founding *really* inspired? In the early years of the republic, it was common for Americans to assert that it was, and

Figure 15.11 Detail of the Creation of Adam *by Michelangelo. How might the idea of a divinely inspired Founding affect the American spirit?*

there was more to the claim than mere chauvinism. The idea was that God had inspired the American Founding not solely for the benefit of Americans but for the benefit of all mankind, and that it marked a turning point in human history.

Several of the Founders appeared to believe this themselves, and these men had no illusions about the bare-knuckle politics, the serial compromises, the appeals to power and interest that ran through their labors from beginning to end. Here, for consideration, is a short list of items supporting the divine inspiration thesis.

First, by any measure, the founding generation was a remarkable one. It would have been remarkable in Enlightenment Europe, and how much more so out on the Atlantic frontier. Take, for example, a roster of those who served in the Continental Congress, of those who guided the first state governments, of those who met in Philadelphia to forge the Constitution. Think of George Washington, Alexander Hamilton, Thomas Jefferson, James Madison, John Adams, Benjamin Franklin, not as marble statues but as men who knew one another, liked one another (for the most part), swapped ideas and hammered out agreements. There has never been such a generation in U.S. history, before or since.

Second, the peculiar mix of personalities in the Founding is remarkable in another way. At every crucial point in the story, forceful characters played off against receptive characters, idealists against pragmatists, visionaries and innovators against compromisers and facilitators. At the Grand Convention, to name a single instance, there were energetic, take-charge individuals and there were subtle, manipulative individuals. Fifty-five of either variety would have spelled disaster. As it was, the personal chemistry was nearly perfect.

Third, there were several junctures during the Founding where a squall far out at sea—or a decision not to attack a stronger foe—might have changed the entire outcome of the Revolution. What if the attack at Princeton had failed—as it very nearly did? What if the French Admiral de Grasse had been delayed a few hours in arriving off Yorktown? What if Daniel Jenifer hadn't missed that particular roll call in Philadelphia, or Abraham Baldwin hadn't abruptly changed his vote? What if Edmund Randolph hadn't supported the document in Williamsburg that he had previously refused to sign in Philadelphia? What if a youthful George Washington had been killed by any of those six bullets that pierced his coat in the 1755 Battle of the Wilderness?

Figure 15.12 Portraits and autographs of the signers of the Declaration of Independence. Lithograph by Ole Erekson, 1876. Will there ever again be such a generation?

Fourth, there was a curious serendipity attending the American Founding. One saw it clearly at the Philadelphia Convention. Compromises that seemed ill-starred not only turned out for the best, they turned out as if touched by genius. Federalism wasn't brilliant political theory—it was a bizarre accommodation. Separation of powers didn't fall out of the pages of Montesquieu—it emerged from a weary committee. We look back on the Constitution as an achievement of Western Civilization, but the men who wrote it saw it as a mishmash of concessions.

Finally, when we consider how surprisingly (and often disappointingly) structural models often work in practice, it is little short of amazing that the Constitution worked at all—much less worked the way it did. Several of the document's carefully contrived mechanisms utterly failed and had to be scrapped. (For example, the method of counting votes in the Electoral College early on produced deadlock ties and eventually resulted in a constitutional crisis.) What if in practice the Constitution simply hadn't functioned or, more probably, had functioned in a limp-along fashion like the Articles of Confederation? Would forces ever have mustered for another try? Unlikely.

None of these points proves the work of the Founders to have been divinely inspired. But when they are listed together they create just enough awe and mystery to bathe the Founding in an aura it has never lost. Americans came to believe that their republic was more than just a sum of its parts, more than any one of the Founders could have created by himself, more than all of them could have created together. And there was more to its Founding than lucky breaks.

Then and Now

Viewing their work as the Founders did, we have a way of assessing the changes that have taken place between their world and ours. Suppose we step into our imaginary time machine once again and bid the Founders to pay us a visit at the dawn of the 21st century. How would they like what they saw?

They would undoubtedly be impressed by many things. The size and power of the United States. The marvels of its technology. The life of abundance enjoyed by most of its citizens. The realization of many founding ideals.

They would also undoubtedly feel some concerns. As a simple thought problem, let us draw our study to a close by sketching out what some of those concerns might be, framing them as questions.

Does the United States still maintain a structural balance among the three branches of government?

This question naturally breaks down into three parts. First, has the executive branch become too powerful? Has the President's colossal role in war-making and foreign affairs indeed created the "imperial Presidency?" Has the President's massive bureaucracy become, for all intents and purposes, *the* government?

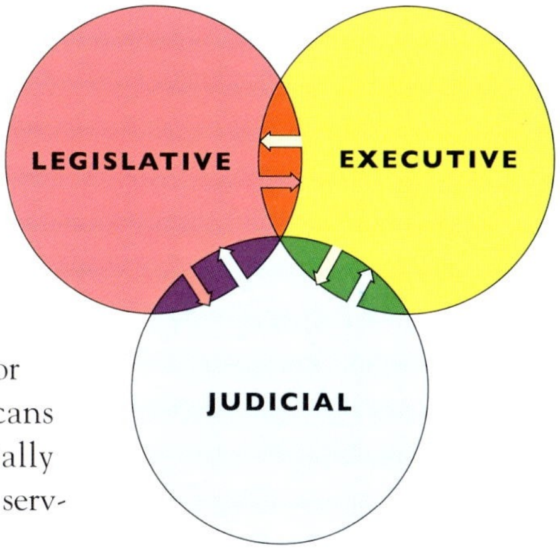

Second, has the stature and power of Congress withered? Is Congress still responsible for the rules by which Americans live? Or has it essentially devolved into a constituent service provider?

Finally, are there sufficient checks on the Supreme Court? Has judicial review honored the bounds of popular sovereignty? Or has it morphed into some quasi-legislative function responsible not to all the people but to an elite class of citizens?

In sum, does the structure of government still operate as a guardian of freedom?

Is the Founding's emphasis on liberty still relevant?

Liberty, for the Founders, depended on specific traits of national character, what they called civic virtue, and on the willingness of individuals to provide most of their own governance from within. Liberty was an American passion—a fire.

Has the fire burned itself out? Has deep change created a world in which freedom is less important than, say, security? Has the American character changed in some fundamental way, such that Americans will now trade personal liberty for a state-guaranteed well-being?

Has the size and power of government suffocated individualism?

The Founding grew out of a world that stressed private action, individual initiative, and personal accountability. It was necessarily a simple world, one that ordinary citizens could comprehend and grapple with.

In the booming, buzzing, ultra-sophisticated society of the 21st century, many activities seem to require the "adult supervision" that government alone can provide. Has modern society simply outgrown the Founding? Or, put another way, is individualism still conceivable in a world where government touches nearly every aspect of our lives?

Do Americans still enjoy the rule of law?

Law has become extraordinarily complex and intricate, addressing itself to a myriad of particular activities. Most of the rules that govern such activities are not laws at all but regulations, rulings, administrative orders, negotiated exceptions, and the like.

The laws themselves are often lost in convolution and beyond the ken of the ordinary citizen. At the time of the Founding, court action was comparatively rare. Now the United States has become the most litigious society in the world, and the number of lawyers per capita is more than ten times that of Japan. Some court cases now drag on for decades.

Are ordinary Americans able to run their lives according to such rules?

Is federalism still viable?

With no mechanisms for maintaining the federal balance, and with the constant drift of power toward Washington, do the individual states continue to provide a counterweight?

The question has a second part. Democracy supposes that citizens have access to the public arena, that they are empowered and ennobled by meaningful political activity. If all the important activity occurs within the Washington Beltway, does the word *democracy* still apply?

Has the American role in world affairs undermined national integrity?

Until well into the 20th century, most Americans believed that their institutions were inconsistent with a proactive role on the world stage. If there was truly such a thing as virtue in human affairs, could it ever be applied beyond the *polis?*

Americans came to confront a hard truth during the Cold War. In order to compete in war's brinksmanship games, they had to play by the same rules their enemies did. And they had to curtail some of their cherished freedoms at home, such as the freedom to embrace communism. Neutrals caught in

Figure 15.13 As we step into the future as a major influence in world affairs, what will we bring with us from the past? How will an understanding of the Founding affect the decisions we will make as a nation?

the crossfire between East and West said they could see no difference in the morals of either side.

In what sense could America be a beacon to the world if it could not stand apart from the world?

Have postmodern values outmoded the worldview of the Founding?

The Founders saw themselves as modernizers, eliminating the follies of long-established political precedent and crafting an entirely new and self-cleaning political engine. The world they envisioned was a Lockean world of rational purpose, materialistic outlook, and personal freedom.

What has come to be called postmodern thought rejects rationalism and materialism, and puts a radical spin on freedom. The liberty extolled by the Founders has been replaced by "liberation," which is not the same thing, while the rights listed in the Bill of Rights have become the basis of a whole new approach to politics. Freedom, in the postmodern world, is mainly for the marginalized and the disadvantaged, not for the spoiled and complacent mainstream—and certainly not for business.

In a world that is coming to think so differently, do the ideas of the Founding still make sense?

We cannot give reassuring answers to any of these questions, but we can point once again to the fact of the Founding's notable success. It was designed to adapt and to survive. That it has already accommodated changes of such epic scale tells us much about its inner strength and resiliency. But just as the Founders, to a degree, invented our first independent America, each generation has to reinvent it. The hard truth is that some common widespread level of virtue is essential if liberty is to be preserved, and this principle stares directly out at each generation. Beyond this necessity for virtue, wisdom is needed to adapt and apply the founding principles to a future with changes that we can't even imagine. Whether or not this 230-year-old experiment continues in a robust form will remain an open question.

Modernism

A cultural movement embracing human empowerment and rejecting traditionalism as outdated. Rationality, industry, and technology were cornerstones of progress and human achievement.

Postmodernism

A skeptical paradigm that critiques the ideals of modernism such as materialism and rational purpose and questions the true objectivity of viewpoints.

Key Terms

American Exceptionalism
Greek virtue (Areté)
liberty
federalism
polis (poleis, pl.)
Christian virtue
patria
postmodernism
social compact
rule of law
morality
modernism

Key People

Alexis De Tocqueville

Questions

1. Although the framers had many conflicting ideas, they also held many ideals in common. What were those ideals and common viewpoints?
2. What is morality? How are America's founding documents moral documents? How is the Founding a moral Founding?
3. If the Founding was a moral statement, what are your obligations to that morality?
4. What evidence does the text give that the founding was "inspired"? What evidence do you think points to the Founding not being inspired? In your opinion, was the Founding inspired?
5. What present day evidence do you see to each of the questions posed under the heading "Then and Now"?
6. Using evidence from an earlier chapter regarding the growth of government, do you believe that the size and power of government has suffocated individualism? Or has government intervention and growth helped protect peoples' rights?
7. What is postmodernism? How is the idea of freedom in a "modern" world different from the idea of freedom in a "postmodern" world?

Appendix A

The Founding and a Market Economy

Every society wants at least two things from the economic system it chooses to use. First, the economic system should be efficient. That is, it should produce as much of goods A as possible without sacrificing goods B, or make John as happy as possible without reducing the happiness of Sarah, and so on. We live in a dynamic world, so the efficiency of the system should also include economic growth that will provide future generations with the best economy possible without sacrificing consumption today. Second, the economic system should be fair. The distribution of income or other measures of economic rewards should conform with the society's beliefs about fairness. In other words, there should be economic equity. To these two widely-held goals, the Founders would have added a third goal. The chosen economic system should help society preserve liberty or freedom. Most people would accept these three goals as the primary objectives of our economic system—**efficiency**, **equity,** and **freedom**.

Efficiency

A goal for economic systems to produce the most goods for the most people.

Equity

A goal for economic systems to distribute goods and rewards fairly.

Freedom

A goal for economic systems to preserve the liberty of the people.

Economic System

A society's structure for making and distributing goods and services.

Socialism

Where government plans and regulates the economy.

The Founders spent little time discussing the type of **economic system** they wanted for their new country. **Socialism**, with government managing the economy, had not yet gained the status that its most powerful advocate, Karl Marx, would give it in the 19th century. The Founders did, by and large, reject mercantilism as described in Chapter 4 because of its clear costs to the colonies under the British version of that economic plan. So by default the Founders were left with a market economy. The purpose of this appendix is to describe the market system of economic organization and to demonstrate how fortunate the Founders were to connect the Constitution and the new republic with a market economy. The description here will be a non-technical, simple description of how markets work from a political point of view. In this appendix, we are going to assume markets are operating without any serious problems. We will take up the problem cases in the next appendix.

Markets are organized around one simple premise, that voluntary exchange or trade between two parties makes both parties better off. There is not one winner and one loser in exchange; there are two winners. Common sense tells us that two people who make a trade or exchange both benefit from the exchange or they would simply not make this voluntary arrangement. Consequently, a market-organized economy encourages and facilitates free exchange. Indeed, to the extent that exchange is controlled, restricted, or prohibited, the economy has departed from a market system and replaced it with a **command** or authoritarian system of economic organization where government exerts more control over the economy. Cuba and, to a lesser degree, China, still control significant aspects of exchange.

This primary premise of a market economy—that voluntary exchange is beneficial—is consistent with the fundamental principles of the Declaration of Independence and the Founding. Individuals, equal in their right to pursue happiness, are judges of what is in their own best interest. That is, they look at the opportunities before them, make decisions, and involve themselves in economic activity with the goal of making themselves better off. Should government restrict such exchanges, then government or society has substituted its notion of what is good for the individual for that individual's own pursuit of happiness. If pursuit of happiness is an inalienable right, then a market economy goes a long way toward preserving and making practical this right.

If a market economy simply allowed us to make voluntary exchanges, it would not be worth our attention. After all, anarchy allows us to do whatever we want to do, assuming other people operating under the same anarchy don't destroy us first. A market starts with free exchange of goods and services, but it goes well beyond simply allowing exchange. A market facilitates exchanges and makes them viable. A perfectly operating market helps us identify the set of exchanges that will make us the best off we can be. And it does not simply do this for you or for a part of the economy; it identifies the best exchanges for everyone in the economy. In other words, markets push us toward the holy grail of economics—efficiency. An economy is **efficient** if the only way to help one person is by hurting someone else. Logically, we can help someone without harming the economic position of anyone else if there is inefficiency in the economy. We just fix the inefficiency and give the gains to someone. Once all inefficiency has been bled out of an economy, the only way to increase Mary's happiness is to reduce John's economic satisfaction. Well-functioning markets root out inefficiency by rewarding the party that eliminates the inefficiency. This glowing description of markets seems very mysterious so let's start at the beginning and sketch the basic operation of a market economy.

Assumption about Human Nature

There is probably no aspect of our lives more dominated by self-interest than our economic behavior. Most of us expect the other side of an exchange to try to get the best of the deal. Most of us follow a moral code of honesty as well as keeping our word when we enter the economic arena. But we tend to interpret that code pretty strictly for our benefit. For example, society expects us to tell the truth in an economic exchange, but we do not have to volunteer everything we know about what we are selling or why we are buying something or other information that might be damaging to us. We normally live up to our contracts, but such a contract or agreement better be clearly explained on paper or at least clearly stated with a firm hand shake. Fortunately, we will see that a free-market system works just fine with self-interested individuals. The famous quotation from Adam Smith's *The Wealth of Nations* cited on p. 50 of this book emphasizes how well self-interest works in a market economy. By appealing to "self-love" we can all get what we want out of the market economy.

Note that this market assumption of self-interest is fairly consistent with the Founders view of human nature. They anticipated that citizens would act with public virtue at times, but they designed a government that assumed self-interest would dominate our behavior just as a market economy assumes that self-interest will guide our economic behavior. Americans seem to be well-fitted to this assumption. In 1831, **Alexis De Tocqueville** found a United States firmly based on self-interest.

> In the United States hardly anybody talks about the beauty of virtue . . . The Americans, on the other hand, are fond of explaining almost all the actions of their lives by the principle of self-interest . . .

For the market to work well, one more assumption about human nature must be made. We must assume that people are rational in their economic decisions. This assumption of rationality does not mean that people are always wise in decisions nor does it mean they have extraordinary abilities. It simply means that most people are consistent in the way they view the world and they try to do what is best for them and their families.

Figure A.1 Alexis de Tocqueville.

Command system

An economy that is heavily controlled by government and does not readily allow free exchange.

Efficient Economy

An economy in which all benefits are maximized to the point that one party cannot increase their benefits without decreasing the benefits of others.

Alexis de Tocqueville

1805–1859
Frenchman who wrote *Democracy in America* (1835), in which he explored the uniqueness of American character and its sources.

Some Basic Principles of Economic Behavior

Opportunity Cost

Scarcity

Economic condition where resources and supply cannot meet all demands, and therefore decisions must be made about which demands to meet.

Opportunity cost

The value of the best alternative not chosen.

The most basic fact of economic life is **scarcity**. There simply are not enough resources to fill all of the wants of everyone in society. There is not enough time, not enough money, not enough kindness, not enough intelligence, and not enough creativity. Therefore we live in a world of scarcity where we have to make choices. If a college student chooses to major in music, time and money limitations probably will require that student to forego degrees in other disciplines such as physics, history, or business. If the government chooses to increase taxes to pay for an expanded educational system, private citizens will not be able to buy the stereos, cars, or new clothes they intended to purchase with the income the government taxed away. If society chooses to clean up the environment by restricting factory emissions, there could be less manufacturing and fewer goods to enjoy. If government decides to spend a larger proportion of its available funds on health care, less attention and funding will go to defense or space research. Making choices in government spending, like making choices in family and individual spending, is often uncomfortable and difficult, for it means that the interests of some groups and individuals will not be met.

In almost every situation imaginable, we pay for our choices by giving up alternatives. It is the act of choice that brings cost into the world. The real cost of a choice is measured by what is given up when that choice is made. This concept is called **opportunity cost.** *The opportunity cost of a choice is the value of the best foregone alternative.*

Suppose Congress is considering three alternative uses for an appropriation of $50 billion: an ambitious new program of space exploration, an increase of $50 a month in basic social security benefits, or grants-in-aid of $10,000 for five million college students. Congress must rank the three alternatives in order to make a choice. Suppose that the ranking is (1) improvement of social security benefits, (2) increased funding to higher education, and (3) increased space activity. The opportunity cost of improving social security benefits is the value of the next best alternative—in this example, more funding for higher education.

In many instances, the opportunity cost of a choice may be measured directly by the money spent for a particular option. When we buy a loaf of bread for a dollar we do not need to know the exact alternative had we not purchased the bread. Rather, we think of the cost in terms of the dollar paid out and the general value we get for one dollar of expenditures. However, in many situations the opportunity cost cannot be measured by the dollar expenditure.

For example, most individuals use airlines rather than buses for long-distance travel, even though a bus ticket may cost only half as much as an airline ticket. This preference for the airplane seems irrational if we only consider the ticket price in opportunity cost. Obviously, there is another dimension to the opportunity cost of travel—that of time. Since people value their time, or their time has alternative uses, the cost of travel includes the ticket price and the value of time used to travel. Travel by bus involves a lower ticket price but a higher opportunity cost overall because of the value people place on their time. Consequently, people are more likely to travel by airline than by bus. To measure cost accurately, we must include all aspects of the next best alternative foregone.

Choices about whom to marry, whether or not to go to college, or whether we should eat that extra slice of pie, are all examples in which dollar expenditure is not the key element of opportunity cost. The principle of opportunity cost applies in every situation that involves choice, because it is necessary to sacrifice something in order to obtain the alternative we prefer. But we always try to reduce the opportunity cost of our choices, and this interesting trait in human nature motivates us to trade or make exchanges. By specializing in production of certain goods, we can exchange the goods we produce for the goods we would like to consume, and reduce the costs of our choices. This process of **specialization** and exchange is guided by opportunity cost. That is, we produce goods where we have a low opportunity cost of production and we exchange for goods where our own production of those goods would carry a high opportunity cost. This analysis of specialization and exchange is called the **law of comparative advantage.**

The Law of Comparative Advantage

One of the principal implications of free exchange is specialization. In our modern economy, most companies specialize in the production of just a few products, or they offer only a few services. Individuals within those companies (or preparing to become employed in those companies) are even more specialized and usually concentrate on one type of service or one small aspect of the production of a product. We apply the **law of comparative advantage** virtually every day. Suppose Jake decides to have pizza for dinner. He can purchase a pizza for $8 at a local restaurant. If he produces the pizza at home, he calculates the opportunity cost of the pizza to be $10, including the value of time and the cost of ingredients. In this case, Jake would exchange for the pizza because others can produce (probably tastier) pizzas at a lower opportunity cost than he can. His gain from the exchange will be the difference between his opportunity cost of production ($10) and the cost through exchange ($8),

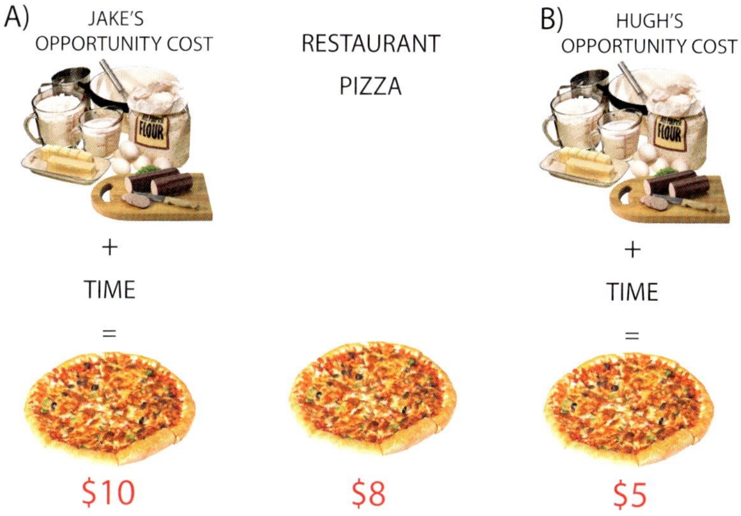

Figure A.2 Opportunity cost

or a benefit of $2. Now lets assume Hugh is a good cook whose uncle is a restaurant supplier. Hugh's opportunity cost to produce the pizza may be only $5, so he should produce the pizza at home. Indeed, he might want to go into the pizza business since his opportunity cost is less than the prevailing price. If Hugh can make pizza for $5 and sell it for $8, your gain from exchange will be $3 on each pizza. We implicitly think through this same sort of calculation for each item we decide to purchase or produce ourselves, whether its pizza or plumbing repairs or a brake job on our car.

It is important to understand that opportunity cost is defined by available alternatives rather than by absolute levels of skill and abilities. A surgeon may be a wonderful chef but rarely cooks because the financial rewards she receives with her medical practice makes the opportunity cost of cooking too high. A university president may be a marvelous teacher who does not teach any university classes because of responsibilities elsewhere. The good news for most of us is that our average or lower abilities (we burn our homemade pizzas, we flooded the downstairs neighbor, and after that unfortunate

Specialization

The economic practice of focusing resources on production of one or a few goods.

Law of comparative advantage

Economic principle in which resources will be specialized in the production of goods where producers have the lowest opportunity cost.

incident on the freeway we no longer install our own brakes) do not eliminate us from participation in the economic game. A person with adequate but not outstanding cooking skills may still find work as a chef. Of course, the wage may only be adequate as well. Another person with rudimentary skills at repairing cars may find employment as a mechanic's helper if he/she does not have better skills in other areas. The law of comparative advantage implies that there is an economic activity that each person, each plot of land, each machine or other resources can perform at a low opportunity cost.

The following example illustrates the importance of available alternatives rather than level of skill in determining opportunity cost. Susan is more capable than John in both accounting and law. Susan can generate an income of $100,000 as an accountant and an income of $90,000 as a lawyer while John can generate an income of $80,000 as either an accountant or a lawyer. Fortunately for John, Susan cannot do both jobs. When she chooses accounting over law because her opportunities are greater in accounting, an opportunity for employment in law then opens for John.

Economists summarize the effect of opportunity cost on specialization and trade in the Law of Comparative Advantage: *Every individual, group, or nation can produce at least one good or service at a lower opportunity cost than others. To maximize their standard of living they should specialize in the production of such goods or services. As long as opportunity costs differ for various individuals or groups, specialization and trade will be beneficial to the parties involved.*

Implications of Specialization and the Law of Comparative Advantage

Full Employment of Resources

First, the law of comparative advantage generates a tendency toward full employment of all resources in their most preferred use in an economy. This tendency toward full employment follows from the principle of opportunity cost. Because everyone is the low-cost producer of something, each specializes in producing the particular good where he or she has the lowest opportunity cost. This principle explains why the engineer who designs a car employs someone else to repair it and why authors of textbooks on typing do not type their own manuscripts. On the other hand, the operation of the law of comparative advantage also explains why there is so much mediocrity in the world; this economic principle means that the best people for the job are only rarely doing the actual work. Most individuals are in their particular jobs because they are low-cost producers and not because they are particularly competent in that area.

The law of comparative advantage leads not only to full employment of resources but to full employment of resources in their best possible use. Planned economies do not necessarily allow the principle of comparative advantage to operate, but they often achieve full employment. Government managers controlling a planned economy have a difficult time allocating and employing resources for their best use.

Widest Possible Gains from Exchange

The second implication of the law of comparative advantage is that it creates the widest possible gains from exchange or trade. Consider the complex process of automobile production. An automobile contains thousands of parts and many different types of materials such as steel, aluminum, chrome, rubber, glass, and plastic. Over 50,000 businesses supply various auto parts to automobile factories and assembly plants in the United States. Automobile production draws on the skills of engineers, computer programmers, tool and die makers, accountants, secretaries, skilled assemblers, and other labor. Yet this very complex object can be purchased for approximately five months of income for a typical middle class American family. The gains from trade and specialization involved in just the manufacture and marketing of an automobile are astounding. The gains are similar for almost all the other goods we consume or use.

In the absence of this specialization and exchange process, we would all be living in caves—being 100 percent self-sufficient and poor. Specialization is the primary feature that distinguishes a modern economy from a more traditional or primitive economy. Specialization dramatically increases the production power of an economy because individuals and firms learn how to do repetitive tasks more efficiently. This process of specialization continually increases the possible gains from exchange and is largely responsible for the modern material civilization we all enjoy.

Economic Interdependence

The third implication of the law of comparative advantage is that it promotes a great deal of economic interdependency. Trade links nearly every individual in the U.S. economy to a vast network of other citizens. Furthermore, we are connected to individuals in virtually every other part of the world. Even very simple manufactured objects involve trade on several continents. Complex objects involve exchanges among thousands of individuals from various parts of the globe. Consequently, we are all completely at the mercy of one another for maintaining our high standard of living. This *interdependency* frightens some people, who equate the word with economic and political vulnerability, making them want to become self-sufficient. However, true self-sufficiency is also self-limiting, and requires those who choose that lifestyle to live relatively hard lives, literally off the sweat of their own brows, without modern conveniences and comforts. Given this fact, most of us are willing to accept interdependency as a necessary evil in return for the high standard of living specialization and exchange offers us. Some argue that the economic interdependency stemming from the law of comparative advantage also links countries together politically, thereby reducing chances of war and violence.

The Role of Economic Competition

The market economy described here works best if there is competition in specialization and exchange. Just as the nation's Founders tried to separate and disperse political power among different branches and levels of government, competition disperses economic power among different buyers and sellers of each good or service, while encouraging those same buyers and sellers to develop new and better products and services. In fact, the best economic outcome occurs when power has been dispersed so widely that no buyers or sellers believe that they have an influence on the price or terms of exchange. This is what economists call **perfect competition**. For example, each wheat farmer accepts the expected price of his next wheat harvest as a given and wastes no time calculating what effect his production might have on the price of all wheat on the market. Similarly, consumers take the price of bread as a given, knowing it would be silly to expect the price of bread to fall because they didn't purchase bread that week.

Perfect competition

When buyers and sellers have no influence on price and terms of exchange.

Competition accomplishes two very important goals for a market economy. It helps the economy attain efficiency. (The harm done by monopolies will be discussed in the next appendix.) Competition helps the economy find and make all of the exchanges that will be beneficial. Competition also disperses the gains from exchange more widely. Imagine that there was just one seller of automobiles. That seller would have an enormous advantage and could capture most of the gains from exchange in automobiles while the consumers would be forced to pay nearly the maximum price where they would have no gain from the exchange. Since there is competition to sell cars, the actual selling price is set much lower than that maximum price so that most consumers can pick and choose, gaining substantially when they purchase an automobile.

Summing up what we have discussed thus far, a market system of economic organization allows people to make any voluntary exchanges they wish to make. An efficient economy requires a producer or consumer to make the set of exchanges that are "right," that will make each participant as well off as possible without harming someone else. Opportunity cost and the law of comparative advantage guide us in making the right set of exchanges and, thereby push the free-market society toward economic efficiency. Competition among many buyers on one side and many sellers on the other disperses the gains from exchange and pushes the economy to the efficient point.

But what determines opportunity costs so the law of comparative advantage can operate? How do we know whether or not we should make our own supper tonight or buy it at the local diner? The answer is prices—the prices of goods and services and wages, the price of time. Prices guide us in our measurement of opportunity cost. To make the right set of exchanges that make us as well off as possible, the prices individuals face must accurately reflect the full opportunity cost of resources. Here lies the real genius of a market economy (if there is genius in it). A market economy is particularly adept at generating the right set of prices on which people make decisions to produce efficiency in the economy. No government or business official, certainly not some smart economist working in a think tank, sets the prices in a market economy. The prices are generated by the actions of millions of households, tens of millions of individual consumers, and many thousands of firms that are all pursuing their own self-interests. Miraculously, out of their independent actions comes a set of efficient prices to guide the economy, thus explaining Adam Smith's metaphor of the "invisible hand" referred to in Chapter 4.

Four Basic Economic Principles for Market Pricing

Four economic principles help us understand how prices are determined in a market economy. They are:

- The Law of Demand
- The Law of Supply
- The Equilibrium Price
- The Role of Profits

These four propositions describe the basic operation of a market system. The combination of these four propositions show us how prices are determined in a market economy, and most importantly, show why those prices are likely to be the prices needed to generate efficiency.

The Law of Demand

There is an inverse relationship between the price of a good or service and the amount people will buy. That is, as the price of a good or service rises, individuals will buy less of that good or service, assuming no other influence on demand has changed. It is important to understand exactly what the law of demand is and also what it is not. The law is a statement about the effect of price changes holding all other factors that might influence demand

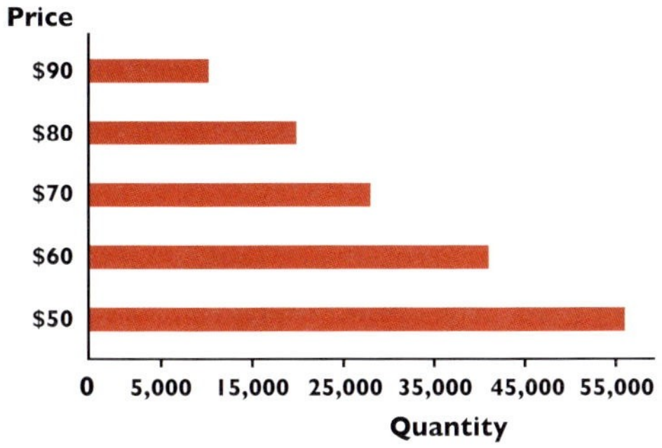

Figure A.3 Law of demand

constant. If the price of running shoes were to rise at the same time that incomes of runners increased, the purchases of running shoes could grow because of the positive effect of larger earnings. If it becomes a fad to wear Nike running shoes, the purchases of Nike shoes could go up in spite of the increased price of running shoes. The law of demand refers to individuals' responses to price changes alone.

The law of demand is the most verifiable law of economics. After literally hundreds of independent tests, no substantial evidence has surfaced that the law of demand is violated with any consistency. An increase in the price of a milk (if all other factors influencing purchases are constant) will tend to discourage individuals from purchasing milk. But what if people still want or even need something to drink? The law of demand is an expression of the power of substitution in our daily behavior. If the cost to the consumer of a product increases, the consumer will find a substitute for that product. If the price of a good falls, we will substitute the now-cheaper good instead of a more expensive alternative. We constantly make small marketplace adjustments, substituting the cheaper for the more expensive.

For some goods such as gasoline, food, and utilities, an increase in price will cause the consumer to make a fairly small downward adjustment in the quantity purchased. This is because substitution for some products (and services) is difficult, particularly for short periods of time. For other goods such as clothing, recreation, and many manufactured goods, a price increase will cause people to buy considerably less or none at all, until prices go down. In many industries, an attempt by one company within an industry to raise its price higher than those of other companies in that industry will cause that particular firm's sales to virtually disappear. The most important point, however, is that in all cases individuals respond to an increase in price by purchasing less of that good and substituting other goods in its place.

The law of demand is so obvious and straightforward that we often forget it is constantly in operation. For example, the embargo on export oil imposed by the OPEC nations in 1973 drastically reduced the supply of oil to the United States, generating what the press quickly called the "energy crisis." As politicians, journalists, and commentators on economic matters discussed the energy crisis, they basically assumed that Americans would consume the same amount of gasoline regardless of the price. Initial commentary centered on the belief that people would not conserve unless forced to by the government. Roderick Cameron, Director of the Environmental Defense Fund, expressed the common view: "We have to start thinking about cutting demand for energy with things like an excise tax on big gas-gulping cars." In other words, he felt, as did many others, that gasoline price increases alone would not reduce gasoline consumption though he did believe an excise tax raising the price of gas-gulping cars would lower the quantity demanded.

Most politicians overlooked the law of demand and insisted that the American public would never make a change in gasoline usage voluntarily. Government initiated programs to force the automobile companies into producing smaller and more efficient engines and smaller cars. Political discussions concerned the need for mandatory rationing because there was no way to induce people to consume less. Presidents Nixon, Ford, and Carter all appeared on television at various times in their administrations, pleading with American citizens to conserve because it was their patriotic duty. The Presidents felt that public virtue rather than self-interest was the best solution for the energy crisis.

However, gasoline prices gradually rose and the law of demand operated predictably. Over time, individuals began to purchase more efficient and smaller cars. Americans drove less and conserved gasoline as it became more expensive. In 1973, the average consumption of gasoline per vehicle in the United States was 850 gallons per year. Gas consumption per automobile fell to 677 gallons by 1990 as more fuel-efficient cars and trucks were developed and marketed. The price of gasoline fell in the 1990s, so consumption increased to 719 gallons in 2003. Now that gasoline prices have increased, we can anticipate that consumption per vehicle will decline as consumers substitute other methods of transportation,

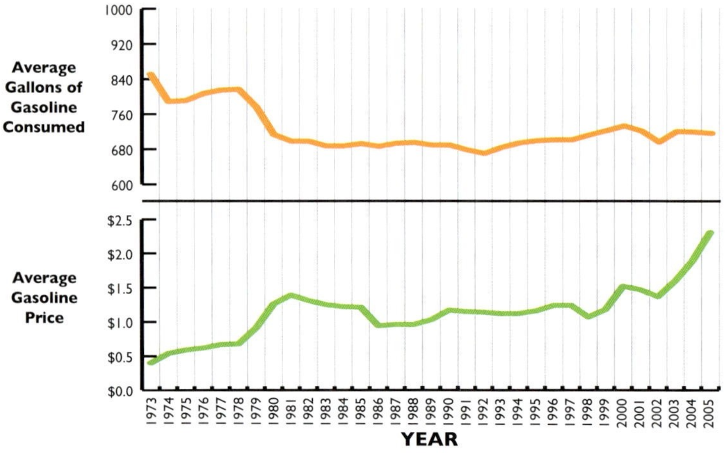

Figure A.4 The price of gasoline and consumption per vehicle. The data is courtesy of www.eia.doe.

and use more efficient and smaller vehicles as manufacturers respond to the profit potential in fuel-efficient cars. Economists studying the relationship between the price of gasoline and the amount purchased have found that a 10 percent rise in the price of gasoline will cause people to buy about three percent less gasoline. Economic events, such as the ebb and flow of energy prices, verify the unceasing operations of the law of demand.

An important property of the law of demand is that the response to a price change will become greater over time. For example, as illustrated by the energy crisis, response to an increase in energy prices is becoming greater over time. The decline in energy and gasoline consumption was not immediate; it took time for people to trade in large cars, insulate homes, buy more energy-efficient appliances, and make other changes to economize on their energy use.

The operation of the law of demand is illustrated clearly by simple economic examples. Yet the law of demand is also at work in more complicated social and political situations. Prices are a powerful social control mechanism as individual behavior can be manipulated through changes in the prices of various goods. For example, a tuition increase lowers the number of applicants to a university, thereby potentially reducing the number of college graduates entering the white-collar job market. An increase in the penalty for a crime (the price of the crime) reduces crime. By contrast, reducing the penalties for a certain crime increases the number of reports of that type of crime. An increase in medical fees causes people to see their doctor less and to use more home remedies. An increase in the costs of having children means people will have fewer children. These social and economic examples are but a few of the countless daily expressions of the law of demand.

The Law of Supply

Price and the quantity supplied are directly related. That is, businesses will produce more goods of a particular type as the price of that particular product rises. The law of supply also plays a critical role in a market economy and applies both when businesses supply goods and services and when households supply labor or savings for business to use. Essentially, prices in a market system act as a reward to sellers. As this reward increases, sellers have an incentive to supply more of their particular product.

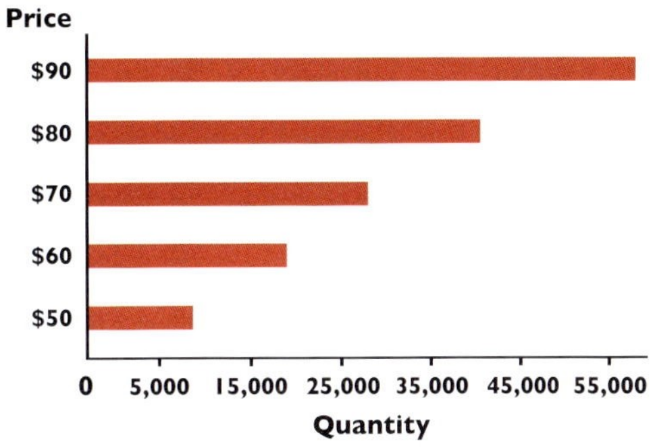

Figure A.5 Law of supply

At the beginning of the summer driving season in 2005, the price of gasoline before taxes was about $1.50 a gallon and refineries were supplying just under 9 million barrels a day. Over the next year, prices rose to about $2.10 a gallon pre-taxation. No new refineries had been built and some refineries were damaged by Hurricane Katrina, but production by U.S. suppliers still rose by approximately 300,000 barrels a day in response to this price increase. As time goes on, supply would likely respond even more.

The law of supply is also easily observable in the labor market. During the 1960s and early 1970s, very few individuals graduated with engineering degrees in the United States. The wages of engineers increased as a shortage developed. Universities and college students responded to the rising financial rewards for engineering degree-holders by producing substantially more engineering graduates. In accordance with the law of supply, 80,000 engineers left universities in 1982 compared to the 46,900 engineering graduates in 1975.

Just as with the law of demand, the effect of a price change on quantity supplied is greater given a longer time period.

The Equilibrium Price

The genius of the market system is that the interaction of the law of supply with the law of demand moves prices to the level at which consumers are willing to buy the same amount producers are offering for sale. In economics, this price is called the **equilibrium price**.

The market system adjusts and moves prices to an equilibrium whenever a particular price is too high or too low. Consider a hypothetical example where the price is very high—say $50—for a compact disc of music featuring a popular singer. The sellers are ecstatic because the singer is extremely "hot" at the moment and the disc has been heavily and successfully promoted before its release. Three times the average number of first release CDs are produced for the expected rush of sales. However, by the time the disc arrives at retail outlets the fickle music consumers have shifted their listening interests almost entirely to another performer, or they are choosing to download music files from the Internet. Consequently, the $50 CD just gathers dust on the retail shelves. Sellers find they are not able to sell the CD at $50 a disc, so they begin cutting the unit price. The surplus acts as a weight that drives the price downward. Now, suppose instead that same music CD was being offered for $.50 an album instead of $50. When fans of that musician see the price tag on the CD they are ecstatic. Everyone wants to buy the CD but few have been produced, and shortages develop. Sellers soon find they can charge a bit more than $.50. In fact, buyers are offering them more money and the CD shortage acts as a lever to move the price upward.

These market forces, producing surpluses when prices are too high and shortages when prices are too low, stop when prices reach the point where buyers and seller desire to exchange the same amount. The length of time necessary to reach the equilibrium price for a particular good or service can vary depending on circumstances. Whether the process is gradual or rapid, the market system will eventually adjust the price of each good to its equilibrium price, where the amount of money buyers are willing to pay for the product precisely equals the amount of money the sellers are willing to accept to move the product.

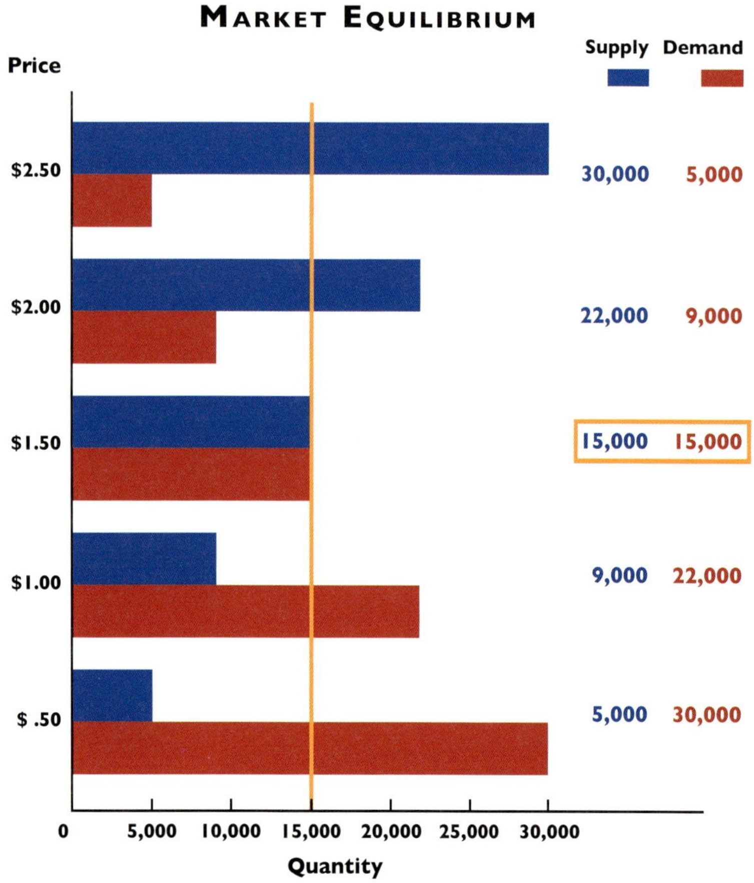

Figure A.6 Market equilibrium graph

The importance of this market system characteristic cannot be exaggerated. The market's ability to find the equilibrium price allows it to ration scarce commodities by moving prices up or down, as dictated by product availability and changes in resource costs. Thus, behavior of both buyers and sellers changes in the desired direction without coercion or governmental edict or control. In contrast to the market system described above, planned or command economic systems go to extensive lengths to try to determine the equilibrium price of every product and to make sure consumers will want to purchase the same amount that sellers hope to sell. On the other hand, the market system uses the equilibrium price and meets the challenge of coordinating buyers and sellers almost effortlessly.

Remember that one of the most important goals of an economy is economic efficiency. The equilibrium price generated by markets for each good plays a crucial role in producing economic efficiency. Efficiency happens when we have found and carried out all of the exchanges that are beneficial. How might we express this notion of all the beneficial exchanges? Exchanges between buyers and sellers are beneficial as long as the value of the good to buyers is greater than the value of that same good to the seller. As shown in Figure A.6, the equilibrium price and quantity divide the possible exchanges into two categories. To the left of the equilibrium quantity are the exchanges where buyers value the product more than sellers or, said differently, the value to the buyers is greater than the sellers' opportunity cost of production. To the right of the equilibrium quantity are exchanges between buyers and sellers that would not be beneficial because the opportunity cost to sellers of producing the good would be more than the value to buyers. A market economy, by generating the right equilibrium price, identifies and encourages beneficial exchanges and eliminates exchanges that are not beneficial.

The Role of Profits

Two of the most difficult problems in any economy are how best to reallocate resources from one sector to another and how to most effectively allocate new investment funds for the economy. **Profits** solve these problems for a market economy. High profits in a particular industry attract new businesses, resources, and investment to that industry while losses cause firms and resources to exit an industry. It is not enough for prices to simply move up and down in a market system. Consumers may respond to prices, but businesses respond to profits. Therefore, it is necessary that—in addition to price changes—profits and losses also change.

New firms enter industries to take advantage of abnormally high profits (or the expected potential for such profits) in a particular sector of the economy. On the other side, as firms incur losses, they depreciate their productive facilities and leave a weak industry to search out more attractive alternatives. For example, lower fares in the airline industry decreased profits for the numerous different carriers. As a result, several airlines got out of the business or merged with other carriers. Other airlines tried to reduce labor and input costs so their profits would compare favorably with the profits they could expect to earn in other industries. Profits and losses dictate when businesses should move resources from one sector of the economy to another.

To illustrate this point, assume the pizza market is working just fine with the price at equilibrium. There are no shortages of pepperoni or surpluses of anchovies, and normal profits are being enjoyed throughout the industry. If the market is disturbed, resources will be reallocated until equilibrium is reestablished in the pizza industry. Suppose, for instance, a medical research team discovers that a steady diet of pepperoni pizza causes significant weight loss. The pizza industry deviates from its equilibrium point. Weight-conscious consumers are overjoyed at the news and now rush out to purchase a lot more pepperoni pizzas. Consequently, a severe shortage develops as sausage makers strain to increase their output of pepperoni, causing the street price of pepperoni pizzas to rise. This rising price encourages pizza sellers to increase their production of pepperoni pizzas (while now paying a slightly higher price for pepperoni—which is also in greater demand—to ensure sufficient quantities to meet the present and anticipated consumer demand). As the pizza price increases, consumers note the increasing bite the higher-prices of much-in-demand pepperoni pizza is now taking out of their budget, and they decide they must begin to economize on their pizza consumption. Because of the law of demand, the amount of pizzas budget-conscious consumers now buy falls as the price rises. The two effects (sellers supplying more and consumers buying less) reduce or eliminate the shortage.

With the initial price increase, firms in the pizza business make abnormally high profits. These profits attract new firms and resources ("It looks like pepperoni! It tastes like pepperoni! But it's BETTER than pepperoni!") into the pizza market. As new firms enter, increasing the supply of pizza, the price falls until the price of pizza once again reaches an equilibrium at some new price (below the initial sharp rise but perhaps slightly above the old equilibrium price). More resources have been developed or accessed to allow increased pizza production, reflecting the change in consumer tastes because of pepperoni pizza's power to take off pounds. Consumers are content because there is no shortage of pizza while producers are satisfied because they are earning normal profits in their pizza businesses. The important point is that the movement of prices and profits prompts changes in consumer behavior during shortages and attracts resources to an area of the economy where they will generate higher benefits. Because of the numerous processes used to produce most of our modern goods, there is frequently a "lag" time between an industry's perceived change in the market (new research, consumer attitudes, etc.) and the ability of that industry to market a product that will take profitable advantage of that change.

Economic Interactions

In the hypothetical example described above (regrettably, pepperoni pizza has been shown to be capable of producing significant weight loss only if it is never removed from the box), the interrelationship of the four economic propositions is illustrated with the following typical chain of economic events:

1. Buyers desire to shift their consumption toward a particular good because of some change in taste.
2. A shortage develops for that product (market out of equilibrium).
3. The price of the product rises.
4. The rise in price causes buyers to reduce the amount of that product they purchase (the law of demand).
5. The rise in price generates profits, stimulating existing firms to produce more (the law of supply).
6. The abnormal profits in this industry cause new firms to enter the production of this product (the role of profits).
7. Output increases until the abnormal profits in this industry disappear.
8. As output increases, surpluses would develop if the price did not fall, inducing consumers to buy more (the law of demand).
9. The market for this particular product returns to equilibrium.

Three important functions summarize the role of prices and profits in a market system. First, **prices ration** scarce goods and valuable resources. If, for some reason, a product or resource becomes scarce, its price will rise, resulting in a natural "rationing" of the available supply. As certain minerals, fuels, or other key elements of the economy diminish in easy availability, the price of those resources will rise, creating incentives to economize on their use. Goods that are fixed in supply, such as paintings by Rembrandt or antique clocks, will be rationed in a market system entirely by the price movements that apportion the limited supply. Tickets to a popular play or sporting event will increase in price until the available number is rationed. Second, **prices and profits act as incentives** to change the behavior of individuals and businesses. A higher price rewards a seller and a penalizes a buyer. Conversely, a lower price hurts a seller and rewards the buyer. Third, **prices and profits send signals** to households and businesses that influence their future plans. If wages for a particular occupation rise, some households and individuals respond to this signal by entering that occupation. When profits change, businesses respond to that signal by shifting resources in or out of that particular area of the economy.

The workings of the free market system are brilliantly simple. To appreciate the full advantages of the market system, consider a world without it—a socialist world. Suppose you were appointed as chief economic planner in charge of shirts for a socialist country. You won this plum position without the benefit of previously observing and learning from market economies. You need to figure out your nation's consumers' preferences for different styles, sizes, and colors of shirts. You also will want to know how consumers will respond to different prices you might decide to charge for various shirts. You must decide what kind of buttons will be best. What combinations of wool, cotton, and polyesters should be used? Your answer to this question must be coordinated with production in agriculture and petrochemicals. Obviously, your shirt production also should be associated closely with the production of pants, ties, jackets, skirts, and related merchandise. In addition, you must determine what wages to pay your fellow socialists working in the shirt factories and sales outlets. The questions go on and on (and most of them will start coming from the glowering socialist committee monitoring you, if your planning goes horribly awry—who knew purple and green shirts would become the next fad?). To summarize, as chief planner in charge of shirts, you need enormous amounts of information, much of which is not available even in a modern, computerized world.

There's a genius within the free market system. It relies on self-interest, opportunity costs, comparative advantages, prices, and profits, and it eliminates the need for one person or group to possess most of this information. Individual decisions of thousands of businesses and millions of households generate the required information in the form of prices, wages, and profits, which is communicated to those directly affected by the decisions on each level. The information transmitted through changes in prices and profits alters consumer behavior and reallocates resources. If households have a sudden shift of preferences toward denim shirts, prices for denim shirts will rise, inducing a response from businesses and households without any individual consciously collecting the information about the change of preferences (a collection that would be essential in a socialist or planned economy). A bumper crop of cotton automatically lowers the price of cotton, leading to a substitution of cotton for other materials in clothing without deliberate collection and analysis of information by planners. In other words, the market system with its central role for prices and profits minimizes the information needed to keep the economic system operating. Actually, most economic planning in the world is done by planners who observe prices operating successfully either in other countries or on the black market. Economic planning in a world totally devoid of markets would be much more difficult than it is in today's socialist economies, where planners can observe how the free market system operates and use elements from that model to help them leap from theory directly to market reality.

Role of Government in a Pure Market System

The argument for a market economy is not an argument for economic anarchy. There are certain functions which government must perform for a market economy to operate efficiently. There are other functions that are likely to fall to government even though it is possible for private institutions to perform those functions.

The first economic function of government in a pure market system is to prevent coercion and fraud in economic matters. The benefits from specialization and exchange can only be certain if all exchanges that take place are voluntary and based on reasonably accurate information. If one party involved in a trade is coerced into the exchange, that exchange only benefits the other party. A mugging or a complex fraud do not fit the definition of free exchange. The first obligation of government in an exchange economy is to prevent such coercion or fraud.

A second function of government in an exchange economy is to provide money, the medium of exchange. It is possible to have money provided by various private organizations; however, there are benefits to having a single form of money provided and controlled by the government. Governments occasionally abuse this function of providing money by creating too much money generating inflation that reduces consumer confidence and the usefulness of money. Nevertheless, the benefits of a single creation source of money seems to outweigh the costs of government mismanagement of the money supply.

A third function that government has often assumed in exchange economies is to subsidize or create the transportation and communication networks, enabling exchange to take place more conveniently and over wider areas. In the 19th century, for example, state and federal governments subsidized the building of canals, better roads, and railroads in order to bring various parts of the country together in an exchange network. Today, the government supports extensive systems of roads, and subsidizes airport construction, port facilities, and space satellites in order to maintain the transportation and communication networks. In principle, private enterprise could provide all transportation and communication, but government has often been heavily involved in this aspect of the economy.

A fourth function of government is to define property rights within the economy. When two parties participate in an exchange, they are actually exchanging goods (or services) with a given set of rights associated with each product. If the property rights associated with the goods are changed, then the value of that particular merchandise also changes. One can see this clearly by considering our laws with respect to automobiles. Under current definitions of property rights, it is illegal for someone to drive an automobile without the permission of the owner. The owner has the right to lock the car. Most cars are fitted with a unique key ignition, and the police enforce the rights of owners over their cars. Auto theft is a serious crime. Consequently, automobiles exchange for significant amounts of money.

Suppose the laws changed, making it illegal to lock automobiles, build cars with key ignitions, or take any precautions to prevent theft. Suppose further that the laws allowed anyone who wished to go somewhere the use of the nearest car to reach his or her destination. Under that set of property rights, the value of the automobile would quickly be reduced to near zero, even though the appearance of the car and its other physical qualities remain unchanged. Thus, an important government function in an exchange economy is the definition and enforcement of property rights relative to different goods and resources within the economy.

A fifth function of government in a market economy is to enforce the exchange agreements that have been made between various parties. In any economic situation where self-interest is important, there may be disputes over the exact terms of the exchange and the compliance of each party to those terms. The government, through its courts and judicial system, plays an important role in arbitrating those disputes so that the exchanges can continue.

Conclusion

The parallels between a free market economic system and the Founders' design of government are striking. Both a market economy and the Constitution are concerned with the control and dispersal of power. Both use human nature and the natural tendency for all of us to pursue self-interest to achieve beneficial outcomes. Both require some level of public virtue to work effectively. (Exchange, the basis of the market economy, would be exciting but hazardous in a society dominated by cheats and liars.) Both the Constitution and the market use competition to restrain power: competition among firms and individuals in a market economy and competition in elections and among branches of government under the Constitution. And above all, the market economy was consistent with the Founders' basic objective—freedom or liberty. What economic system could be more consistent with life, liberty, and the pursuit of happiness than one that allowed you to make any transaction or exchange you wanted to, as long as you could persuade someone else to participate as buyer or seller?

These parallels are not all that surprising. The understanding of a market economy came out of the same intellectual environment that produced key principles of the nation's founding, such as the rule of law, popular sovereignty, and the use of mechanical devices. Some of the Founders undoubtedly had some knowledge of the writings of Adam Smith. They had plenty of first-hand experience with the problems of government planning and intervention into economic matters. They implicitly accepted a market economy as the basis for their commercial lives just as they accepted the need for constitutional government grounded in a realistic view of human nature. Most observers would agree that the Constitution and the market economy have served us well over the past two hundred years.

This praise of the market system is not meant to imply that our free market system is perfect. On the contrary, actual market systems contain a number of weaknesses. However, both in the ideal model and in practice, the market system deserves praise. A market economy accomplishes two of the three basic economic goals—efficiency and freedom—with ease, and does not compare too unfavorably with planning systems in terms of fairness. The market does all this without fanfare, for its genius is in the quiet, day-to-day operation and interplay of prices and profits.

Appendix B

Market Weaknesses and the Growth of Government

A market economy allows government to have a limited economic role in defining a context for the operation of free competitive markets. Government defines and enforces property rights and provides other services to encourage free exchange and competitive markets. In reality, government has gone well beyond the limited role described for it in Appendix A. Our government regulates many markets, sometimes even setting prices. It prohibits some exchange and strongly oversees exchange in other areas. In fact, government today is directly involved in one-third of the U.S. economy, through taxes and expenditures. All businesses find their markets regulated to varying degrees by government, through local zoning laws, state business licensing and regulation, federal oversight of stock markets, labor arrangements and so on. If a market economy is the point of departure of the American Founding and the dominant form of economic organization throughout the world, why is there so much government involvement in modern economies, especially in the United States, where we pride ourselves on our free market system? That question is the focus of this appendix.

Appendix A described a perfectly functioning market economy where the principles of supply, demand, equilibrium price, and role of profits allocated resources efficiently across the economy. Individuals and businesses made exchanges that were beneficial. No better set of exchanges were left on the table. Obviously, markets do not always operate so perfectly; no human institutions do. There are times when the equilibrium prices are the wrong set of prices to produce efficiency. There are times when businesses or individuals have market power and exploit that power for their own gain at the expense of other interests. There are other times when exchanges that would be beneficial (for example, someone looking for work but not able to find any employment) are not made. These market weaknesses can be classified for ease of reference into six general categories.

1. **Imperfect Information**
   Exchanges are made that are not beneficial because they are not based on good information.
2. **Monopoly**
   The monopolist restricts exchange to raise price and profits, leaving some beneficial exchange unmade.
3. **Public Goods**
   Goods consumed in common where consumption by one individual does not diminish the amount available to others. Consequently, individuals do not have an incentive to provide a public good.
4. **Externalities**
   Exchanges where some costs and benefits go to third parties not involved in the exchange. Exchanges are sometimes made that are not beneficial because some costs are ignored. Likewise, some beneficial exchanges are not made because some benefits are ignored.
5. **Recession**
   Resources (labor, raw materials, machinery, etc.) are left unused even though full employment/use of resources would be beneficial.
6. **Economic Injustice**
   Society decides that the distribution of income or other economic benefits generated by the market economy is unfair.

The sixth issue listed above, regarding the large and complex issue of economic fairness or justice, has more to do with social perceptions than actual market exchanges. In other words, sometimes we just don't like the outcome we get from the exchange process. Still, this must be factored into the "big picture" of a nation's economy, because that one issue is central to politics and the economic actions of many governments. The perception of economic injustice, stark fact or cleverly promoted fiction, is a powerful, if often intangible force that can adversely affect a market's effectiveness and the government supporting that market.

Analyzing Market Weaknesses

There are three steps a researcher must take to accurately and adequately analyze a market weakness.

1. *Find the source of the market breakdown.* Are exchanges taking place that should not be taking place? Are there beneficial exchanges that are not occurring? Is the market price too high or too low?
2. *Define the necessary government policy that will correct the market problem.*
3. *Evaluate the success of the government's policy when it is used to correct the market weakness.*

Participants in a debate about which is the best economic system tend to talk about theoretical ideals rather than concrete reality. In reality, our nation's market system is not perfect. Government actions taken to correct market weakness or outright failure are not likely to be perfect, either. Every market breakdown has similarities to previous breakdowns, but each also differs in many ways, so while the ups and downs and calamities of the market in past years should be studied and learned from, every new downturn or disaster in the economy requires close scrutiny because it will be full of new elements that were not in play in previous economic events. Government intervention may sometimes help correct the market's mistakes. At other times, government intervention might actually worsen the problem. The best policy may be acceptance of the market weakness as well as a realistic view of government and the consequences of its intervention to correct market weaknesses.

Common Market Weaknesses

1. Imperfect Information

The Problem

A voluntary exchange makes both parties of the trade better off if those involved are acting on good information. If those taking part in the exchange do not have adequate information about the products involved or the terms of the exchange, it is possible they will miscalculate and the exchange may harm one or both of the parties. For example, a teenager goes into a grocery store to buy a package of cigarettes. Many would argue that this exchange is not mutually beneficial because the teenager has imperfect information; he or she may be unaware of the consequences of cigarette addiction and make the exchange not knowing that a lifetime of smoking wastes money, increases the probability of lung cancer, and reduces the smoker's expectation of a long life.

Information is often difficult and costly to obtain. Which car should we buy? Are computer prices going to fall or rise? Is a new medical procedure safe and effective? There is no question that imperfect information causes exchanges to take place that do not benefit all parties, and prevent truly beneficial exchanges from occurring. Critics of a free exchange system can always point to examples where consumers have been cheated in the marketplace. Popular television shows like "60 Minutes" or "20/20" enjoy their greatest audience response when they uncover some unscrupulous business practice that hurts benighted, uninformed consumers or workers. Obviously, exchanges that are not mutually beneficial do transpire. The real issue is whether such exchanges are so systematic that the market system's usual remedies are not sufficient to solve the problem.

A market economy does have some safeguards against the problem of poor information. For example, many exchanges are repetitive. We purchase gasoline, food, clothing, and personal services such as meals or haircuts on a regular basis. This means that both parties to the exchange have an incentive to treat each other in a manner that makes future exchanges more likely. A grocery store that mistreats its customers will quickly fail and restaurants that provide bad meals or incompetent service soon lose their clientele.

However, some exchanges involve technical information that consumers do not normally have. In these instances, intermediaries develop in the market system to help consumers. We employ the services of a physician to diagnose illness and tell us what medicines and therapy will make us better. While physicians may not always use their information for their patients' full advantage, they generally act as good intermediaries and help their patients make advantageous decisions. If a physician fails to perform this role effectively, the patient has the option of finding another doctor. Customers always have the power to withhold their business in a market system, so businesses, professionals, and individuals performing services have strong incentives to provide customers with good products, good service, and good information.

Suppose a company has a superior product that many consumers would purchase (and thereby gain from the exchange) if they only had information about the product. Obviously, the company has a strong profit incentive to provide consumers with information. This information is conveyed through advertising. Advertising expenditures in the United States total more than $100 billion a year—a lot of money to disseminate information. Even an ardent devotee of the market economy ideal would probably not insist that an ad featuring a Hollywood starlet draped across a sports car provides sufficient information about the comfort, functions, and resale value of that vehicle. Nevertheless, a great deal of information is conveyed through advertising. Consider how much we have learned about personal computers, cellular phones, compact disc players, and digital cameras in the last few years through advertising. Grocery shoppers scrutinize the weekly ads in order to manage their shopping for the week. A vast amount of information about both quality and prices of products is available to us through the media without government sponsorship.

In spite of all the incentives to provide information and the development of intermediaries to help consumers, exchanges based on poor information may still occur. Such exchanges violate the very premise upon which free exchange and the market system is based, because they don't benefit both parties to the exchange. Since these exchanges don't promote efficiency or economic welfare, they should be eliminated, if possible.

Government Policies to Potentially Correct Market Problems

The traditional government method of correcting misinformed exchanges has long been the court system. For centuries, fraud and deception have been grounds under common law for invalidating contracts and providing compensation to victims. Civil courts provide a way to retain freedom of exchange while still protecting individuals from deception or fraud.

Some harmful exchanges do not directly involve fraud or deception. Rather, consumers simply do not have all the information relevant to making decisions, and no one in the system has an interest in providing such information. Smokers need to know that there is strong statistical evidence linking cigarettes smoking to lung cancer. However, no private individual or group has any self-interested motive for providing this information. Consequently, the government has taken responsibility to disseminate such information. Labeling on food and other packages warning about health risks of some products, and television ads encouraging young people not to use drugs are both examples of government intervention to correct the market weakness of imperfect information. The government tries to provide information directly with labels and warnings and indirectly through laws requiring private individuals and businesses who profit directly or indirectly from marketing particular products to provide such information to consumers and potential consumers.

In recent years, government has gone beyond providing information to intervene directly in the exchange process on the grounds that consumers and workers are sometimes incapable of making sound exchanges. Laws mandating safety devices such as air bags and seat belts for automobiles, and laws prohibiting trade in certain drugs such as cocaine or heroin, reflect an increasing social impulse for government to control the exchange process. Whether this interventionist impulse will remain limited, as it is now, or whether it represents a significant movement toward more planning in the economy is unclear. Some find it ironic that modern democracies entrust important electoral decisions affecting life, death, and taxes to ordinary citizens, while these same governments worry that citizens are incapable of buying pajamas without guidance.

Success of Government Intervention

Few criticize the government's efforts to provide information through labeling, government pamphlets, or television messages. Government involvement that simply increases information appears to be quite successful. Consumers seem to gain from the information that is provided at a relatively low cost. The criticisms have come when government has taken a more active role in prohibiting trade because it feels the exchange is based on poor information. The problem is a simple one: under the guise of preventing bad exchanges, the government also prevents beneficial exchanges. Consequently, the harm done through regulation must be compared to the good that is accomplished. For example, government's licensing of occupations such as physicians, cosmetologists, accountants, etc., is designed to prevent harmful exchange. Unfortunately, the licensing agencies are often used to restrict entry and exchange in order to raise the price charged by the licensed group. Government requires drug companies to send their new prescription drugs through elaborate tests and meet stringent regulations before those compounds reach the general public. If the drug proves to be beneficial, then delay causes harm just as a drug offering no consumer benefit causes harm.

To summarize, poor information may lead to exchange that is not beneficial to the parties involved or prevent beneficial exchanges from taking place. While the market system has some strong incentives to minimize this problem, there is a role for government in providing information that will not ordinarily be provided by businesses or individuals following their self-interest. In recent years, government has moved beyond just providing information to controlling quality and the price of some goods on the grounds that participants in the market do not have the capabilities to act in their own best interest. How far the government will go with this intervention policy remains to be seen.

2. Monopoly Power

The Problem

Adam Smith wrote in *The Wealth of Nations*, "People of the same trade seldom meet together, even for merriment and diversion, but the conversation ends in a conspiracy against the public, or in some contrivance to raise prices." One of the most serious potential weaknesses of a market economy is the possibility that economic power will be concentrated in the hands of a few individuals or businesses.

The fear of economic power being so concentrated has been heightened by the development of large corporations employing thousands of workers and with billions of dollars in assets. ExxonMobil rode the climbing price of oil to over $30 billion in profits in 2005. Wal-Mart employs 1.8 million workers worldwide. General Motors has nearly $500 billion in assets or capital. Perhaps some of these huge corporations should apply to join the United Nations, because their sales are larger than the gross domestic product of many U.N. member nations.

Monopoly

When one person or group is the sole seller in a market with many buyers; the lack of competition in a market.

While the size of these large corporations may create some fear and distrust in a citizen who is confused by such enormous organizations' complexity, the real issue is the extent of their economic power and the implications of that power. Economic power is associated with the term **monopoly**, which literally means *one seller*. True monopolies are quite rare and are normally confined to utilities such as local phone service, electricity, or natural gas. More often, there are a few large producers or sellers that appear to have considerable economic power, such as several powerful corporations in the automobile and oil industry.

Cartel

A group of firms that act together like a monopoly.

The quote above from *The Wealth of Nations* indicates that businesses have strong interests in conspiring to set prices at high levels. A group of firms that come together to make such agreements and act together like a monopoly is referred to as a **cartel**. The Organization of Petroleum Exporting Countries (OPEC) has been a cartel since 1973. Using an oil embargo and production quotas, the cartel has restricted production and dramatically increased the price of crude oil to U.S refineries, causing increased prices for gasoline at the corner gas station. The countries involved in OPEC have reaped enormous profits from the cartel's actions.

While there are incentives to form cartels to raise prices, there are also strong incentives for any cartel member to cheat on the agreement by lowering the price of its product, which will multiply its own profits by increasing sales. Cartels and price-fixing agreements are constantly being formed, but their life span is usually quite short. By the 1980s, OPEC started to lose its power. Member governments constantly exhorted each other to maintain the high price, but most governments were producing more than their allotted quota. By 1985, the price of oil had fallen to $25 per barrel, down dramatically from its previous high of $35 per barrel. In early 1986 crude oil was being sold to American refineries for less than $15 per barrel. Even OPEC, one of the most famous cartels, seems to be unstable because of the incentives of member nations to cheat. The spectacular increases in oil prices in 2005 and 2006 appear to be due to increases in demand and some unusual supply effects rather than the actions of the cartel.

In a competitive market, no individual or business has significant power over the price of the good or service being exchanged. Businesses may appear to have the power to set prices, but a price set above the market equilibrium will eliminate virtually all sales for that business. Consequently, when there is economic competition, the true power of a business over price is limited. On the other hand, a monopoly does have real power over the price because of limited competition. This power to control price is a serious threat to a market economy. It is important to understand the nature of that threat.

Exchange continues to take place even though one party has economic power because exchange continues to benefit the parties involved. Economic power does not imply that monopolists are able to force consumers to buy their product. Businesses are always constrained by the law of demand. However, businesses or individuals with economic power can dramatically increase their own gains at the expense of others by setting a high price for their product. Thus, one of the most important effects of monopoly is to redistribute the gains from trade so that the monopolist or party with power captures nearly all of the gain.

If the monopolist only redistributed the gains from exchange, economic power would be a problem of equity or economic injustice. Unfortunately, the method monopolists use to capture those gains from exchange also creates inefficiency. Market inefficiency occurs because beneficial exchanges are lost as the monopolist restricts production to raise the price.

For example, the pizza market is fairly competitive in most cities. Suppose Luigi's Happy Pizzeria's cost of making a pizza with everything on it, and then some, is $10. The forces of competition will keep the selling price close to that pizza's $10 production cost. A hungry consumer who values that particular pizza more than the $10 he has in his pocket is able to purchase one, and Luigi and the consumer both make a beneficial exchange in the pizza market. But now suppose the Amalgamated Pizza Corporation across the street from Luigi's gains full monopoly power in the pizza market by controlling the supply of mozzarella cheese. The strategists in the Amalgamated boardroom decide to reduce pizza production by half and charge $20 for their pizza to maximize the profits (or gains) from each exchange. They do this even though their pizzas still cost only $10 each to produce. Individuals who previously always bought pizzas that they valued at between $20 and $10 now pass by both the closed Luigi's and the Amalgamated pizzerias on their way to buy hamburgers or chicken, or they just warm up a can of chili and eat at home. Too few resources are employed in pizza production while too many resources are being used elsewhere because the monopolist has created an artificial scarcity of pizza. For this reason, economic power creates inefficiency in a market economy.

Government Policies to Potentially Correct Problem

Government has two strategies available for controlling economic power:

- Government may regulate economically powerful firms, forcing them to price their products and behave like competitive firms.
- Government may take action to *foster competition* in order to eliminate the economic power of firms.

Regulation is most often used to control industries that naturally tend toward monopoly. Utilities such as natural gas, electricity, and local telephone service are the most important contemporary examples. Most states have regulatory commissions that control the economic power of utilities by setting the rates for their services. The federal government also has set up agencies such as the ICC (Interstate Commerce Commission) to regulate businesses.

For nearly a century, the government has tried to inhibit monopolies and promote competition through laws that have been grouped together under the title of "antitrust policy." Monopolies (and groups of firms joined together to act like monopolies) were called trusts in the late 19th century. Antitrust policy began with the passage of the Sherman Antitrust Act in 1890, which made conspiracy to restrain trade or the creation of monopoly power illegal. This law gave government the power to use courts to prosecute conspiracies, set prices, and break up companies that had monopoly power. Later laws, such as the Clayton Act (1914) and legislation passed during the New Deal, strengthened the power of government over business. Antitrust legislation essentially requires the government to prosecute businesses it charges with monopoly practices or conspiracy to set prices. This legislation also allows other businesses or consumers to sue for damages caused by monopolies. If a business proves it was damaged by another firm's violation of the antitrust laws, it is allowed to collect triple the amount of damages actually incurred.

Success of Government Intervention

The record of government regulation of monopolies or businesses with economic power is mixed at best. In fact, the businesses being regulated often control the regulatory agencies. Individual consumers have relatively small stakes in the actions of any regulatory agency, so they pay little attention to the actions of such agencies. However, businesses being regulated have enormous stakes in the actions of the regulatory agency. Consequently, and regardless of the ethics involved, they lobby the agency intensely and often persuade it to take action in their favor. Indeed, the government regulatory agencies created to control monopoly power have instead been used by business to create additional monopoly power by setting high prices and controlling entry into the regulated business. Most policymakers and citizens still would not want natural monopolies such as the public utilities to go unregulated.

The implementation of antitrust laws has not been entirely successful either. The use of the courts to control economic power through antitrust laws is expensive and time consuming. For example, the government sued IBM for restraint of trade. The litigation took 10 years, after which the government dropped the case because all parties recognized that technological changes in the industry IBM had controlled had eliminated the monopoly IBM enjoyed earlier. However, antitrust policy remains an important tool to control monopoly. Businesses may change their economic behavior in anticipation of possible antitrust action against them. Thus, antitrust laws may reduce the number of business conspiracies to control prices. The government may be encouraging more economic competition simply by fostering the threat of antitrust action against monopolies and economic power.

International trade may be the most effective method of promoting competition to reduce monopoly power. Three automobile companies, General Motors, Ford, and Daimler-Chrysler, account for nearly all domestic automobile production, with one firm, General Motors, producing the major share. (General Motors itself is a conglomerate of numerous automobile companies that were gathered together under the GM banner, starting in the 1930s.) Clearly, the potential for abuse of economic power exists in this situation. A few casual meetings by GM, Ford, and Chrysler executives on golf courses in Bloomfield Hills or Grosse Point, Michigan, would be sufficient to set prices at a monopoly level. Fortunately, international competition is intense in the automobile industry. Pressure from Nissan, Toyota, Mitsubishi, Volkswagen, Fiat, Honda, and others effectively removes the threat of monopoly as long as government leaves international trade unfettered. International trade has grown over the past few decades through inexpensive ocean shipping coupled with wide-spread economic development, so the threat of monopoly has been significantly reduced.

Governments occasionally have prevented international trade under the guise of job protection and fair play. Indeed, much economic power exercised today is caused by government actions that create domestic monopolies through protectionism, or international government cartels such as OPEC, which act like monopolies. Government has opposed monopoly power, but that opposition has been fainthearted at best. Fortunately, true monopolies, a concentration of economic powers, and the resulting market inefficiencies are all rare.

3. Public Goods

The Problem

Some market imperfections, such as inadequate information, allow markets to continue to operate even though they do so in a flawed manner. However, in the case of public goods, the market often fails to operate at all. For this reason, almost all economists accept the necessity of government intervention to provide public goods that cannot be provided effectively by the markets.

The term **public goods** has a narrow and technical definition. It does not refer to all goods provided by government. *A public good is a good or service that, if consumed by one individual, does not diminish the amount of that good or service that remains available to others. Furthermore, it is difficult or impossible to prevent individuals from using a public good.* Public goods have the remarkable property of being available for all to consume jointly if anyone should happen to purchase or provide them. If one shipping company constructs a lighthouse to aid in navigation, that service would be available for every ship in the vicinity of the lighthouse. Furthermore, use by one ship does not reduce the amount of light available for other ships to use. Not all government-provided goods are public goods. The U.S. government provides most elementary and secondary education, but education, while it is a service, is not a public good. Many goods have some element of a public good in them. Police protection has an element of a public good in the sense that service by police may reduce crime in a city, providing protection for all. However, if the police are busy in one area of the city, they are not available for others. In this sense, police protection fails the test of a public service because consumption by one individual does diminish the amount of police service available to other consumers. Radio and television signals, navigational aids of all kinds and, most significantly, national defense, are examples of pure public goods.

The failure of markets to provide public goods lies in the nature of such goods. If one individual provides the good, others can reap the benefits without cost. Therefore, each individual has an incentive to let others provide public goods. In other words, no one goes down to the friendly store and buys public goods because it pays to get someone else to make the purchase. This incentive problem is called the **free rider problem**. Consider the lighthouse example. Every ship owner finds the services of the lighthouse useful. However, each owner knows the lighthouse is available to everyone. Therefore, each ship owner hopes that other ship owners will build and maintain the lighthouse so that he or she may be a free rider, consuming the services of the lighthouse without cost. If there were few ship owners, they could probably provide the public good privately in a collective manner. As the number of consumers grow, a public good will probably have to be provided by the government. Even though each business or household would be better off if it made an

Public goods

A good or service that, if consumed by one individual, does not diminish the amount of that good or service that remains available to others.

Free rider program

A good or service provided by one individual or company that benefits or is consumed at no charge by another individual or company.

exchange for the good, each has an incentive to refrain from making the exchange. If public goods were left to the free market, there would be almost no production or consumption of these particular goods, and many beneficial exchanges would be missed.

Radio and television signals represent unique public goods because communications technology may either create or eliminate the free rider problem. Everyone with an antenna may receive a radio or television signal. Reception and use by one party does not diminish the potential use by someone else. Yet, in many circumstances, the market for these signals operates efficiently. For ordinary radio and television reception, the free rider problem is eliminated by the use of commercial advertising. Commercials are the "price" we pay for reception of the signal. Since we are forced to listen to the commercials in order to receive the signal, it is impossible to be a free rider in this case. This combination of commercials with the program eliminates the usual public good problem. Similarly, satellite scramblers and receivers eliminate the free rider problem with dish and cable TV.

The most important public good is national defense. Each individual within the borders of a particular country is protected regardless of the amount each has paid for defense. The protection that missiles give citizens in New York does not diminish the amount of protection available for citizens in California. But each individual has an incentive to be a free rider and let others pay for the defense provided. Moreover, there is no way to make people pay for the benefits of national defense as there is with the radio or TV signal. In economic terms, there will be no revealed demand for national defense even though individuals would benefit from exchanging income or resources or the protection of national defense.

Could national defense be provided by a market system? Probably not. Even if private firms were created to provide the product, each of us would encourage our neighbors to purchase national defense goods in the hope that we could get a free ride on their purchase. The market would underproduce these goods, if they were produced at all. The only solution to the problem appears to be government provision of these public goods, levying taxes to pay for them.

The war against terrorism is an international public good. The countries threatened by terrorist actions have an incentive to prosecute the war on terror. However, each country also has an incentive to encourage other countries (i.e., the United States) to provide the bulk of the public good. The U.S. for some reason seems willing to provide the public good and let most other threatened countries be free riders.

Government Policies to Potentially Correct the Problem

Since markets for public goods either collapse or are poorly developed, governments have assumed the role of providing public goods, especially national defense, throughout history. Indeed, governments are almost synonymous with public goods. The primary government solution is to directly provide the public good. There are a few examples of contracting the production out to private firms, but most public goods are provided directly by government. Public goods are no small matter—most modern societies spend a significant fraction of their income on national defense and police protection. In 2007, the U.S. government budget for national defense is about $440 billion.

Success of Government Intervention

The fact that government has assumed the responsibility to provide public goods does not mean that the problems surrounding public goods have been totally solved. Policymakers have no clear guide from the public as to the exact quantity to provide. With other goods, consumers reveal their preferences for the good through their purchases or consumption. Shortages or surpluses develop if the wrong amount of a good or service has been produced. However, there will be no observed shortages or surpluses of public goods because users do not reveal their demand for the product. Consequently, the quantity of national defense or any other public good provided by government must be determined through reason and debate. Nothing in this process ensures that government will determine the correct quantity of public goods. Clearly, turning public goods production over to the government does not necessarily solve the problem. But the government solution is superior to the market alternative.

4. Externalities

The Problem

In the ideal market world, exchanges only affect the two parties involved in the trade. Everyone else in the economy is left unaffected by any particular exchange. Problems develop when parties not directly involved in the transaction are affected by the exchange. To illustrate, suppose that a student entrepreneur rents a house and stages a party each weekend. Local college students gladly pay the party admission price of $10 per person. The music is played at 100 decibels so that all may enjoy dancing to the beat of The Black Eyed Peas and other artists. However, a quiet professor lives next door, working each weekend on his magnum opus—*The Economics of Banking in Siena in the Thirteenth Century*. The exchange is beneficial to both the entrepreneur and his customers, but harmful to the professor. The professor is a third party adversely affected by the exchange.

These third party effects in exchanges are labeled externalities because they are costs or benefits that are *external* to the two parties directly involved in the exchange. An externality occurs whenever a third party not directly involved in an exchange either benefits or suffers cost because of that exchange. The exchange between the student promoter and his customers generates an external cost for the professor. Examples of external costs would include factory pollution damaging nearby homes, contamination of drinking water by farmers' pesticides, upstream manufacturing from a city, offshore oil drilling which harms fish and mammals along a coast, blood donations by persons with diseases such as AIDS or hepatitis, and so forth. Other transactions may generate external benefits. In fact, if the professor were writing his one great book on the musical tastes of contemporary college students, studying the parties next door could generate external benefits to him. Examples of external benefits might include inoculations that reduce the chance that others will contract a particular disease, honey production providing pollination for certain agricultural crops, citizenship education because more enlightened voting will benefit others, and scientific discoveries available for general use.

Externalities do not eliminate exchange, yet they cause problems in a market economy. Perhaps the problems are best explained by two examples—one involving external costs and the other involving external benefits.

Return to the student entrepreneur with his weekend parties specializing in loud rock music. Both the entrepreneur and his customers are better off because of the party. To illustrate, assume the gain from the exchange after all costs of the entrepreneur are considered is $60. If this is the case, the parties will continue, since everyone directly involved in the exchange benefits. Suppose that the professor suffers a loss valued at $100 because the party impedes his great intellectual endeavor. From the viewpoint of the whole society, the exchange should not take place since the benefits only total $60 while the costs total $100. However, the market will ignore the costs incurred by the professor because he is not directly involved in the exchange.

External costs allow exchanges to take place that should not because no net benefits are created. If certain goods in the economy create external costs, those goods will be overproduced in a market system as the external costs are ignored. The market price for such goods—too low from the point of view of the whole society—induces consumers to buy more than the efficient amount. Since producers do not pay these external costs, they are willing to supply such goods at lower prices because their private costs are covered.

External benefits create as many problems as external costs for the efficient operation of a market economy, although society tends to focus on the problem of external costs. A vaccination provides a clear example of an external benefit. If an individual goes to a physician and pays a fee to receive a vaccination against a particular disease such as polio, both the doctor and the individual benefit and are better off because of the exchange. However, third parties also benefit because the treated individual is less likely to communicate the disease to others. Suppose that a polio vaccination costs $50 in physician's fee and patient's time and pain. Only those individuals who value the vaccination more than $50 will be vaccinated. Suppose, also, that each vaccination generates $40 in benefits to the rest of society in terms of increased prevention. If George Jones only values a polio vaccine shot at $20 in personal benefits, he will not get the shot even though the benefits to society as a whole are valued at $60. In other words, *goods or services involving external benefits will be under-produced by a market economy.* Notice that George Jones would purchase the vaccination if it cost $20 or less. The price of goods involving external benefits is not an efficient price in a market economy. Just as the price to consumers of goods involving external costs is too low for consumers in a market economy, the price of goods involving external benefits is too high.

Government Policies to Potentially Correct the Problem

One of the basic functions of government in a market economy is to define and enforce property rights. Externalities occur because property rights are not fully defined. Some resources, such as air and water, are either not owned or are owned in common and misused. If ownership of all resources could be fully and explicitly defined, externalities would disappear. Persons suffering external costs would be able to sue for damage to their property, forcing the two exchanging parties to bear all costs involved in the exchange. If the professor had a clear property right to the air surrounding his home, he could sue the student entrepreneur for violating his air with the sounds of loud music. This lawsuit would force the college students and entrepreneur to pay the total costs for their party. Unfortunately, definition of resource ownership for air and water is somewhere between difficult and impossible. Consequently, the problems caused by externalities can't be solved easily by enforcing property rights.

Property rights are vague, but some externalities can be eliminated by private arrangement, significantly reducing the problem. In the case of the professor and the student entrepreneur, for example, the gains of the rock party were $60 for the entrepreneur and his customers, while the losses to the professor were valued at $100. Because the loss to the professor exceeded the gain to the students generating the external cost, there is room for an exchange to eliminate the externality. If the professor offers more than $60, the student entrepreneur may be willing to stop offering the parties and the externalities. Externalities that involve just a few people can often be eliminated though private arrangement.

More serious problems with external costs or benefits occur when the costs or benefits involve large numbers of people or businesses in situations where property rights are difficult to define. For example, air pollution of major metropolitan centers can't be fixed by a simple private arrangement or by specification of property rights. Furthermore, technological change sometimes creates new situations where it is difficult to protect property rights. The copying of motion pictures or computer programs generates an externality that requires some government actions for solution, because it is otherwise impossible to enforce the property right of the owner and private agreements are unlikely.

When property rights can't be defined, the government can reduce the externality problem with taxes, fees, or subsidies that make individuals or businesses that generate external costs pay for those costs, or make those receiving external benefits compensate the producers of the external benefits. For example, drivers in Los Angeles and surrounding counties could be taxed according to the amount they drive and pollute the air. (The tax could be collected on gasoline and other fuels.) Such a tax turns the *external cost* into an *internal cost,* which would be paid by the correct parties in the economy. Similarly, general taxes might be levied on the segment of the population that benefits from the polio vaccination program. These taxes may then be used to subsidize vaccinations so that the cost to the individuals obtaining the shots only reflects their personal benefits, with society paying for the external benefits.

Success of Government Intervention

Government's record in dealing with externalities is mixed. Certainly, the definition and adjudication of property rights by government has helped reduce the problem of externalities and promote efficiency. Yet, direct control of external costs by government has not been notably successful. Government's attempts to internalize the costs of pollution, for example, have been muddled because it is difficult to determine what level of pollution is acceptable to the whole society. Environmentalists argue for no pollution, which would mean ending the production of some goods as well as fewer jobs in certain industries. Advocates for the polluting industries and the associated jobs argue for leaving the control of industrial pollution to the public virtue of the industry doing the polluting, but that would result in a great deal of pollution with the associated threats to public health and degradation of the environment. There is actually an optimal level of pollution (in nature, as well as in human activities) where the increment to the external costs just equals the net benefits of the increment to production. It is difficult to calculate this optimal level of pollution. Consequently, government sometimes regulates too much and at other times too little. Getting things just right in the environment is not government's comparative advantage.

While government has generally done a mediocre job in dealing with external costs, it has usually done a poor job when it deals with external benefits. With the possible exception of education, most external benefits in medicine, the arts, and research are ignored or under-funded by government. Consequently, goods and services that have external benefits are under-produced in most economies.

External costs and benefits are likely to remain serious problems in some sectors of a market economy. Fortunately, most economic activity has quite limited external effects. Government policies to correct externalities are actually quite easy to design, but political implementation of such policies is often difficult because the individual losses from those policies can be substantial.

5. Recession or Depression

The Problem

Market economies have always had up and down cycles of output and employment. This economic instability poses a profound challenge because we still don't have sufficient economic knowledge to prevent cyclical movements in a market economy. Whenever the actual level of output in the economy falls significantly below the output that the economy can produce with full employment of resources, the economy is in a **recession**. A depression is a severe recession. We usually refer to an economic downturn in which unemployment is less than 10 percent as a recession.

Figure B.1 shows the variability or "cycles" in unemployment over time. Some unemployment is inevitable because people are moving to new jobs or have recently entered the labor force. If most unemployment is voluntary or short term (so that exchange in the labor market is working smoothly), the economy is not in a recession. When large numbers of people are involuntarily unemployed, indicating that exchange in the labor market is being frustrated, the economy is in a recession.

The causes and remedies of recession are still only partially understood, but it is possible to see that the exchange process does not work properly during either a recession or a depression. In the ideal market economy, prices, wages, and interest rates adjust to economic conditions so that all exchange possibilities are realized. For some reason (or a variety of reasons) these adjustments are not complete during recessions. Workers want to exchange labor for goods but are unable to do so. Merchants want to sell their accumulating inventories but can't at prevailing prices. Capital is not invested because of the downturn in the economy. A recession is a direct contradiction of the law of comparative advantage, which is normally confirmed by economic experience.

Recessions and depressions represent the most serious weaknesses of the market system of economic organization. The problem of recession or depression is of vital concern to our evaluation of a market economy. There is little doubt that it is the most serious market inefficiency in terms of costs. Monopolies or externalities may cost an economy like the United States tens of billions of dollars. The costs of recession and unemployment are often measured in hundreds of billions of dollars.

Recessions and depressions appear to be the result of random shocks to the economy that cannot be readily blamed on government action. Explaining recessions as natural market phenomena does not totally absolve the government from responsibility. Evidence strongly suggests that a number of recessions, such as the Great Depression of the 1930s, have been worsened by government action.

Nevertheless, market economies have a pattern of cyclical economic behavior—a period of full employment and rapid economic growth followed by a period of unemployment and economic stagnation.

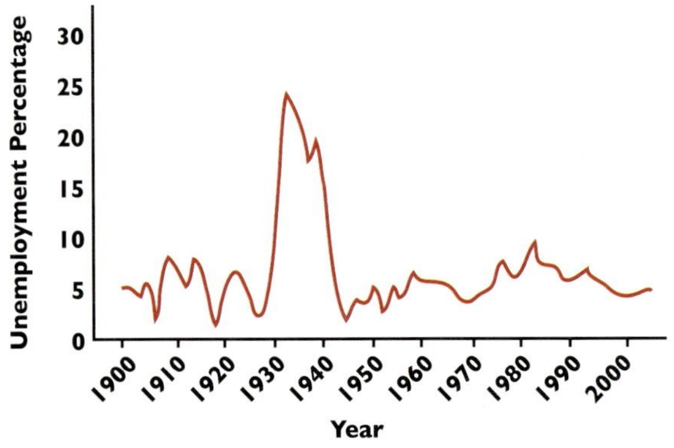

Figure B.1 Unemployment rates.

The basic puzzle of recession concerns the persistence of *involuntary* unemployment. Why can't people find jobs during a recession? If a market is working properly, a surplus should make the price fall until the surplus is eliminated. Unemployed labor, which is a surplus in the labor market, theoretically should make the price of labor (wages) fall until all unemployment is eliminated. Furthermore, the law of comparative advantage implies that the opportunity cost of unemployed labor should be very low, which should lead to the employment of that labor in some specialty. Basic economic principles suggest that unemployed labor should not persist. But it does. A puzzle must be solved: why doesn't the wage fall until everyone who wants to exchange some labor for income is accommodated? The price and wage flexibility that controls exchange, the genius of the market economic system, fails during recession. The labor market is by far the most important market in the economy, and it also may be the least effective.

Current economic analysis reveals that wages do adjust to control labor exchange and eliminate unemployment, but the process is very slow. This sluggish response in the labor market means that the economy can't respond well to economic shocks such as the dramatic increase in oil prices in 1973 and again in 2005–06, sudden changes in governmental tax policy, or a large reduction in the money supply, such as the decline from 1929 to 1933. The best market response to these shocks would be wage and price changes that leave the economy at full employment. However, prices and wages do not change as rapidly as they should, and this causes unemployment as products sit on shelves and output falls.

Because the economy does not adjust quickly or properly, most of these unexpected events that shock the economy reduce the total or aggregate demand. As demand falls in the economy, prices and wages should also adjust by falling. Unfortunately, prices and wages tend to be "sticky" or inflexible because of contracts and business plans. Instead of prices and wages adjusting, people are laid off from their jobs and some businesses close or suspend production. Thus, the combination of a decline in demand and inflexible wages and prices produces a recession. In other words, output is forced downward in response to the decline in aggregate demand because prices and wages are slow to adjust to the new economic conditions. Recessions occur because the response to economic shocks is in terms of *quality* adjustments rather than *price* adjustments—sales, output employment, and plant utilization all slow down. If prices and wages would adjust quickly, some individuals or companies would be hurt by the adjustments, but the economy would stay at full employment of resources.

Not all recessions are caused by declines in aggregate or total demand. There can also be supply-side shocks to the economy, such as a steep increase in the price or changes in the availability of a key resource like oil or bad weather that limits agricultural production. Agriculture-based economies often have supply-side recessions that are caused by bad weather (droughts, floods, unusually long winters), reducing or actually eliminating the amount of soybeans, wheat, etc. that can be sent to market. Industry-based or service-oriented economies are more likely to have demand-side recessions.

Modern economies are also susceptible to recession because of growing international interdependence. For example, a sudden fall in the demand for the exports of an economy can cause a recession. During the Great Depression of the 1930s, the United States passed a high tariff (to protect the price levels of its own products against cheaper-produced foreign goods) that sent a shudder through world trade and helped spread the internal problems of the American depression worldwide. As economies of today become increasingly interdependent, recession in one country has an adverse effect on the economies of others. Alternatively, an increase in the international demand for the products of a country can pull it out of recession.

Government Policies to Potentially Correct the Problem

Government policies designed to reduce the effect of recession or depression must attempt to offset the reduction in spending by household consumption or business investment that causes a recession to begin. The government has two types of economic policies to stabilize total demand or total spending—fiscal policy, which involves government expenditures and taxes, and monetary policy, which involves the money supply and interest rates.

Fiscal Policy

First, let's assume the government knows that aggregate demand is going to fall in the coming year if it does not change its policies. Government policymakers can use public expenditures and taxes to offset or eliminate the expected decline. Government spending can be increased so that total demand will not fall, even though consumer spending or business investment might fall. If policymakers do not want to increase government expenditures, they can reduce taxes on household income to stimulate consumption, or cut taxes on business investment to encourage more investment in the economy.

The fiscal policy prescribed for recession requires the government to spend more and tax less—in other words, run a deficit in the federal budget. The graphs in Figure B.2 show surpluses and deficits in the federal budget for the past three decades and the accumulated government debt from budget deficits. Politicians seem to enjoy running deficits in the budget; they have chosen to do so for all but two years since 1970. Most economists recommend that government officials apply fiscal policy consistently over the business cycle—deficits in recessions, balanced budgets in normal years, and surpluses in the budget in years with increased inflation. Politicians find this diet to be low in political sugar (i.e., benefits to their geographical constituency, in the form of funds to build a new bridge, widen a highway, install a dam, etc.) and have chosen to ignore it in favor of getting votes from those benefiting from new jobs, etc.

Monetary Policy

The other potentially stabilizing government policy is monetary policy, which involves the action of the Federal Reserve System upon the money supply and interest rates. Most economists would recommend that the Federal Reserve increase the money supply and push down interest rates in a recession. The value of the U.S. dollar is no longer based on a gold or silver standard, but on the strength of the economy as well as the demand and supply of dollars. An act of Congress in 1913 gave the Federal Reserve, which is a corporation of private banking interests, the power to increase or decrease the amount of money in the economy, simply by printing or creating currency or encouraging increases in money holdings. It can also temporarily influence, although not completely control, interest rates. For example, the Fed might be able to temporarily reduce interest rates by injecting large quantities of newly printed money into the economy. The lower interest rates would encourage individual consumers and businesses to react with increased spending. Furthermore, households and businesses would try to spend the increased money supply, which would stimulate the economy.

During 1981 and the first nine months of 1982, the Fed concentrated on controlling inflation. Consequently, the money supply was restricted tightly during this period. This action by the Fed probably initiated the recession of 1981 and 1982. The restrictive monetary policy of the Fed, combined with inflationary expectations, caused interest rates to rise abnormally high during this period. High interest rates and worries about the economy caused investment to fall between the third quarter of 1981 and the fourth quarter of 1982, pushing the economy into a recession.

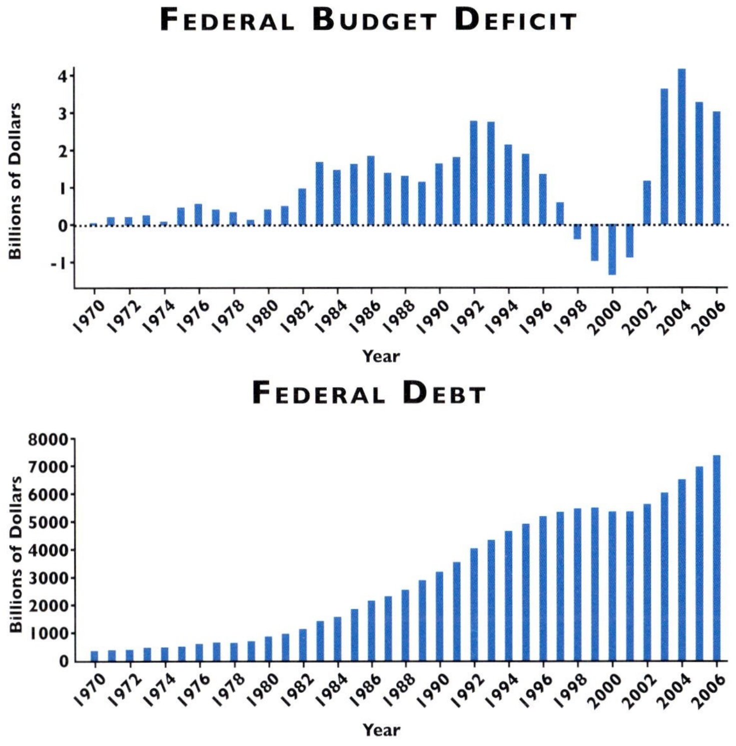

Figure B.2 Federal budget deficits and debts.

Success of Government Intervention

Though these measures to combat recession appear to be simple, government has been unable to eliminate recession from the market economy. Complicating factors make the implementation of these prescriptions much more difficult than first appearances suggest. Policies designed to thwart recession are not immediately effective; rather, they take time to work through the economy. Government planning, like that of business and—hopefully—student budgets, must always be done with an eye toward the future rather than the present. Therefore, the government encounters significant difficulties as it collects information, forecasts the future, and prescribes changes today to intervene and potentially correct any projected economic ups or downs. The difficulties in applying economic stabilizing policies are best understood by examining the problematic steps in the policy process.

Collecting Information

The first step in applying stabilizing policy is to obtain data about the current state of the economy and then to forecast the future, based on what is known about current circumstances, recent changes in production and demand, and how those changes might alter future circumstances. It takes government officials in the Departments of Labor and Commerce from 30 to 45 days to develop a hopefully accurate picture of the economy as it exists at a given point of time. For example, due to the lag in accumulating and cataloging information, it will be February of 2007 before government knows the true state of the economy as it existed in December, 2006.

Forecasting the Economic Future

The second step is to use this delayed information to forecast the future. Any forecasting for a time period beyond 18 months, however, is really no better than guesses. All forecasts have a degree of error, since the forecast must accurately predict the aggregate behavior of millions of households and businesses. Economists are sure that GNP will be positive in the next year. They are reasonably sure that GNP will be within five percent of the previous year's GNP. Beyond this expert knowledge, determining the course of the economy is really guesswork.

Our ability to forecast the economic future is comparable to our ability to accurately forecast the weather a week from now.

Deciding on a Policy

Step three requires the President and the Federal Reserve to decide on the appropriate economic policy. Cooperative government decisions take an excruciatingly long time even when economic events are moving rapidly.

Suppose the President decides the economy is headed into recession in a year. He determines that a tax cut or an increase in government spending is the appropriate fiscal policy. He then makes these recommendations to Congress. A tax proposal goes to the House Ways and Means Committee, which takes it under deliberation. If things go well, the Senate Finance Committee will act at the same time, but all tax measures must originate in the House Ways and Means Committee. In most instances, it takes Congress a long time to deliberate tax policy. For example, the tax cut proposed by President John F. Kennedy as a remedy for the mild recession of 1960 was not enacted until 1964, when the recession had already passed.

Obviously, Congress usually takes so much time responding to recommendations by the President that the policies may be ineffective, or even adversely influence the economy as they fuel the next inflation cycle. Belated implementation of policies does not fix the economic conditions that the President had in mind when making the request. Furthermore, deciding to create a government spending policy is a long process, but deciding to undo the spending policy once the recession is over is just as time-consuming and problematic.

The Federal Reserve can change the rate of growth of the money supply in the economy more quickly since its actions do not require congressional approval. However, because the Federal Reserve is an independent corporation, the policies of the Fed, Congress, and the President often conflict. The difficulties in deciding on and coordinating policies to combat recession are exemplified because of the independence of the actors; monetary policy may be directed at controlling inflation while fiscal policy is battling recession.

Implementing the Policy

Once the decisions have been made, the programs must be implemented. An increase in government expenditures may be implemented very quickly if the program simply involves transfers to households. However, government construction projects, such as highways and dams, require years for planning and construction. The problems of implementing changes in the money supply are even greater, for a change in the money supply today will have its effect on the economy anywhere from a year to two years into the future. These types of lags and imperfections in the implementation of economic policy create enormous difficulties for the government in its attempt to eliminate the cycle of recessions in a market economy.

Inconsistency

Another difficulty in reducing the effect of recession on the economy is the short time span of most political decisions. In the United States, politicians have two-year intervals between elections. They tend to be impatient with policies that promise results three, four, or five years into the future. Consequently, they constantly search for quick and painless fixes to the country's economic problems. The remedy for inflation involves slow rates of growth of the money supply, which may cause high interest rates for a time. The monetary prescription for pulling an economy out of a recession is just the opposite. The money supply should grow quickly, interest rates should be low, and government should spend more than it takes in. All too often the political system must choose between controlling inflation and fighting off recessionary tendencies in the economy, a choice politicians find extremely difficult. Each administration comes into office promising to combat recession and inflation at the same time, finds it is unable to do so, and eventually chooses economic policies that contain inconsistencies. One phase of the policy may be anti-inflationary while the other phase increases demand.

Political Pressures

Recession and inflation have significant political consequences. Presidents are swept into office because of the poor economic situations confronting their predecessors. Franklin D. Roosevelt, John F. Kennedy, and Ronald W. Reagan won their elections largely because they promised to conduct vigorous campaigns against weaknesses in the economy.

In 1932, as Roosevelt was running against Hoover, the unemployment rate in the economy was about 24 percent. The depth of the Great Depression made the winning Presidential candidate in the 1932 election a foregone conclusion. Roosevelt won with 472 electoral votes compared to Hoover's 59 electoral votes.

Almost three decades later, Kennedy campaigned against the unemployment and slow growth caused by the recession of 1960—a period which, for obvious reasons, Eisenhower and Nixon refused to call a recession, preferring instead to call it an "economic pause."

In 1980, Reagan campaigned successfully against the results of Carter's economic policies. Unemployment in that election year stood at 7.1 percent, while inflation crested 13.5 percent. Reagan was very critical of Carter's economic policies, and his criticism was readily accepted by most of the voters.

Once in office and faced with political reality, most administrations cannot actually deliver what they promised. For example, if the problem is perceived to be inflation, the remedy calls for some combination of reducing the money supply, increasing interest rates, lowering government spending, or raising taxes. There is no policy on the menu that does not negatively impact an important part of the economy. No one likes tight money and high interest rates, least of all the housing, construction, and timber industries. While it may be popular to propose reducing government spending, the consequences of that proposal are losses of thousands of high-tech and low-tech jobs in the space industry, or eliminate the water and reclamation projects in the growing and thirsty western states, and so on. Most members of Congress support reducing government waste, only to balk at any cutbacks that would affect their own districts. Government waste often seems to be defined as "all government spending outside my district." And finally, no President or member of Congress wants to return to the voters having just gone on record approving raising taxes by 10 percent. Much of the economic policy debate in Washington is simply an elegant dance without music in the search for nonexistent answers to very difficult questions.

Responses of Households and Business

Even if all the political problems of implementing monetary and fiscal policy could be swept away, these policies still may not be very effective. The purpose of fiscal and monetary policies is to control or at least influence aggregate demand. Yet, it is possible that households and businesses offset the government actions in very rational ways from their individual points of view. If government spends more, households and businesses may spend less so that demand is left unchanged in the economy. If government cuts current taxes to simulate consumption, households, believing that future taxes will rise to pay for the interest and debt incurred by the government, may not increase consumption even though their take-home pay has increased. If the Fed increases the paper money supply to reduce interest rates, households and businesses may revise their expectations about inflation and reduce savings, thereby forcing the interest rates up instead of down. In other words, if households and businesses are as smart as government policymakers (a reasonable assumption), then neither monetary nor fiscal policy may be able to influence aggregate demand very much.

Recessions are clearly a weakness of the market economy that exact a high cost from time to time. As societies throughout history came to expect government to solve more and more problems, they likewise passed the responsibility for eliminating recession on to government. However, government has not been very successful in controlling recession because of the problems with intervention discussed above. We may now be at a point where we can avoid disastrous policies, such as the government's actions in the 1930s that turned a recession into the Great Depression. However, the unemployment rate of 10.5 percent in 1983 should be ample evidence that absolute control of recession still eludes market economies.

The search for control of recession and depression continues. The market economy does too many things well to be rejected because of its cyclical nature. Yet, recession imposes unacceptable costs in the form of unemployment, bankruptcy, and vanished dreams. Economists will continue to search for economic understanding of recessions. A Nobel Prize awaits anyone who can significantly further our knowledge about the problem—with at least one solution to that same problem. Politicians will continue to promise economic panaceas and experiment with new policies because reelection and political power are waiting for the political party that deals successfully with recession—the most costly market weakness.

6. Economic Injustice

Perhaps the most important and contentious intersection of government and economics is on the issue of economic injustice. Reform of the tax system or social security or publicly administered health care will quickly move to a discussion of the morality of the distribution of income and other economic benefits in society. Rock stars and movie celebrities beat a now well-worn path to Africa to draw attention to the plight of the world's poorest (and most politically oppressed) groups. Political campaigns tend to center on candidates' plans to redistribute the wealth of society. ("A chicken in every pot." "Something for everybody.") Economic injustice is, in some sense, not an economic issue, but a moral one. Economic injustice is in the eye of the beholder and is really an expression of our view of human nature and our view of the operation of our basic institutions. For those who see individuals as largely responsible for their own position in society, economic injustice recedes as an issue. For those who see individuals as being ground down into poverty by the corporate juggernaut, government redistribution is the only answer. Most people are in the middle, but the debate about economic justice goes on and on without conclusion or much progress.

The Problem—The Case for Redistribution

A market economy has no built-in sense of morality about the distribution of income or economic rewards. If particular individuals own resources or possess talents that are in high demand, they will earn a high income. The forces of competition do tend to push down excessive returns to resources that can be supplied, but competition cannot change the returns to resources or talents that cannot respond to price. To illustrate, consider market responses to the popularity of a sporting event like professional basketball. If basketball becomes very popular, makers of basketballs, or basketball backboards and rims, or even basketball arenas will not earn particularly high returns or incomes because it is easy for others to enter and supply those products. Also, referees are not likely to receive much higher incomes because other individuals can easily learn the skills needed to officiate. Players, on the other hand, do often contain talents or characteristics (height, athleticism, ability to shoot baskets from long distances) that are difficult for others to supply even though the rewards are high. In 2006, *Forbes* magazine estimated that LeBron James, a very talented 22-year old basketball player, had an income of $26 million because of the high demand for his unusual talents. Other people have unusual talents that they have spent years developing and which are hard to do (street mimes and owners of flea circuses, for instance), but in our economy are not likely to earn high incomes because the rest of the economy is only marginally interested in those talents. Therefore, the beginning of the case for government to redistribute income is the recognition that a market economy has no moral compass when it comes to the distribution of income.

Figure B.3 reflects the next point: there is a great deal of inequality in our world. The distribution of income in most market economies is very unequal. Half of all income goes to the 20 percent of households earning the highest incomes. The top 5 percent of households have incomes that are 25 times larger than the incomes of households in the bottom 20 percent. There is a high degree of inequality in this distribution. Wealth is even more unequally distributed than income. The lifestyles of the wealthy, whether they are beautiful, handsome, and/or talented, or just born with that proverbial silver spoon in their mouths, are favorite topics for the media. Window shopping on Rodeo Drive in Los Angeles or along Fifth Avenue in New York City quickly assures most Americans that their income is inadequate to live the Good Life. On the other hand, a few miles from either shopping mecca are neighborhoods filled with people living on much less that the average American family. Although incomes within the United States are unequal, probably unfair in some cases, and likely to remain both unequal and unfair, worldwide financial inequality and unfairness is even greater. A thoughtful observer cannot help

Group	Poorest Fifth	2nd Poorest Fifth	Middle Fifth	2nd Richest Fifth	Richest Fifth	Richest 5%
Income Limit	<$18,000	<$33,000	<$53,000	<$83,500	—	>$154,120
Average Income	$10,000	$25,500	$43,500	$69,000	$147,078	$253,000
Share of total income	3.5%	8.8%	14.8%	23.3%	49.7%	21.7%

Figure B.3 Distribution of household income in 2001.

Group	Share of Income	Average Tax Rate	Share of Taxes
Top 1%	21%	28%	37%
Top 5%	35%	24%	56%
Top 10%	46%	22%	67%
Top 25%	67%	19%	84%
Top 50%	87%	17%	96%
Bottom 50%	13%	5%	4%

Figure B.4 Distribution of Federal Income Tax in 2000.

but wonder whether or not the concept of "fairness" has ever been, or can ever be factored into the distribution of wealth and income in a free market society, where its best chance of survival should surely exist.

We next observe that income seems to be correlated with traits or characteristics that our moral sense tells us should not be allowed to affect income. Many characteristics over which an individual has no control may affect that individual's income. There are systematic relationships between income and race, gender, attractiveness, and family wealth. This basic, observable, and undeniable fact of life strikes many if not most individuals as unfair. If you were setting up an ideal world of economic justice with no knowledge of anyone's individual characteristics, what traits would you permit to influence the distribution of income? Chances are that many of the traits that currently have an effect would not appear on your list of what is fair. But such mind games quickly become complicated. Suppose you decided that inheritance from parents to children was unfair, but you discovered that such inheritance caused parents to work harder and make everyone better off. Would you allow a source of inequality that raised the incomes of everyone, but some people much more than others?

For most people the market economy carries no moral authority concerning economic justice, that the distribution of income resulting from the market economy is very unequal and, to many people, there are unfair aspects to that distribution. People who are either affected negatively by the unfairness, or who empathize with those who are stuck on the bottom rung of the ladder to success often put pressure on the government to change the distribution or do more to redress some, if not all, of the inequities. A minority of American households receives more than half of the total economic gain in this nation, and so there are often political pressures in our representative democracy to redistribute income. In fact, James Madison in Federalist 10 said these two factions would be among the most durable. "But the most common and durable source of factions has been the various and unequal distribution of property. Those who hold and those who are without property have ever formed distinct interests in society." It would seem that proposals to redistribute away from the rich to the poor have always been made. But the wealthy have always had more political clout than the poor. This is a nation governed by people who every few years must prove their right to govern others, and provide proof through expensive campaign advertising and other costly devices. Politicians usually try to position themselves as champions of the "little guy," the unwealthy, but they readily gravitate to the wealthiest voters for campaign funds, and it is naïve to imagine that sizable campaign donations don't come with promises of post-election benefits to the large contributors. The wealthiest donors often use their behind-the-scenes influence (i.e., handfuls of saved IOUs) to encourage the elected politician to stop redistributional proposals that will adversely affect the wealth of the donor.

Government Efforts toward Economic Justice

Government efforts concerning economic justice fall into two categories—equality of opportunity and equality of result. Government efforts at education and anti-discrimination laws are examples of government programs to create more equality of opportunity. Social security, Medicare, Medicaid, welfare programs, food stamps, and subsidized housing are examples of programs aimed at equality of result.

Tax policy has also been used to try to redistribute after-tax income. A tax that increases percentage of income taken in taxes as the total income increases is called a "progressive tax." A tax that takes relatively more from the poor than the rich is called a "regressive tax." Figure B.4 shows that the Internal Revenue Service's federal income tax is somewhat progressive. However, other taxes, such as social security taxes, are regressive.

Success of Government Efforts

Government efforts aimed at achieving economic justice always seem to confront a dilemma. That is, government efforts to divide GDP more fairly create inefficiencies that reduce the size of GDP or its rate of growth. One economist used the analogy of the leaky bucket. The government often tries to redistribute income from one group to another. But the government programs or tax policies used to redistribute have leaks of inefficiencies in them. So as the government dips its bucket in the "haves" pool to carry water to the "have nots," that water is spilling out and being wasted. How much water do you allow to leak out of the bucket before you decide that the whole effort of transferring from one pool to the other is useless? That question captures the basic problem government faces as it tries to redistribute. Nevertheless, the desire for someone's definition of economic justice is an attractive political goal. Whether the discussion is about health care, education, energy prices, social security, or tax changes, economic justice is part of the agenda.

Conclusion

These six weaknesses or perceived weaknesses of a market economy motivate government's efforts to intervene in the economy. Figure B.6 and the pie chart in Figure B.5 show the relationship of each of the weaknesses to the growth of government. Much of the impetus for government to solve these market problems has roots in the Progressive Period (1880 to 1920). The New Deal following the Great Depression expanded government programs in all areas except national defense, which has been a

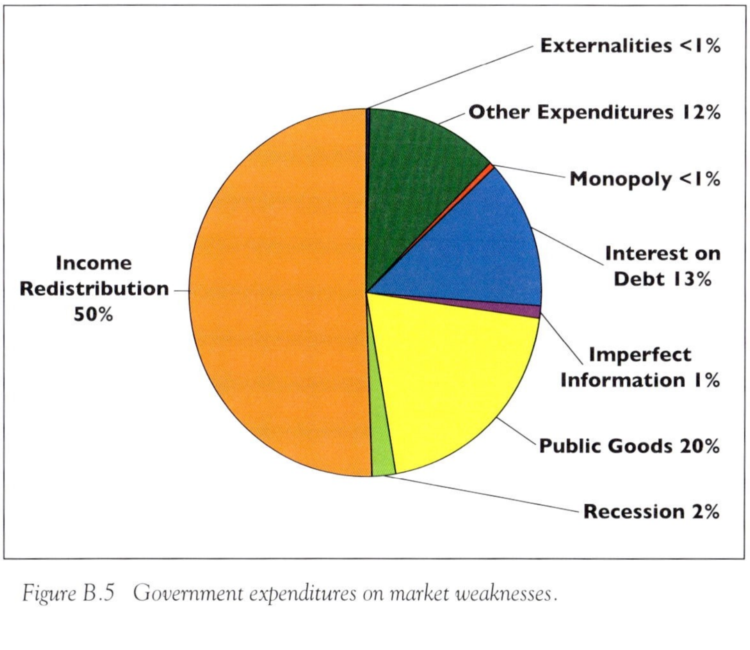

Figure B.5 Government expenditures on market weaknesses.

primary focus since World War II, continuing on into the Cold War years and beyond, in the current War on Terror. If government maintains its commitments to social security, Medicare, welfare, and education, government will continue to grow faster than the economy as a whole, causing government's share of output to increase. Whether this growth is good or bad is left for each person to decide. There is plenty of room for disagreement about the severity of the various market weaknesses and the effectiveness of government intervention to correct those weaknesses. We cannot answer whether or not the size of government is what voters want. Nonetheless, it is clear that government is a much larger part of our lives today than it was 100 or 200 years ago. We leave it to each individual to decide whether the benefits are worth the costs.

Market Weakness	Historical Examples of Government Intervention	Current Examples
MISINFORMATION	Food and Drug Administration (FDA) Prohibition	FDA, FTC, Prohibited and Restricted Goods, Securities and Exchange Commission
MONOPOLY A major focus of the progressive period	Interstate Commerce Commission (1887) Sherman Antitrust Act (1890)	Antitrust laws implemented by the Justice Department, utility regulation
EXTERNALITIES A major focus of government since the 1970s	Public Sanitation (1880s)	Public Sanitation and health, Environmental Protection agency, National Institutes of Health, National Science Foundation
PUBLIC GOODS National defense always a role of government	National Defense (1776) Lighthouses	National Defense Navigation Systems
RECESSION Major focus since the great depression and new deal	Federal Reserve Act (1913)	Unemployment insurance, monetary and fiscal policies
ECONOMIC INJUSTICE Major focus of the New Deal and The Great Society in the 1960s.	Local poor relief laws (1600s) Progressive income tax, (1913) Federal relief laws (1930s) Social Security (1935)	Social Security, Medicare, welfare, tax policy

Figure B.6 Market weaknesses and their relationship to the growth of government.

Appendix C

Founding Documents

Contents

The Declaration of Independence

IN CONGRESS, July 4, 1776.

The unanimous Declaration of the thirteen united States of America,

When in the Course of human events, it becomes necessary for one people to dissolve the political bands which have connected them with another, and to assume among the powers of the earth, the separate and equal station to which the Laws of Nature and of Nature's God entitle them, a decent respect to the opinions of mankind requires that they should declare the causes which impel them to the separation.

We hold these truths to be self-evident, that all men are created equal, that they are endowed by their Creator with certain unalienable Rights, that among these are Life, Liberty and the pursuit of Happiness.—That to secure these rights, Governments are instituted among Men, deriving their just powers from the consent of the governed, —That whenever any Form of Government becomes destructive of these ends, it is the Right of the People to alter or to abolish it, and to institute new Government, laying its foundation on such principles and organizing its powers in such form, as to them shall seem most likely to effect their Safety and Happiness. Prudence, indeed, will dictate that Governments long established should not be changed for light and transient causes; and accordingly all experience hath shewn, that mankind are more disposed to suffer, while evils are sufferable, than to right themselves by abolishing the forms to which they are accustomed. But when a long train of abuses and usurpations, pursuing invariably the same Object evinces a design to reduce them under absolute Despotism, it is their right, it is their duty, to throw off such Government, and to provide new Guards for their future security.—Such has been the patient sufferance of these Colonies; and such is now the necessity which constrains them to alter their former Systems of Government. The history of the present King of Great Britain is a history of repeated injuries and usurpations, all having in direct object the establishment of an absolute Tyranny over these States. To prove this, let Facts be submitted to a candid world.

He has refused his Assent to Laws, the most wholesome and necessary for the public good.

He has forbidden his Governors to pass Laws of immediate and pressing importance, unless suspended in their operation till his Assent should be obtained; and when so suspended, he has utterly neglected to attend to them.

He has refused to pass other Laws for the accommodation of large districts of people, unless those people would relinquish the right of Representation in the Legislature, a right inestimable to them and formidable to tyrants only.

He has called together legislative bodies at places unusual, uncomfortable, and distant from the depository of their public Records, for the sole purpose of fatiguing them into compliance with his measures.

He has dissolved Representative Houses repeatedly, for opposing with manly firmness his invasions on the rights of the people.

He has refused for a long time, after such dissolutions, to cause others to be elected;

whereby the Legislative powers, incapable of Annihilation, have returned to the People at large for their exercise; the State remaining in the mean time exposed to all the dangers of invasion from without, and convulsions within.

He has endeavoured to prevent the population of these States; for that purpose obstructing the Laws for Naturalization of Foreigners; refusing to pass others to encourage their migrations hither, and raising the conditions of new Appropriations of Lands.

He has obstructed the Administration of Justice, by refusing his Assent to Laws for establishing Judiciary powers.

He has made Judges dependent on his Will alone, for the tenure of their offices, and the amount and payment of their salaries.

He has erected a multitude of New Offices, and sent hither swarms of Officers to harrass our people, and eat out their substance.

He has kept among us, in times of peace, Standing Armies without the Consent of our legislatures.

He has affected to render the Military independent of and superior to the Civil power.

He has combined with others to subject us to a jurisdiction foreign to our constitution, and unacknowledged by our laws; giving his Assent to their Acts of pretended Legislation:

For Quartering large bodies of armed troops among us:

For protecting them, by a mock Trial, from punishment for any Murders which they should commit on the Inhabitants of these States:

For cutting off our Trade with all parts of the world:

For imposing Taxes on us without our Consent:

For depriving us in many cases, of the benefits of Trial by Jury:

For transporting us beyond Seas to be tried for pretended offences

For abolishing the free System of English Laws in a neighbouring Province, establishing therein an Arbitrary government, and enlarging its Boundaries so as to render it at once an example and fit instrument for introducing the same absolute rule into these Colonies:

For taking away our Charters, abolishing our most valuable Laws, and altering fundamentally the Forms of our Governments:

For suspending our own Legislatures, and declaring themselves invested with power to legislate for us in all cases whatsoever.

He has abdicated Government here, by declaring us out of his Protection and waging War against us.

He has plundered our seas, ravaged our Coasts, burnt our towns, and destroyed the lives of our people.

He is at this time transporting large Armies of foreign Mercenaries to compleat the works of death, desolation and tyranny, already begun with circumstances of Cruelty & perfidy scarcely paralleled in the most barbarous ages, and totally unworthy the Head of a civilized nation.

He has constrained our fellow Citizens taken Captive on the high Seas to bear Arms against their Country, to become the executioners of their friends and Brethren, or to fall themselves by their Hands.

He has excited domestic insurrections amongst us, and has endeavoured to bring on the inhabitants of our frontiers, the merciless Indian Savages, whose known rule of warfare, is an undistinguished destruction of all ages, sexes and conditions.

In every stage of these Oppressions We have Petitioned for Redress in the most humble terms: Our repeated Petitions have been answered only by repeated injury. A Prince whose character is thus marked by every act which may define a Tyrant, is unfit to be the ruler of a free people.

Nor have We been wanting in attentions to our Brittish brethren. We have warned them from time to time of attempts by their legislature to extend an unwarrantable jurisdiction over us. We have reminded them of the circumstances of our emigration and settlement here. We have appealed to their native justice and magnanimity, and we have conjured them by the ties of our common kindred to disavow these usurpations, which, would inevitably interrupt our connections and correspondence. They too have been deaf to the voice of justice and of consanguinity. We must, therefore, acquiesce in the necessity, which denounces our Separation, and hold them, as we hold the rest of mankind, Enemies in War, in Peace Friends.

We, therefore, the Representatives of the united States of America, in General Congress, Assembled, appealing to the Supreme Judge of the world for the rectitude of our intentions, do, in the Name, and by Authority of the good People of these Colonies, solemnly publish and declare, That these United Colonies are, and of Right ought to be Free and Independent States; that they are Absolved from all Allegiance to the British Crown, and that all political connection between them and the State of Great Britain, is and ought to be totally dissolved; and that as Free and Independent States, they have full Power to levy War, conclude Peace, contract Alliances, establish Commerce, and to do all other Acts and Things which Independent States may of right do. And for the support of this Declaration, with a firm reliance on the protection of divine Providence, we mutually pledge to each other our Lives, our Fortunes and our sacred Honor.

John Hancock
Button Gwinnett
Lyman Hall
George Walton
William Hooper
Joseph Hewes
John Penn
Edward Rutledge
Thomas Heyward, Jr.
Thomas Lynch, Jr.
Arthur Middleton
Samuel Chase
William Paca
Thomas Stone
Charles Carroll of Carrollton
George Wythe
Richard Henry Lee
Thomas Jefferson
Benjamin Harrison
Thomas Nelson, Jr.
Francis Lightfoot Lee
Carter Braxton
Robert Morris
Benjamin Rush
Benjamin Franklin
John Morton
George Clymer
James Smith
George Taylor
James Wilson
George Ross
Caesar Rodney
George Read
Thomas McKean
William Floyd
Philip Livingston
Francis Lewis
Lewis Morris
Richard Stockton
John Witherspoon
Francis Hopkinson
John Hart
Abraham Clark
Josiah Bartlett
William Whipple
Samuel Adams
John Adams
Robert Treat Paine
Elbridge Gerry
Stephen Hopkins
William Ellery
Roger Sherman
Samuel Huntington
William Williams
Oliver Wolcott
Matthew Thornton

THE UNITED STATES CONSTITUTION

We the People of the United States, in Order to form a more perfect Union, establish Justice, insure domestic Tranquility, provide for the common defence, promote the general Welfare, and secure the Blessings of Liberty to ourselves and our Posterity, do ordain and establish this Constitution for the United States of America.

Article. I.

Section. 1.

All legislative Powers herein granted shall be vested in a Congress of the United States, which shall consist of a Senate and House of Representatives.

Section. 2.

The House of Representatives shall be composed of Members chosen every second Year by the People of the several States, and the Electors in each State shall have the Qualifications requisite for Electors of the most numerous Branch of the State Legislature.

No Person shall be a Representative who shall not have attained to the Age of twenty five Years, and been seven Years a Citizen of the United States, and who shall not, when elected, be an Inhabitant of that State in which he shall be chosen.

Representatives and direct Taxes shall be apportioned among the several States which may be included within this Union, according to their respective Numbers, which shall be determined by adding to the whole Number of free Persons, including those bound to Service for a Term of Years, and excluding Indians not taxed, three fifths of all other Persons. The actual Enumeration shall be made within three Years after the first Meeting of the Congress of the United States, and within every subsequent Term of ten Years, in such Manner as they shall by Law direct. The Number of Representatives shall not exceed one for every thirty Thousand, but each State shall have at Least one Representative; and until such enumeration shall be made, the State of New Hampshire shall be entitled to chuse three, Massachusetts eight, Rhode-Island and Providence Plantations one, Connecticut five, New-York six, New Jersey four, Pennsylvania eight, Delaware one, Maryland six, Virginia ten, North Carolina five, South Carolina five, and Georgia three.

When vacancies happen in the Representation from any State, the Executive Authority thereof shall issue Writs of Election to fill such Vacancies.

The House of Representatives shall chuse their Speaker and other Officers; and shall have the sole Power of Impeachment.

Section. 3.

The Senate of the United States shall be composed of two Senators from each State, chosen by the Legislature thereof for six Years; and each Senator shall have one Vote.

Immediately after they shall be assembled in Consequence of the first Election, they shall be divided as equally as may be into three Classes. The Seats of the Senators of the first Class shall be vacated at the Expiration of the second Year, of the second Class at the

Expiration of the fourth Year, and of the third Class at the Expiration of the sixth Year, so that one third may be chosen every second Year; and if Vacancies happen by Resignation, or otherwise, during the Recess of the Legislature of any State, the Executive thereof may make temporary Appointments until the next Meeting of the Legislature, which shall then fill such Vacancies.

No Person shall be a Senator who shall not have attained to the Age of thirty Years, and been nine Years a Citizen of the United States, and who shall not, when elected, be an Inhabitant of that State for which he shall be chosen.

The Vice President of the United States shall be President of the Senate, but shall have no Vote, unless they be equally divided.

The Senate shall chuse their other Officers, and also a President pro tempore, in the Absence of the Vice President, or when he shall exercise the Office of President of the United States.

The Senate shall have the sole Power to try all Impeachments. When sitting for that Purpose, they shall be on Oath or Affirmation. When the President of the United States is tried, the Chief Justice shall preside: And no Person shall be convicted without the Concurrence of two thirds of the Members present.

Judgment in Cases of Impeachment shall not extend further than to removal from Office, and disqualification to hold and enjoy any Office of honor, Trust or Profit under the United States: but the Party convicted shall nevertheless be liable and subject to Indictment, Trial, Judgment and Punishment, according to Law.

Section. 4.

The Times, Places and Manner of holding Elections for Senators and Representatives, shall be prescribed in each State by the Legislature thereof; but the Congress may at any time by Law make or alter such Regulations, except as to the Places of chusing Senators.

The Congress shall assemble at least once in every Year, and such Meeting shall be on the first Monday in December, unless they shall by Law appoint a different Day.

Section. 5.

Each House shall be the Judge of the Elections, Returns and Qualifications of its own Members, and a Majority of each shall constitute a Quorum to do Business; but a smaller Number may adjourn from day to day, and may be authorized to compel the Attendance of absent Members, in such Manner, and under such Penalties as each House may provide.

Each House may determine the Rules of its Proceedings, punish its Members for disorderly Behaviour, and, with the Concurrence of two thirds, expel a Member.

Each House shall keep a Journal of its Proceedings, and from time to time publish the same, excepting such Parts as may in their Judgment require Secrecy; and the Yeas and Nays of the Members of either House on any question shall, at the Desire of one fifth of those Present, be entered on the Journal.

Neither House, during the Session of Congress, shall, without the Consent of the other, adjourn for more than three days, nor to any other Place than that in which the two Houses shall be sitting.

Section. 6.

The Senators and Representatives shall receive a Compensation for their Services, to be ascertained by Law, and paid out of the Treasury of the United States. They shall in all Cases, except Treason, Felony and Breach of the Peace, be privileged from Arrest during their Attendance at the Session of their respective Houses, and in going to and returning from the same; and for any Speech or Debate in either House, they shall not be questioned in any other Place.

No Senator or Representative shall, during the Time for which he was elected, be appointed to any civil Office under the Authority of the United States, which shall have been created, or the Emoluments whereof shall have been encreased during such time; and no Person holding any Office under the United States, shall be a Member of either House during his Continuance in Office.

Section. 7.

All Bills for raising Revenue shall originate in the House of Representatives; but the Senate may propose or concur with Amendments as on other Bills.

Every Bill which shall have passed the House of Representatives and the Senate, shall, before it become a Law, be presented to the President of the United States: If he approve he shall sign it, but if not he shall return it, with his Objections to that House in which it shall have originated, who shall enter the Objections at large on their Journal, and proceed to reconsider it. If after such Reconsideration two thirds of that House shall agree to pass the Bill, it shall be sent, together with the Objections, to the other House, by which it shall likewise be reconsidered, and if approved by two thirds of that House, it shall become a Law. But in all such Cases the Votes of both Houses shall be determined by yeas and Nays, and the Names of the Persons voting for and against the Bill shall be entered on the Journal of each House respectively. If any Bill shall not be returned by the President within ten Days (Sundays excepted) after it shall have been presented to him, the Same shall be a Law, in like Manner as if he had signed it, unless the Congress by their Adjournment prevent its Return, in which Case it shall not be a Law.

Every Order, Resolution, or Vote to which the Concurrence of the Senate and House of Representatives may be necessary (except on a question of Adjournment) shall be presented to the President of the United States; and before the Same shall take Effect, shall be approved by him, or being disapproved by him, shall be repassed by two thirds of the Senate and House of Representatives, according to the Rules and Limitations prescribed in the Case of a Bill.

Section. 8.

The Congress shall have Power To lay and collect Taxes, Duties, Imposts and Excises, to pay the Debts and provide for the common Defence and general Welfare of the United States; but all Duties, Imposts and Excises shall be uniform throughout the United States;

To borrow Money on the credit of the United States;

To regulate Commerce with foreign Nations, and among the several States, and with the Indian Tribes;

To establish an uniform Rule of Naturalization, and uniform Laws on the subject of Bankruptcies throughout the United States;

To coin Money, regulate the Value thereof, and of foreign Coin, and fix the Standard of Weights and Measures;

To provide for the Punishment of counterfeiting the Securities and current Coin of the United States;

To establish Post Offices and post Roads;

To promote the Progress of Science and useful Arts, by securing for limited Times to Authors and Inventors the exclusive Right to their respective Writings and Discoveries;

To constitute Tribunals inferior to the supreme Court;

To define and punish Piracies and Felonies committed on the high Seas, and Offences against the Law of Nations;

To declare War, grant Letters of Marque and Reprisal, and make Rules concerning Captures on Land and Water;

To raise and support Armies, but no Appropriation of Money to that Use shall be for a longer Term than two Years;

To provide and maintain a Navy;

To make Rules for the Government and Regulation of the land and naval Forces;

To provide for calling forth the Militia to execute the Laws of the Union, suppress Insurrections and repel Invasions;

To provide for organizing, arming, and disciplining, the Militia, and for governing such Part of them as may be employed in the Service of the United States, reserving to the States respectively, the Appointment of the Officers, and the Authority of training the Militia according to the discipline prescribed by Congress;

To exercise exclusive Legislation in all Cases whatsoever, over such District (not exceeding ten Miles square) as may, by Cession of particular States, and the Acceptance of Congress, become the Seat of the Government of the United States, and to exercise like Authority over all Places purchased by the Consent of the Legislature of the State in which the Same shall be, for the Erection of Forts, Magazines, Arsenals, dock-Yards, and other needful Buildings;—And

To make all Laws which shall be necessary and proper for carrying into Execution the foregoing Powers, and all other Powers vested by this Constitution in the Government of the United States, or in any Department or Officer thereof.

Section. 9.

The Migration or Importation of such Persons as any of the States now existing shall think proper to admit, shall not be prohibited by the Congress prior to the Year one thousand eight hundred and eight, but a Tax or duty may be imposed on such Importation, not exceeding ten dollars for each Person.

The Privilege of the Writ of Habeas Corpus shall not be suspended, unless when in Cases of Rebellion or Invasion the public Safety may require it.

No Bill of Attainder or ex post facto Law shall be passed.

No Capitation, or other direct, Tax shall be laid, unless in Proportion to the Census or enumeration herein before directed to be taken.

No Tax or Duty shall be laid on Articles exported from any State.

No Preference shall be given by any Regulation of Commerce or Revenue to the Ports of one State over those of another; nor shall Vessels bound to, or from, one State, be obliged to enter, clear, or pay Duties in another.

No Money shall be drawn from the Treasury, but in Consequence of Appropriations made by Law; and a regular Statement and Account of the Receipts and Expenditures of all public Money shall be published from time to time.

No Title of Nobility shall be granted by the United States: And no Person holding any

Office of Profit or Trust under them, shall, without the Consent of the Congress, accept of any present, Emolument, Office, or Title, of any kind whatever, from any King, Prince, or foreign State.

Section. 10.

No State shall enter into any Treaty, Alliance, or Confederation; grant Letters of Marque and Reprisal; coin Money; emit Bills of Credit; make any Thing but gold and silver Coin a Tender in Payment of Debts; pass any Bill of Attainder, ex post facto Law, or Law impairing the Obligation of Contracts, or grant any Title of Nobility.

No State shall, without the Consent of the Congress, lay any Imposts or Duties on Imports or Exports, except what may be absolutely necessary for executing it's inspection Laws: and the net Produce of all Duties and Imposts, laid by any State on Imports or Exports, shall be for the Use of the Treasury of the United States; and all such Laws shall be subject to the Revision and Controul of the Congress.

No State shall, without the Consent of Congress, lay any Duty of Tonnage, keep Troops, or Ships of War in time of Peace, enter into any Agreement or Compact with another State, or with a foreign Power, or engage in War, unless actually invaded, or in such imminent Danger as will not admit of delay.

Article. II.

Section. 1.

The executive Power shall be vested in a President of the United States of America. He shall hold his Office during the Term of four Years, and, together with the Vice President, chosen for the same Term, be elected, as follows:

Each State shall appoint, in such Manner as the Legislature thereof may direct, a Number of Electors, equal to the whole Number of Senators and Representatives to which the State may be entitled in the Congress: but no Senator or Representative, or Person holding an Office of Trust or Profit under the United States, shall be appointed an Elector.

The Electors shall meet in their respective States, and vote by Ballot for two Persons, of whom one at least shall not be an Inhabitant of the same State with themselves. And they shall make a List of all the Persons voted for, and of the Number of Votes for each; which List they shall sign and certify, and transmit sealed to the Seat of the Government of the United States, directed to the President of the Senate. The President of the Senate shall, in the Presence of the Senate and House of Representatives, open all the Certificates, and the Votes shall then be counted. The Person having the greatest Number of Votes shall be the President, if such Number be a Majority of the whole Number of Electors appointed; and if there be more than one who have such Majority, and have an equal Number of Votes, then the House of Representatives shall immediately chuse by Ballot one of them for President; and if no Person have a Majority, then from the five highest on the List the said House shall in like Manner chuse the President. But in chusing the President, the Votes shall be taken by States, the Representation from each State having one Vote; A quorum for this purpose shall consist of a Member or Members from two thirds of the States, and a Majority of all the States shall be necessary to a Choice. In every Case, after the Choice of the President, the Person having the greatest Number of Votes of the Electors shall be the Vice President. But if there should remain two or more who have equal Votes, the Senate shall chuse from them by Ballot the Vice President.

The Congress may determine the Time of chusing the Electors, and the Day on which they shall give their Votes; which Day shall be the same throughout the United States.

No Person except a natural born Citizen, or a Citizen of the United States, at the time of the Adoption of this Constitution, shall be eligible to the Office of President; neither shall any Person be eligible to that Office who shall not have attained to the Age of thirty five Years, and been fourteen Years a Resident within the United States.

In Case of the Removal of the President from Office, or of his Death, Resignation, or Inability to discharge the Powers and Duties of the said Office, the Same shall devolve on the Vice President, and the Congress may by Law provide for the Case of Removal, Death, Resignation or Inability, both of the President and Vice President, declaring what Officer shall then act as President, and such Officer shall act accordingly, until the Disability be removed, or a President shall be elected.

The President shall, at stated Times, receive for his Services, a Compensation, which shall neither be increased nor diminished during the Period for which he shall have been elected, and he shall not receive within that Period any other Emolument from the United States, or any of them.

Before he enter on the Execution of his Office, he shall take the following Oath or Affirmation:—"I do solemnly swear (or affirm) that I will faithfully execute the Office of President of the United States, and will to the best of my Ability, preserve, protect and defend the Constitution of the United States."

Section. 2.

The President shall be Commander in Chief of the Army and Navy of the United States, and of the Militia of the several States, when called into the actual Service of the United States; he may require the Opinion, in writing, of the principal Officer in each of the executive Departments, upon any Subject relating to the Duties of their respective Offices, and he shall have Power to grant Reprieves and Pardons for Offences against the United States, except in Cases of Impeachment.

He shall have Power, by and with the Advice and Consent of the Senate, to make Treaties, provided two thirds of the Senators present concur; and he shall nominate, and by and with the Advice and Consent of the Senate, shall appoint Ambassadors, other public Ministers and Consuls, Judges of the supreme Court, and all other Officers of the United States, whose Appointments are not herein otherwise provided for, and which shall be established by Law: but the Congress may by Law vest the Appointment of such inferior Officers, as they think proper, in the President alone, in the Courts of Law, or in the Heads of Departments.

The President shall have Power to fill up all Vacancies that may happen during the Recess of the Senate, by granting Commissions which shall expire at the End of their next Session.

Section. 3.

He shall from time to time give to the Congress Information of the State of the Union, and recommend to their Consideration such Measures as he shall judge necessary and expedient; he may, on extraordinary Occasions, convene both Houses, or either of them, and in Case of Disagreement between them, with Respect to the Time of Adjournment, he may adjourn them to such Time as he shall think proper; he shall receive Ambassadors and other public Ministers; he shall take Care that the Laws be faithfully executed, and shall Commission all the Officers of the United States.

Section. 4.

The President, Vice President and all civil Officers of the United States, shall be removed from Office on Impeachment for, and Conviction of, Treason, Bribery, or other high Crimes and Misdemeanors.

Article III.

Section. 1.

The judicial Power of the United States shall be vested in one supreme Court, and in such inferior Courts as the Congress may from time to time ordain and establish. The Judges, both of the supreme and inferior Courts, shall hold their Offices during good Behaviour, and shall, at stated Times, receive for their Services a Compensation, which shall not be diminished during their Continuance in Office.

Section. 2.

The judicial Power shall extend to all Cases, in Law and Equity, arising under this Constitution, the Laws of the United States, and Treaties made, or which shall be made, under their Authority;—to all Cases affecting Ambassadors, other public Ministers and Consuls;—to all Cases of admiralty and maritime Jurisdiction;—to Controversies to which the United States shall be a Party;—to Controversies between two or more States;— between a State and Citizens of another State;—between Citizens of different States;—between Citizens of the same State claiming Lands under Grants of different States, and between a State, or the Citizens thereof, and foreign States, Citizens or Subjects.

In all Cases affecting Ambassadors, other public Ministers and Consuls, and those in which a State shall be Party, the supreme Court shall have original Jurisdiction. In all the other Cases before mentioned, the supreme Court shall have appellate Jurisdiction, both as to Law and Fact, with such Exceptions, and under such Regulations as the Congress shall make.

The Trial of all Crimes, except in Cases of Impeachment, shall be by Jury; and such Trial shall be held in the State where the said Crimes shall have been committed; but when not committed within any State, the Trial shall be at such Place or Places as the Congress may by Law have directed.

Section. 3.

Treason against the United States, shall consist only in levying War against them, or in adhering to their Enemies, giving them Aid and Comfort. No Person shall be convicted of Treason unless on the Testimony of two Witnesses to the same overt Act, or on Confession in open Court.

The Congress shall have Power to declare the Punishment of Treason, but no Attainder of Treason shall work Corruption of Blood, or Forfeiture except during the Life of the Person attainted.

Article. IV.

Section. 1.

Full Faith and Credit shall be given in each State to the public Acts, Records, and judicial Proceedings of every other State. And the Congress may by general Laws prescribe the Manner in which such Acts, Records and Proceedings shall be proved, and the Effect thereof.

Section. 2.

The Citizens of each State shall be entitled to all Privileges and Immunities of Citizens in the several States.

A Person charged in any State with Treason, Felony, or other Crime, who shall flee from Justice, and be found in another State, shall on Demand of the executive Authority of the State from which he fled, be delivered up, to be removed to the State having Jurisdiction of the Crime.

No Person held to Service or Labour in one State, under the Laws thereof, escaping into another, shall, in Consequence of any Law or Regulation therein, be discharged from such Service or Labour, but shall be delivered up on Claim of the Party to whom such Service or Labour may be due.

Section. 3.

New States may be admitted by the Congress into this Union; but no new State shall be formed or erected within the Jurisdiction of any other State; nor any State be formed by the Junction of two or more States, or Parts of States, without the Consent of the Legislatures of the States concerned as well as of the Congress.

The Congress shall have Power to dispose of and make all needful Rules and Regulations respecting the Territory or other Property belonging to the United States; and nothing in this Constitution shall be so construed as to Prejudice any Claims of the United States, or of any particular State.

Section. 4.

The United States shall guarantee to every State in this Union a Republican Form of Government, and shall protect each of them against Invasion; and on Application of the Legislature, or of the Executive (when the Legislature cannot be convened), against domestic Violence.

Article. V.

The Congress, whenever two thirds of both Houses shall deem it necessary, shall propose Amendments to this Constitution, or, on the Application of the Legislatures of two thirds of the several States, shall call a Convention for proposing Amendments, which, in either Case, shall be valid to all Intents and Purposes, as Part of this Constitution, when ratified by the Legislatures of three fourths of the several States, or by Conventions in three fourths thereof, as the one or the other Mode of Ratification may be proposed by the Congress; Provided that no Amendment which may be made prior to the Year One thousand eight hundred and eight shall in any Manner affect the first and fourth Clauses in the Ninth Section of the first Article; and that no State, without its Consent, shall be deprived of its equal Suffrage in the Senate.

Article. VI.

All Debts contracted and Engagements entered into, before the Adoption of this Constitution, shall be as valid against the United States under this Constitution, as under the Confederation.

This Constitution, and the Laws of the United States which shall be made in Pursuance thereof; and all Treaties made, or which shall be made, under the Authority of the United States, shall be the supreme Law of the Land; and the Judges in every State shall be bound thereby, any Thing in the Constitution or Laws of any State to the Contrary notwithstanding.

The Senators and Representatives before mentioned, and the Members of the several State Legislatures, and all executive and judicial Officers, both of the United States and of the several States, shall be bound by Oath or Affirmation, to support this Constitution; but no religious Test shall ever be required as a Qualification to any Office or public Trust under the United States.

Article. VII.

The Ratification of the Conventions of nine States, shall be sufficient for the Establishment of this Constitution between the States so ratifying the Same.

Done in Convention by the Unanimous Consent of the States present the Seventeenth Day of September in the Year of our Lord one thousand seven hundred and Eighty seven and of the Independence of the United States of America the Twelfth In witness whereof We have hereunto subscribed our Names,

G°. Washington—Presidt and deputy from Virginia

Delaware

Geo: Read
Gunning Bedford jun
John Dickinson
Richard Bassett
Jaco: Broom

Maryland

James McHenry
Dan of St Thos. Jenifer
Danl. Carroll

Virginia

John Blair
James Madison Jr.

North Carolina

Wm. Blount
Richd. Dobbs Spaight
Hu Williamson

South Carolina

J. Rutledge
Charles Cotesworth Pinckney
Charles Pinckney
Pierce Butler

Georgia

William Few
Abr Baldwin

New Hampshire

John Langdon
Nicholas Gilman

Massachusetts

Nathaniel Gorham
Rufus King

Connecticut

Wm. Saml. Johnson
Roger Sherman

New York

Alexander Hamilton

New Jersey

Wil: Livingston
David Brearley
Wm. Paterson
Jona: Dayton

Pennsylvania

B Franklin
Thomas Mifflin
Robt. Morris
Geo. Clymer
Thos. FitzSimons
Jared Ingersoll
James Wilson
Gouv Morris

Attest William Jackson
Secretary

THE BILL OF RIGHTS AND AMENDMENTS TO THE CONSTITUTION

Congress of the United States
begun and held at the City of New-York, on Wednesday the fourth of March, one thousand seven hundred and eighty nine.

THE Conventions of a number of the States, having at the time of their adopting the Constitution, expressed a desire, in order to prevent misconstruction or abuse of its powers, that further declaratory and restrictive clauses should be added: And as extending the ground of public confidence in the Government, will best ensure the beneficent ends of its institution.

RESOLVED by the Senate and House of Representatives of the United States of America, in Congress assembled, two thirds of both Houses concurring, that the following Articles be proposed to the Legislatures of the several States, as amendments to the Constitution of the United States, all, or any of which Articles, when ratified by three fourths of the said Legislatures, to be valid to all intents and purposes, as part of the said Constitution; viz.

ARTICLES in addition to, and Amendment of the Constitution of the United States of America, proposed by Congress, and ratified by the Legislatures of the several States, pursuant to the fifth Article of the original Constitution.

Amendment I

Congress shall make no law respecting an establishment of religion, or prohibiting the free exercise thereof; or abridging the freedom of speech, or of the press; or the right of the people peaceably to assemble, and to petition the Government for a redress of grievances.

Amendment II

A well regulated Militia, being necessary to the security of a free State, the right of the people to keep and bear Arms, shall not be infringed.

Amendment III

No Soldier shall, in time of peace be quartered in any house, without the consent of the Owner, nor in time of war, but in a manner to be prescribed by law.

Amendment IV

The right of the people to be secure in their persons, houses, papers, and effects, against unreasonable searches and seizures, shall not be violated, and no Warrants shall issue, but upon probable cause, supported by Oath or affirmation, and particularly describing the place to be searched, and the persons or things to be seized.

Amendment V

No person shall be held to answer for a capital, or otherwise infamous crime, unless on a presentment or indictment of a Grand Jury, except in cases arising in the land or naval forces, or in the Militia, when in actual service in time of War or public danger; nor shall any person be subject for the same offence to be twice put in jeopardy of life or limb; nor shall be compelled in any criminal case to be a witness against himself, nor be deprived of life, liberty, or property, without due process of law; nor shall private property be taken for public use, without just compensation.

Amendment VI

In all criminal prosecutions, the accused shall enjoy the right to a speedy and public trial, by an impartial jury of the State and district wherein the crime shall have been committed, which district shall have been previously ascertained by law, and to be informed of the nature and cause of the accusation; to be confronted with the witnesses against him; to have compulsory process for obtaining witnesses in his favor, and to have the Assistance of Counsel for his defence.

Amendment VII

In Suits at common law, where the value in controversy shall exceed twenty dollars, the right of trial by jury shall be preserved, and no fact tried by a jury, shall be otherwise re-examined in any Court of the United States, than according to the rules of the common law.

Amendment VIII

Excessive bail shall not be required, nor excessive fines imposed, nor cruel and unusual punishments inflicted.

Amendment IX

The enumeration in the Constitution, of certain rights, shall not be construed to deny or disparage others retained by the people.

Amendment X

The powers not delegated to the United States by the Constitution, nor prohibited by it to the States, are reserved to the States respectively, or to the people.

Amendment XI

Passed by Congress March 4, 1794. Ratified February 7, 1795.

[Note: Article III, section 2, of the Constitution was modified by amendment 11.]

The Judicial power of the United States shall not be construed to extend to any suit in law or equity, commenced or prosecuted against one of the United States by Citizens of another State, or by Citizens or Subjects of any Foreign State.

Amendment XII

Passed by Congress December 9, 1803. Ratified June 15, 1804.

[Note: A portion of Article II, section 1 of the Constitution was superseded by the 12th amendment.]

The Electors shall meet in their respective states and vote by ballot for President and

Vice-President, one of whom, at least, shall not be an inhabitant of the same state with themselves; they shall name in their ballots the person voted for as President, and in distinct ballots the person voted for as Vice-President, and they shall make distinct lists of all persons voted for as President, and of all persons voted for as Vice-President, and of the number of votes for each, which lists they shall sign and certify, and transmit sealed to the seat of the government of the United States, directed to the President of the Senate; — the President of the Senate shall, in the presence of the Senate and House of Representatives, open all the certificates and the votes shall then be counted; — The person having the greatest number of votes for President, shall be the President, if such number be a majority of the whole number of Electors appointed; and if no person have such majority, then from the persons having the highest numbers not exceeding three on the list of those voted for as President, the House of Representatives shall choose immediately, by ballot, the President. But in choosing the President, the votes shall be taken by states, the representation from each state having one vote; a quorum for this purpose shall consist of a member or members from two-thirds of the states, and a majority of all the states shall be necessary to a choice. [And if the House of Representatives shall not choose a President whenever the right of choice shall devolve upon them, before the fourth day of March next following, then the Vice-President shall act as President, as in case of the death or other constitutional disability of the President. —]* The person having the greatest number of votes as Vice-President, shall be the Vice-President, if such number be a majority of the whole number of Electors appointed, and if no person have a majority, then from the two highest numbers on the list, the Senate shall choose the Vice-President; a quorum for the purpose shall consist of two-thirds of the whole number of Senators, and a majority of the whole number shall be necessary to a choice. But no person constitutionally ineligible to the office of President shall be eligible to that of Vice-President of the United States.

*Superseded by section 3 of the 20th amendment.

Amendment XIII

Passed by Congress January 31, 1865. Ratified December 6, 1865.

[Note: A portion of Article IV, section 2, of the Constitution was superseded by the 13th amendment.]

Section 1.

Neither slavery nor involuntary servitude, except as a punishment for crime whereof the party shall have been duly convicted, shall exist within the United States, or any place subject to their jurisdiction.

Section 2.

Congress shall have power to enforce this article by appropriate legislation.

Amendment XIV

Passed by Congress June 13, 1866. Ratified July 9, 1868.

[Note: Article I, section 2, of the Constitution was modified by section 2 of the 14th amendment.]

Section 1.

All persons born or naturalized in the United States, and subject to the jurisdiction thereof, are citizens of the United States and of the State wherein they reside. No State shall make or enforce any law which shall abridge the privileges or immunities of citizens of the United States; nor shall any State deprive any person of life, liberty, or property, without due process of law; nor deny to any person within its jurisdiction the equal protection of the laws.

Section 2.

Representatives shall be apportioned among the several States according to their respective numbers, counting the whole number of persons in each State, excluding Indians not taxed. But when the right to vote at any election for the choice of electors for President and Vice-President of the United States, Representatives in Congress, the Executive and Judicial officers of a State, or the members of the Legislature thereof, is denied to any of the male inhabitants of such State, being twenty-one years of age,* and citizens of the United States, or in any way abridged, except for participation in rebellion, or other crime, the basis of representation therein shall be reduced in the proportion which the number of such male citizens shall bear to the whole number of male citizens twenty-one years of age in such State.

Section 3.

No person shall be a Senator or Representative in Congress, or elector of President and Vice-President, or hold any office, civil or military, under the United States, or under any State, who, having previously taken an oath, as a member of Congress, or as an officer of the United States, or as a member of any State legislature, or as an executive or judicial officer of any State, to support the Constitution of the United States, shall have engaged in insurrection or rebellion against the same, or given aid or comfort to the enemies thereof. But Congress may by a vote of two-thirds of each House, remove such disability.

Section 4.

The validity of the public debt of the United States, authorized by law, including debts incurred for payment of pensions and bounties for services in suppressing insurrection or rebellion, shall not be questioned. But neither the United States nor any State shall assume or pay any debt or obligation incurred in aid of insurrection or rebellion against the United States, or any claim for the loss or emancipation of any slave; but all such debts, obligations and claims shall be held illegal and void.

Section 5.

The Congress shall have the power to enforce, by appropriate legislation, the provisions of this article.

*Changed by section 1 of the 26th amendment.

Amendment XV

Passed by Congress February 26, 1869. Ratified February 3, 1870.

Section 1.

The right of citizens of the United States to vote shall not be denied or abridged by the United States or by any State on account of race, color, or previous condition of servitude—

Section 2.

The Congress shall have the power to enforce this article by appropriate legislation.

Amendment XVI

Passed by Congress July 2, 1909. Ratified February 3, 1913.

[Note: Article I, section 9, of the Constitution was modified by amendment 16.]

The Congress shall have power to lay and collect taxes on incomes, from whatever source derived, without apportionment among the several States, and without regard to any census or enumeration.

Amendment XVII

Passed by Congress May 13, 1912. Ratified April 8, 1913.

[Note: Article I, section 3, of the Constitution was modified by the 17th amendment.]

The Senate of the United States shall be composed of two Senators from each State, elected by the people thereof, for six years; and each Senator shall have one vote. The electors in each State shall have the qualifications requisite for electors of the most numerous branch of the State legislatures.

When vacancies happen in the representation of any State in the Senate, the executive authority of such State shall issue writs of election to fill such vacancies: Provided, That the legislature of any State may empower the executive thereof to make temporary appointments until the people fill the vacancies by election as the legislature may direct.

This amendment shall not be so construed as to affect the election or term of any Senator chosen before it becomes valid as part of the Constitution.

Amendment XVIII

Passed by Congress December 18, 1917. Ratified January 16, 1919. Repealed by amendment 21.

Section 1.

After one year from the ratification of this article the manufacture, sale, or transportation of intoxicating liquors within, the importation thereof into, or the exportation thereof from the United States and all territory subject to the jurisdiction thereof for beverage purposes is hereby prohibited.

Section 2.

The Congress and the several States shall have concurrent power to enforce this article by appropriate legislation.

Section 3.

This article shall be inoperative unless it shall have been ratified as an amendment to the Constitution by the legislatures of the several States, as provided in the Constitution, within seven years from the date of the submission hereof to the States by the Congress.

Amendment XIX

Passed by Congress June 4, 1919. Ratified August 18, 1920.

The right of citizens of the United States to vote shall not be denied or abridged by the United States or by any State on account of sex.

Congress shall have power to enforce this article by appropriate legislation.

Amendment XX

Passed by Congress March 2, 1932. Ratified January 23, 1933.

[Note: Article I, section 4, of the Constitution was modified by section 2 of this amendment. In addition, a portion of the 12th amendment was superseded by section 3.]

Section 1.

The terms of the President and the Vice President shall end at noon on the 20th day of January, and the terms of Senators and Representatives at noon on the 3d day of January, of the years in which such terms would have ended if this article had not been ratified; and the terms of their successors shall then begin.

Section 2.

The Congress shall assemble at least once in every year, and such meeting shall begin at noon on the 3d day of January, unless they shall by law appoint a different day.

Section 3.

If, at the time fixed for the beginning of the term of the President, the President elect shall have died, the Vice President elect shall become President. If a President shall not have been chosen before the time fixed for the beginning of his term, or if the President elect shall have failed to qualify, then the Vice President elect shall act as President until a President shall have qualified; and the Congress may by law provide for the case wherein neither a President elect nor a Vice President shall have qualified, declaring who shall then act as President, or the manner in which one who is to act shall be selected, and such person shall act accordingly until a President or Vice President shall have qualified.

Section 4.

The Congress may by law provide for the case of the death of any of the persons from whom the House of Representatives may choose a President whenever the right of choice shall have devolved upon them, and for the case of the death of any of the persons from whom the Senate may choose a Vice President whenever the right of choice shall have devolved upon them.

Section 5.

Sections 1 and 2 shall take effect on the 15th day of October following the ratification of this article.

Section 6.

This article shall be inoperative unless it shall have been ratified as an amendment to the Constitution by the legislatures of three-fourths of the several States within seven years from the date of its submission.

Amendment XXI

Passed by Congress February 20, 1933. Ratified December 5, 1933.

Section 1.

The eighteenth article of amendment to the Constitution of the United States is hereby repealed.

Section 2.

The transportation or importation into any State, Territory, or Possession of the United States for delivery or use therein of intoxicating liquors, in violation of the laws thereof, is hereby prohibited.

Section 3.

This article shall be inoperative unless it shall have been ratified as an amendment to the Constitution by conventions in the several States, as provided in the Constitution, within seven years from the date of the submission hereof to the States by the Congress.

Amendment XXII

Passed by Congress March 21, 1947. Ratified February 27, 1951.

Section 1.

No person shall be elected to the office of the President more than twice, and no person who has held the office of President, or acted as President, for more than two years of a term to which some other person was elected President shall be elected to the office of President more than once. But this Article shall not apply to any person holding the office of President when this Article was proposed by Congress, and shall not prevent any person who may be holding the office of President, or acting as President, during the term within which this Article becomes operative from holding the office of President or acting as President during the remainder of such term.

Section 2.

This article shall be inoperative unless it shall have been ratified as an amendment to the Constitution by the legislatures of three-fourths of the several States within seven years from the date of its submission to the States by the Congress.

Amendment XXIII

Passed by Congress June 16, 1960. Ratified March 29, 1961.

Section 1.

The District constituting the seat of Government of the United States shall appoint in such manner as Congress may direct:

A number of electors of President and Vice President equal to the whole number of Senators and Representatives in Congress to which the District would be entitled if it were a State, but in no event more than the least populous State; they shall be in addition to those appointed by the States, but they shall be considered, for the purposes of the election of President and Vice President, to be electors appointed by a State; and they shall meet in the District and perform such duties as provided by the twelfth article of amendment.

Section 2.

The Congress shall have power to enforce this article by appropriate legislation.

Amendment XXIV

Passed by Congress August 27, 1962. Ratified January 23, 1964.

Section 1.

The right of citizens of the United States to vote in any primary or other election for President or Vice President, for electors for President or Vice President, or for Senator or Representative in Congress, shall not be denied or abridged by the United States or any State by reason of failure to pay poll tax or other tax.

Section 2.

The Congress shall have power to enforce this article by appropriate legislation.

Amendment XXV

Passed by Congress July 6, 1965. Ratified February 10, 1967.

[Note: Article II, section 1, of the Constitution was affected by the 25th amendment.]

Section 1.

In case of the removal of the President from office or of his death or resignation, the Vice President shall become President.

Section 2.

Whenever there is a vacancy in the office of the Vice President, the President shall nominate a Vice President who shall take office upon confirmation by a majority vote of both Houses of Congress.

Section 3.

Whenever the President transmits to the President pro tempore of the Senate and the Speaker of the House of Representatives his written declaration that he is unable to discharge the powers and duties of his office, and until he transmits to them a written declaration to the contrary, such powers and duties shall be discharged by the Vice President as Acting President.

Section 4.

Whenever the Vice President and a majority of either the principal officers of the executive departments or of such other body as Congress may by law provide, transmit to the President pro tempore of the Senate and the Speaker of the House of Representatives their written declaration that the President is unable to discharge the powers and duties of his office, the Vice President shall immediately assume the powers and duties of the office as Acting President.

Thereafter, when the President transmits to the President pro tempore of the Senate and the Speaker of the House of Representatives his written declaration that no inability exists, he shall resume the powers and duties of his office unless the Vice President and a majority of either the principal officers of the executive department or of such other body as Congress may by law provide, transmit within four days to the President pro tempore of the Senate and the Speaker of the House of Representatives their written declaration that the President is unable to discharge the powers and duties of his office. Thereupon Congress shall decide the issue, assembling within forty-eight hours for that purpose if not in session. If the Congress, within twenty-one days after receipt of the latter written declaration, or, if Congress is not in session, within twenty-one days after Congress is required to assemble, determines by two-thirds vote of both Houses that the President is unable to discharge the powers and duties of his office, the Vice President shall continue to discharge the same as Acting President; otherwise, the President shall resume the powers and duties of his office.

Amendment XXVI

Passed by Congress March 23, 1971. Ratified July 1, 1971.

[Note: Amendment 14, section 2, of the Constitution was modified by section 1 of the 26th amendment.]

Section 1.

The right of citizens of the United States, who are eighteen years of age or older, to vote shall not be denied or abridged by the United States or by any State on account of age.

Section 2.

The Congress shall have power to enforce this article by appropriate legislation.

Amendment XXVII

Originally proposed Sept. 25, 1789. Ratified May 7, 1992.

No law, varying the compensation for the services of the Senators and Representatives, shall take effect, until an election of representatives shall have intervened.

THE FEDERALIST NO. 10

The Utility of the Union as a Safeguard Against Domestic Faction and Insurrection (continued)

To the People of the State of New York:

AMONG the numerous advantages promised by a well constructed Union, none deserves to be more accurately developed than its tendency to break and control the violence of faction. The friend of popular governments never finds himself so much alarmed for their character and fate, as when he contemplates their propensity to this dangerous vice. He will not fail, therefore, to set a due value on any plan which, without violating the principles to which he is attached, provides a proper cure for it. The instability, injustice, and confusion introduced into the public councils, have, in truth, been the mortal diseases under which popular governments have everywhere perished; as they continue to be the favorite and fruitful topics from which the adversaries to liberty derive their most specious declamations. The valuable improvements made by the American constitutions on the popular models, both ancient and modern, cannot certainly be too much admired; but it would be an unwarrantable partiality, to contend that they have as effectually obviated the danger on this side, as was wished and expected. Complaints are everywhere heard from our most considerate and virtuous citizens, equally the friends of public and private faith, and of public and personal liberty, that our governments are too unstable, that the public good is disregarded in the conflicts of rival parties, and that measures are too often decided, not according to the rules of justice and the rights of the minor party, but by the superior force of an interested and overbearing majority. However anxiously we may wish that these complaints had no foundation, the evidence, of known facts will not permit us to deny that they are in some degree true. It will be found, indeed, on a candid review of our situation, that some of the distresses under which we labor have been erroneously charged on the operation of our governments; but it will be found, at the same time, that other causes will not alone account for many of our heaviest misfortunes; and, particularly, for that prevailing and increasing distrust of public engagements, and alarm for private rights, which are echoed from one end of the continent to the other. These must be chiefly, if not wholly, effects of the unsteadiness and injustice with which a factious spirit has tainted our public administrations.

By a faction, I understand a number of citizens, whether amounting to a majority or a minority of the whole, who are united and actuated by some common impulse of passion, or of interest, adversed to the rights of other citizens, or to the permanent and aggregate interests of the community.

There are two methods of curing the mischiefs of faction: the one, by removing its causes; the other, by controlling its effects.

There are again two methods of removing the causes of faction: the one, by destroying the liberty which is essential to its existence; the other, by giving to every citizen the same opinions, the same passions, and the same interests.

It could never be more truly said than of the first remedy, that it was worse than the disease. Liberty is to faction what air is to fire, an aliment without which it instantly expires. But it could not be less folly to abolish liberty, which is essential to political life, because it nourishes faction, than it would be to wish the annihilation of air, which is essential to animal life, because it imparts to fire its destructive agency.

The second expedient is as impracticable as the first would be unwise. As long as the reason of man continues fallible, and he is at liberty to exercise it, different opinions will be formed. As long as the connection subsists between his reason and his self-love, his opinions and his passions will have a reciprocal influence on each other; and the former will be objects to which the latter will attach themselves. The diversity in the faculties of men, from which the rights of property originate, is not less an insuperable obstacle to a uniformity of interests. The protection of these faculties is the first object of government. From the protection of different and unequal faculties of acquiring property, the possession of different degrees and kinds of property immediately results; and from the influence of these on the sentiments and views of the respective proprietors, ensues a division of the society into different interests and parties.

The latent causes of faction are thus sown in the nature of man; and we see them everywhere brought into different degrees of activity, according to the different circumstances of civil society. A zeal for different opinions concerning religion, concerning government, and many other points, as well of speculation as of practice; an attachment to different leaders ambitiously contending for pre-eminence and power; or to persons of other descriptions whose fortunes have been interesting to the human passions, have, in turn, divided mankind into parties, inflamed them with mutual animosity, and rendered them much more disposed to vex and oppress each other than to co-operate for their common good. So strong is this propensity of mankind to fall into mutual animosities, that where no substantial occasion presents itself, the most frivolous and fanciful distinctions have been sufficient to kindle their unfriendly passions and excite their most violent conflicts. But the most common and durable source of factions has been the various and unequal distribution of property. Those who hold and those who are without property have ever formed distinct interests in society. Those who are creditors, and those who are debtors, fall under a like discrimination. A landed interest, a manufacturing interest, a mercantile interest, a moneyed interest, with many lesser interests, grow up of necessity in civilized nations, and divide them into different classes, actuated by different sentiments and views. The regulation of these various and interfering interests forms the principal task of modern legislation, and involves the spirit of party and faction in the necessary and ordinary operations of the government.

No man is allowed to be a judge in his own cause, because his interest would certainly bias his judgment, and, not improbably, corrupt his integrity. With equal, nay with greater reason, a body of men are unfit to be both judges and parties at the same time; yet what are many of the most important acts of legislation, but so many judicial determinations, not indeed concerning the rights of single persons, but concerning the rights of large bodies of citizens? And what are the different classes of legislators but advocates and parties to the causes which they determine? Is a law proposed concerning private debts? It is a question to which the creditors are parties on one side and the debtors on the other. Justice ought to hold the balance between them. Yet the parties are, and must be, themselves the judges; and the most numerous party, or, in other words, the most powerful faction must be expected to prevail. Shall domestic manufactures be encouraged, and in what degree, by restrictions on foreign manufactures? are questions which would be differently decided by the landed and the manufacturing classes, and probably by neither with a sole regard to justice and the public good. The apportionment of taxes on the various descriptions of property is an act which seems to require the most exact impartiality; yet there is, perhaps, no legislative act in which greater opportunity and temptation are given to a predominant party to trample on the rules of justice. Every shilling with which they overburden the inferior number, is a shilling saved to their own pockets.

It is in vain to say that enlightened statesmen will be able to adjust these clashing interests, and render them all subservient to the public good. Enlightened statesmen will not always be at the helm. Nor, in many cases, can such an adjustment be made at all without taking into view indirect and remote considerations, which will rarely prevail over the immediate interest which one party may find in disregarding the rights of another or the good of the whole.

The inference to which we are brought is, that the causes of faction cannot be removed, and that relief is only to be sought in the means of controlling its effects.

If a faction consists of less than a majority, relief is supplied by the republican principle, which enables the majority to defeat its sinister views by regular vote. It may clog the administration, it may convulse the society; but it will be unable to execute and mask its violence under the forms of the Constitution. When a majority is included in a faction, the form of popular government, on the other hand, enables it to sacrifice to its ruling passion or interest both the public good and the rights of other citizens. To secure the public good and private rights against the danger of such a faction, and at the same time to preserve the spirit and the form of popular government, is then the great object to which our inquiries are directed. Let me add that it is the great desideratum by which this form of government can be rescued from the opprobrium under which it has so long labored, and be recommended to the esteem and adoption of mankind.

By what means is this object attainable? Evidently by one of two only. Either the existence of the same passion or interest in a majority at the same time must be prevented, or the majority, having such coexistent passion or interest, must be rendered, by their number and local situation, unable to concert and carry into effect schemes of oppression. If the impulse and the opportunity be suffered to coincide, we well know that neither moral nor religious motives can be relied on as an adequate control. They are not found to be such on the injustice and violence of individuals, and lose their efficacy in proportion to the number combined together, that is, in proportion as their efficacy becomes needful.

From this view of the subject it may be concluded that a pure democracy, by which I mean a society consisting of a small number of citizens, who assemble and administer the government in person, can admit of no cure for the mischiefs of faction. A common passion or interest will, in almost every case, be felt by a majority of the whole; a communication and concert result from the form of government itself; and there is nothing to check the inducements to sacrifice the weaker party or an obnoxious individual. Hence it is that such democracies have ever been spectacles of turbulence and contention; have ever been found incompatible with personal security or the rights of property; and have in general been as short in their lives as they have been violent in their deaths. Theoretic politicians, who have patronized this species of government, have erroneously supposed that by reducing mankind to a perfect equality in their political rights, they would, at the same time, be perfectly equalized and assimilated in their possessions, their opinions, and their passions.

A republic, by which I mean a government in which the scheme of representation takes place, opens a different prospect, and promises the cure for which we are seeking. Let us examine the points in which it varies from pure democracy, and we shall comprehend both the nature of the cure and the efficacy which it must derive from the Union.

The two great points of difference between a democracy and a republic are: first, the delegation of the government, in the latter, to a small number of citizens elected by the rest; secondly, the greater number of citizens, and greater sphere of country, over which the latter may be extended.

The effect of the first difference is, on the one hand, to refine and enlarge the public views, by passing them through the medium of a chosen body of citizens, whose wisdom may best discern the true interest of their country, and whose patriotism and love of justice will be least likely to sacrifice it to temporary or partial considerations. Under such a regulation, it may well happen that the public voice, pronounced by the representatives of the people, will be more consonant to the public good than if pronounced by the people themselves, convened for the purpose. On the other hand, the effect may be inverted. Men of factious tempers, of local prejudices, or of sinister designs, may, by intrigue, by corruption, or by other means, first obtain the suffrages, and then betray the interests, of the people. The question resulting is, whether small or extensive republics are more favorable to the election of proper guardians of the public weal; and it is clearly decided in favor of the latter by two obvious considerations:

In the first place, it is to be remarked that, however small the republic may be, the representatives must be raised to a certain number, in order to guard against the cabals of a few; and that, however large it may be, they must be limited to a certain number, in order to guard against the confusion of a multitude. Hence, the number of representatives in the two cases not being in proportion to that of the two constituents, and being proportionally greater in the small republic, it follows that, if the proportion of fit characters be not less in the large than in the small republic, the former will present a greater option, and consequently a greater probability of a fit choice.

In the next place, as each representative will be chosen by a greater number of citizens in the large than in the small republic, it will be more difficult for unworthy candidates to practice with success the vicious arts by which elections are too often carried; and the suffrages of the people being more free, will be more likely to centre in men who possess the most attractive merit and the most diffusive and established characters.

It must be confessed that in this, as in most other cases, there is a mean, on both sides of which inconveniences will be found to lie. By enlarging too much the number of electors, you render the representatives too little acquainted with all their local circumstances and lesser interests; as by reducing it too much, you render him unduly attached to these, and too little fit to comprehend and pursue great and national objects. The federal Constitution forms a happy combination in this respect; the great and aggregate interests being referred to the national, the local and particular to the State legislatures.

The other point of difference is, the greater number of citizens and extent of territory which may be brought within the compass of republican than of democratic government; and it is this circumstance principally which renders factious combinations less to be dreaded in the former than in the latter. The smaller the society, the fewer probably will be the distinct parties and interests composing it; the fewer the distinct parties and interests, the more frequently will a majority be found of the same party; and the smaller the number of individuals composing a majority, and the smaller the compass within which they are placed, the more easily will they concert and execute their plans of oppression. Extend the sphere, and you take in a greater variety of parties and interests; you make it less probable that a majority of the whole will have a common motive to invade the rights of other citizens; or if such a common motive exists, it will be more difficult for all who feel it to discover their own strength, and to act in unison with each other. Besides other impediments, it may be remarked that, where there is a consciousness of unjust or dishonorable purposes, communication is always checked by distrust in proportion to the number whose concurrence is necessary.

Hence, it clearly appears, that the same advantage which a republic has over a democracy, in controlling the effects of faction, is enjoyed by a large over a small republic, -- is enjoyed by the Union over the States composing it. Does the advantage consist in the substitution of representatives whose enlightened views and virtuous sentiments render them superior to local prejudices and schemes of injustice? It will not be denied that the representation of the Union will be most likely to possess these requisite endowments. Does it consist in the greater security afforded by a greater variety of parties, against the event of any one party being able to outnumber and oppress the rest? In an equal degree does the increased variety of parties comprised within the Union, increase this security. Does it, in fine, consist in the greater obstacles opposed to the concert and accomplishment of the secret wishes of an unjust and interested majority? Here, again, the extent of the Union gives it the most palpable advantage.

The influence of factious leaders may kindle a flame within their particular States, but will be unable to spread a general conflagration through the other States. A religious sect may degenerate into a political faction in a part of the Confederacy; but the variety of sects dispersed over the entire face of it must secure the national councils against any danger from that source. A rage for paper money, for an abolition of debts, for an equal division of property, or for any other improper or wicked project, will be less apt to pervade the whole body of the Union than a particular member of it; in the same proportion as such a malady is more likely to taint a particular county or district, than an entire State.

In the extent and proper structure of the Union, therefore, we behold a republican remedy for the diseases most incident to republican government. And according to the degree of pleasure and pride we feel in being republicans, ought to be our zeal in cherishing the spirit and supporting the character of Federalists.

Publius
[James Madison]
Daily Advertiser
Thursday, November 22, 1787

THE FEDERALIST No. 51

The Structure of the Government Must Furnish the Proper Checks and Balances Between the Different Departments

To the People of the State of New York:

TO WHAT expedient, then, shall we finally resort, for maintaining in practice the necessary partition of power among the several departments, as laid down in the Constitution? The only answer that can be given is, that as all these exterior provisions are found to be inadequate, the defect must be supplied, by so contriving the interior structure of the government as that its several constituent parts may, by their mutual relations, be the means of keeping each other in their proper places. Without presuming to undertake a full development of this important idea, I will hazard a few general observations, which may perhaps place it in a clearer light, and enable us to form a more correct judgment of the principles and structure of the government planned by the convention.

In order to lay a due foundation for that separate and distinct exercise of the different powers of government, which to a certain extent is admitted on all hands to be essential to the preservation of liberty, it is evident that each department should have a will of its own; and consequently should be so constituted that the members of each should have as little agency as possible in the appointment of the members of the others. Were this principle rigorously adhered to, it would require that all the appointments for the supreme executive, legislative, and judiciary magistracies should be drawn from the same fountain of authority, the people, through channels having no communication whatever with one another. Perhaps such a plan of constructing the several departments would be less difficult in practice than it may in contemplation appear. Some difficulties, however, and some additional expense would attend the execution of it. Some deviations, therefore, from the principle must be admitted. In the constitution of the judiciary department in particular, it might be inexpedient to insist rigorously on the principle: first, because peculiar qualifications being essential in the members, the primary consideration ought to be to select that mode of choice which best secures these qualifications; secondly, because the permanent tenure by which the appointments are held in that department, must soon destroy all sense of dependence on the authority conferring them.

It is equally evident, that the members of each department should be as little dependent as possible on those of the others, for the emoluments annexed to their offices. Were the executive magistrate, or the judges, not independent of the legislature in this particular, their independence in every other would be merely nominal.

But the great security against a gradual concentration of the several powers in the same department, consists in giving to those who administer each department the necessary constitutional means and personal motives to resist encroachments of the others. The provision for defense must in this, as in all other cases, be made commensurate to the danger of attack.

Ambition must be made to counteract ambition. The interest of the man must be connected with the constitutional rights of the place. It may be a reflection on human nature, that such devices should be necessary to control the abuses of government. But what is government itself, but the greatest of all reflections on human nature? If men were angels, no government would be necessary. If angels were to govern men, neither external nor internal controls on government would be necessary. In framing a government which is to be administered by men over men, the great difficulty lies in this: you must first enable the government to control the governed; and in the next place oblige it to control itself. A dependence on the people is, no doubt, the primary control on the government; but experience has taught mankind the necessity of auxiliary precautions.

This policy of supplying, by opposite and rival interests, the defect of better motives, might be traced through the whole system of human affairs, private as well as public. We see it particularly displayed in all the subordinate distributions of power, where the constant aim is to divide and arrange the several offices in such a manner as that each may be a check on the other -- that the private interest of every individual may be a sentinel over the public rights. These inventions of prudence cannot be less requisite in the distribution of the supreme powers of the State.

But it is not possible to give to each department an equal power of self-defense. In republican government, the legislative authority necessarily predominates. The remedy for this inconveniency is to divide the legislature into different branches; and to render them, by different modes of election and different principles of action, as little connected with each other as the nature of their common functions and their common dependence on the society will admit. It may even be necessary to guard against dangerous encroachments by still further precautions. As the weight of the legislative authority requires that it should be thus divided, the weakness of the executive may require, on the other hand, that it should be fortified. An absolute negative on the legislature appears, at first view, to be the natural defense with which the executive magistrate should be armed. But perhaps it would be neither altogether safe nor alone sufficient. On ordinary occasions it might not be exerted with the requisite firmness, and on extraordinary occasions it might be perfidiously abused. May not this defect of an absolute negative be supplied by some qualified connection between this weaker department and the weaker branch of the stronger department, by which the latter may be led to support the constitutional rights of the former, without being too much detached from the rights of its own department?

If the principles on which these observations are founded be just, as I persuade myself they are, and they be applied as a criterion to the several State constitutions, and to the federal Constitution it will be found that if the latter does not perfectly correspond with them, the former are infinitely less able to bear such a test.

There are, moreover, two considerations particularly applicable to the federal system of America, which place that system in a very interesting point of view.

First. In a single republic, all the power surrendered by the people is submitted to the administration of a single government; and the usurpations are guarded against by a division of the government into distinct and separate departments. In the compound republic of America, the power surrendered by the people is first divided between two distinct governments, and then the portion allotted to each subdivided among distinct and separate departments. Hence a double security arises to the rights of the people. The different governments will control each other, at the same time that each will be controlled by itself.

Second. It is of great importance in a republic not only to guard the society against the oppression of its rulers, but to guard one part of the society against the injustice of the other part. Different interests necessarily exist in different classes of citizens. If a majority be united by a common interest, the rights of the minority will be insecure. There are but two methods of providing against this evil: the one by creating a will in the community independent of the majority -- that is, of the society itself; the other, by comprehending in the society so many separate descriptions of citizens as will render an unjust combination of a majority of the whole very improbable, if not impracticable. The first method prevails in all governments possessing an hereditary or self-appointed authority. This, at best, is but a precarious security; because a power independent of the society may as well espouse the unjust views of the major, as the rightful interests of the minor party, and may possibly be turned against both parties. The second method will be exemplified in the federal republic of the United States. Whilst all authority in it will be derived from and dependent on the society, the society itself will be broken into so many parts, interests, and classes of citizens, that the rights of individuals, or of the minority, will be in little danger from interested combinations of the majority. In a free government the security for civil rights must be the same as that for religious rights. It consists in the one case in the multiplicity of interests, and in the other in the multiplicity of sects. The degree of security in both cases will depend on the number of interests and sects; and this may be presumed to depend on the extent of country and number of people comprehended under the same government. This view of the subject must particularly recommend a proper federal system to all the sincere and considerate friends of republican government, since it shows that in exact proportion as the territory of the Union may be formed into more circumscribed Confederacies, or States oppressive combinations of a majority will be facilitated: the best security, under the republican forms, for the rights of every class of citizens, will be diminished: and consequently the stability and independence of some member of the government, the only other security, must be proportionately increased. Justice is the end of government. It is the end of civil society. It ever has been and ever will be pursued until it be obtained, or until liberty be lost in the pursuit. In a society under the forms of which the stronger faction can readily unite and oppress the weaker, anarchy may as truly be said to reign as in a state of nature, where the weaker individual is not secured against the violence of the stronger; and as, in the latter state, even the stronger individuals are prompted, by the uncertainty of their condition, to submit to a government which may protect the weak as well as themselves; so, in the former state, will the more powerful factions or parties be gradnally induced, by a like motive, to wish for a government which will protect all parties, the weaker as well as the more powerful. It can be little doubted that if the State of Rhode Island was separated from the Confederacy and left to itself, the insecurity of rights under the popular form of government within such narrow limits would be displayed by such reiterated oppressions of factious majorities that some power altogether independent of the people would soon be called for by the voice of the very factions whose misrule had proved the necessity of it. In the extended republic of the United States, and among the great variety of interests, parties, and sects which it embraces, a coalition of a majority of the whole society could seldom take place on any other principles than those of justice and the general good; whilst there being thus less danger to a minor from the will of a major party, there must be less pretext, also, to provide for the security of the former, by introducing into the government a will not dependent on the latter, or, in other words, a will independent of the society itself. It is no less certain than it is important, notwithstanding the contrary opinions which have been entertained, that the larger the society, provided it lie within a practical sphere, the more duly capable it will be of self-government. And happily for the republican cause, the practicable sphere may be carried to a very great extent, by a judicious modification and mixture of the federal principle.

Publius
[James Madison]
Independent Journal
Wednesday, February 6, 1788

George Washington's Farewell Address

Friends, And Fellow Citizens

The period for a new election of a citizen, to administer the executive government of the United States, being not far distant, and the time actually arrived, when your thoughts must be employed designating the person, who is to be clothed with that important trust, it appears to me proper, especially as it may conduce to a more distinct expression of the public voice, that I should now apprize you of the resolution I have formed, to decline being considered among the number of those out of whom a choice is to be made.

I beg you at the same time to do me the justice to be assured that this resolution has not been taken without a strict regard to all the considerations appertaining to the relation which binds a dutiful citizen to his country; and that in withdrawing the tender of service, which silence in my situation might imply, I am influenced by no diminution of zeal for your future interest, no deficiency of grateful respect for your past kindness, but am supported by a full conviction that the step is compatible with both.

The acceptance of, and continuance hitherto in, the office to which your suffrages have twice called me, have been a uniform sacrifice of inclination to the opinion of duty, and to a deference for what appeared to be your desire. I constantly hoped, that it would have been much earlier in my power, consistently with motives, which I was not at liberty to disregard, to return to that retirement, from which I had been reluctantly drawn. The strength of my inclination to do this, previous to the last election, had even led to the preparation of an address to declare it to you; but mature reflection on the then perplexed and critical posture of our affairs with foreign nations, and the unanimous advice of persons entitled to my confidence impelled me to abandon the idea.

I rejoice, that the state of your concerns, external as well as internal, no longer renders the pursuit of inclination incompatible with the sentiment of duty, or propriety; and am persuaded, whatever partiality may be retained for my services, that, in the present circumstances of our country, you will not disapprove my determination to retire.

The impressions, with which I first undertook the arduous trust, were explained on the proper occasion. In the discharge of this trust, I will only say, that I have, with good intentions, contributed towards the organization and administration of the government the best exertions of which a very fallible judgment was capable. Not unconscious, in the outset, of the inferiority of my qualifications, experience in my own eyes, perhaps still more in the eyes of others, has strengthened the motives to diffidence of myself; and every day the increasing weight of years admonishes me more and more, that the shade of retirement is as necessary to me as it will be welcome. Satisfied, that, if any circumstances have given peculiar value to my services, they were temporary, I have the consolation to believe, that, while choice and prudence invite me to quit the political scene, patriotism does not forbid it.

In looking forward to the moment, which is intended to terminate the career of my public life, my feelings do not permit me to suspend the deep acknowledgment of that debt of gratitude, which I owe to my beloved country for the many honors it has conferred upon me; still more for the steadfast confidence with which it has supported me; and for the opportunities I have thence enjoyed of manifesting my inviolable attachment, by services faithful and persevering, though in usefulness unequal to my zeal. If benefits have resulted to our country from these services, let it always be remembered to your praise, and as an instructive example in our annals, that under circumstances in which the passions, agitated in every direction, were liable to mislead, amidst appearances sometimes dubious, vicissitudes of fortune often discouraging, in situations in which not unfrequently want of success has countenanced the spirit of criticism, the constancy of your support was the essential prop of the efforts, and a guarantee of the plans by which they were effected. Profoundly penetrated with this idea, I shall carry it with me to my grave, as a strong incitement to unceasing vows that Heaven may continue to you the choicest tokens of its beneficence; that your union and brotherly affection may be perpetual; that the free constitution, which is the work of your hands, may be sacredly maintained; that its administration in every department may be stamped with wisdom and virtue; than, in fine, the happiness of the people of these States, under the auspices of liberty, may be made complete, by so careful a preservation and so prudent a use of this blessing, as will acquire to them the glory of recommending it to the applause, the affection, and adoption of every nation, which is yet a stranger to it.

Here, perhaps I ought to stop. But a solicitude for your welfare which cannot end but with my life, and the apprehension of danger, natural to that solicitude, urge me, on an occasion like the present, to offer to your solemn contemplation, and to recommend to your frequent review, some sentiments which are the result of much reflection, of no inconsiderable observation, and which appear to me all-important to the permanency of your felicity as a people. These will be offered to you with the more freedom, as you can only see in them the disinterested warnings of a parting friend, who can possibly have no personal motive to bias his counsel. Nor can I forget, as an encouragement to it, your indulgent reception of my sentiments on a former and not dissimilar occasion.

Interwoven as is the love of liberty with every ligament of your hearts, no recommendation of mine is necessary to fortify or confirm the attachment.

The unity of Government, which constitutes you one people, is also now dear to you. It is justly so; for it is a main pillar in the edifice of your real independence, the support of your tranquillity at home, your peace abroad; of your safety; of your prosperity; of that very Liberty, which you so highly prize. But as it is easy to foresee, that, from different causes and from different quarters, much pains will be taken, many artifices employed, to weaken in your minds the conviction of this truth; as this is the point in your political fortress against which the batteries of internal and external enemies will be most constantly and actively (though often covertly and insidiously) directed, it is of infinite moment, that you should properly estimate the immense value of your national Union to your collective and individual happiness; that you should cherish a cordial, habitual, and immovable attachment to it; accustoming yourselves to think and speak of it as of the Palladium of your political safety and prosperity; watching for its preservation with jealous anxiety; discountenancing whatever may suggest even a suspicion, that it can in any event be abandoned; and indignantly frowning upon the first dawning of every attempt to alienate any portion of our country from the rest, or to enfeeble the sacred ties which now link together the various parts.

For this you have every inducement of sympathy and interest. Citizens, by birth or choice, of a common country, that country has a right to concentrate your affections. The name of

AMERICAN, which belongs to you, in your national capacity, must always exalt the just pride of Patriotism, more than any appellation derived from local discriminations. With slight shades of difference, you have the same religion, manners, habits, and political principles. You have in a common cause fought and triumphed together; the Independence and Liberty you possess are the work of joint counsels, and joint efforts, of common dangers, sufferings, and successes.

But these considerations, however powerfully they address themselves to your sensibility, are greatly outweighed by those, which apply more immediately to your interest. Here every portion of our country finds the most commanding motives for carefully guarding and preserving the Union of the whole.

The North, in an unrestrained intercourse with the South, protected by the equal laws of a common government, finds, in the productions of the latter, great additional resources of maritime and commercial enterprise and precious materials of manufacturing industry. The South, in the same intercourse, benefiting by the agency of the North, sees its agriculture grow and its commerce expand. Turning partly into its own channels the seamen of the North, it finds its particular navigation invigorated; and, while it contributes, in different ways, to nourish and increase the general mass of the national navigation, it looks forward to the protection of a maritime strength, to which itself is unequally adapted. The East, in a like intercourse with the West, already finds, and in the progressive improvement of interior communications by land and water, will more and more find, a valuable vent for the commodities which it brings from abroad, or manufactures at home. The West derives from the East supplies requisite to its growth and comfort, and, what is perhaps of still greater consequence, it must of necessity owe the secure enjoyment of indispensable outlets for its own productions to the weight, influence, and the future maritime strength of the Atlantic side of the Union, directed by an indissoluble community of interest as one nation. Any other tenure by which the West can hold this essential advantage, whether derived from its own separate strength, or from an apostate and unnatural connection with any foreign power, must be intrinsically precarious.

While, then, every part of our country thus feels an immediate and particular interest in Union, all the parts combined cannot fail to find in the united mass of means and efforts greater strength, greater resource, proportionably greater security from external danger, a less frequent interruption of their peace by foreign nations; and, what is of inestimable value, they must derive from Union an exemption from those broils and wars between themselves, which so frequently afflict neighbouring countries not tied together by the same governments, which their own rivalships alone would be sufficient to produce, but which opposite foreign alliances, attachments, and intrigues would stimulate and embitter. Hence, likewise, they will avoid the necessity of those overgrown military establishments, which, under any form of government, are inauspicious to liberty, and which are to be regarded as particularly hostile to Republican Liberty. In this sense it is, that your Union ought to be considered as a main prop of your liberty, and that the love of the one ought to endear to you the preservation of the other.

These considerations speak a persuasive language to every reflecting and virtuous mind, and exhibit the continuance of the union as a primary object of Patriotic desire. Is there a doubt, whether a common government can embrace so large a sphere? Let experience solve it. To listen to mere speculation in such a case were criminal. We are authorized to hope, that a proper organization of the whole, with the auxiliary agency of governments for the respective subdivisions, will afford a happy issue to the experiment. It is well worth a fair and full experiment. With such powerful and obvious motives to Union, affecting all parts of our country, while experience shall not have demonstrated its impracticability, there will always be reason to distrust the patriotism of those, who in any quarter may endeavour to weaken its bands.

In contemplating the causes, which may disturb our Union, it occurs as matter of serious concern, that any ground should have been furnished for characterizing parties by Geographical discriminations, Northern and Southern, Atlantic and Western; whence designing men may endeavour to excite a belief, that there is a real difference of local interests and views. One of the expedients of party to acquire influence, within particular districts, is to misrepresent the opinions and aims of other districts. You cannot shield yourselves too much against the jealousies and heart-burnings, which spring from these misrepresentations; they tend to render alien to each other those, who ought to be bound together by fraternal affection. The inhabitants of our western country have lately had a useful lesson on this head; they have seen, in the negotiation by the Executive, and in the unanimous ratification by the Senate, of the treaty with Spain, and in the universal satisfaction at that event, throughout the United States, a decisive proof how unfounded were the suspicions propagated among them of a policy in the General Government and in the Atlantic States unfriendly to their interests in regard to the Mississippi; they have been witnesses to the formation of two treaties, that with Great Britain, and that with Spain, which secure to them every thing they could desire, in respect to our foreign relations, towards confirming their prosperity. Will it not be their wisdom to rely for the preservation of these advantages on the union by which they were procured? Will they not henceforth be deaf to those advisers, if such there are, who would sever them from their brethren, and connect them with aliens?

To the efficacy and permanency of your Union, a Government for the whole is indispensable. No alliances, however strict, between the parts can be an adequate substitute; they must inevitably experience the infractions and interruptions, which all alliances in all times have experienced. Sensible of this momentous truth, you have improved upon your first essay, by the adoption of a Constitution of Government better calculated than your former for an intimate Union, and for the efficacious management of your common concerns. This Government, the offspring of our own choice, uninfluenced and unawed, adopted upon full investigation and mature deliberation, completely free in its principles, in the distribution of its powers, uniting security with energy, and containing within itself a provision for its own amendment, has a just claim to your confidence and your support. Respect for its authority, compliance with its laws, acquiescence in its measures, are duties enjoined by the fundamental maxims of true Liberty. The basis of our political systems is the right of the people to make and to alter their Constitutions of Government. But the Constitution which at any time exists, till changed by an explicit and authentic act of the whole people, is sacredly obligatory upon all. The very idea of the power and the right of the people to establish Government presupposes the duty of every individual to obey the established Government.

All obstructions to the execution of the Laws, all combinations and associations, under whatever plausible character, with the real design to direct, control, counteract, or awe the regular deliberation and action of the constituted authorities, are destructive of this fundamental principle, and of fatal tendency. They serve to organize faction, to give it an artificial and extraordinary force; to put, in the place of the delegated will of the nation, the will of a party, often a small but artful and enterprising minority of the community; and, according to the alternate triumphs of different parties, to make the public administration the mirror of the ill-concerted and incongruous projects of faction, rather than the organ of consistent and wholesome plans digested by common counsels, and modified by mutual interests.

However combinations or associations of the above description may now and then answer popular ends, they are likely, in the course of time and things, to become potent engines, by which cunning, ambitious, and unprincipled men will be enabled to subvert the power of the people, and to usurp for themselves the reins of government; destroying afterwards the very engines, which have lifted them to unjust dominion.

Towards the preservation of your government, and the permanency of your present happy state, it is requisite, not only that you steadily discountenance irregular oppositions to its acknowledged authority, but also that you resist with care the spirit of innovation upon its principles, however specious the pretexts. One method of assault may be to effect, in the forms of the constitution, alterations, which will impair the energy of the system, and thus to undermine what cannot be directly overthrown. In all the changes to which you may be invited, remember that time and habit are at least as necessary to fix the true character of governments, as of other human institutions; that experience is the surest standard, by which to test the real tendency of the existing constitution of a country; that facility in changes, upon the credit of mere hypothesis and opinion, exposes to perpetual change, from the endless variety of hypothesis and opinion; and remember, especially, that, for the efficient management of our common interests, in a country so extensive as ours, a government of as much vigor as is consistent with the perfect security of liberty is indispensable. Liberty itself will find in such a government, with powers properly distributed and adjusted, its surest guardian. It is, indeed, little else than a name, where the government is too feeble to withstand the enterprises of faction, to confine each member of the society within the limits prescribed by the laws, and to maintain all in the secure and tranquil enjoyment of the rights of person and property.

I have already intimated to you the danger of parties in the state, with particular reference to the founding of them on geographical discriminations. Let me now take a more comprehensive view, and warn you in the most solemn manner against the baneful effects of the spirit of party, generally.

This spirit, unfortunately, is inseparable from our nature, having its root in the strongest passions of the human mind. It exists under different shapes in all governments, more or less stifled, controlled, or repressed; but, in those of the popular form, it is seen in its greatest rankness, and is truly their worst enemy.

The alternate domination of one faction over another, sharpened by the spirit of revenge, natural to party dissension, which in different ages and countries has perpetrated the most horrid enormities, is itself a frightful despotism. But this leads at length to a more formal and permanent despotism. The disorders and miseries, which result, gradually incline the minds of men to seek security and repose in the absolute power of an individual; and sooner or later the chief of some prevailing faction, more able or more fortunate than his competitors, turns this disposition to the purposes of his own elevation, on the ruins of Public Liberty.

Without looking forward to an extremity of this kind, (which nevertheless ought not to be entirely out of sight,) the common and continual mischiefs of the spirit of party are sufficient to make it the interest and duty of a wise people to discourage and restrain it.

It serves always to distract the Public Councils, and enfeeble the Public Administration. It agitates the Community with ill-founded jealousies and false alarms; kindles the animosity of one part against another, foments occasionally riot and insurrection. It opens the door to foreign influence and corruption, which find a facilitated access to the government itself through the channels of party passions. Thus the policy and the will of one country are subjected to the policy and will of another.

There is an opinion, that parties in free countries are useful checks upon the administration of the Government, and serve to keep alive the spirit of Liberty. This within certain limits is probably true; and in Governments of a Monarchical cast, Patriotism may look with indulgence, if not with favor, upon the spirit of party. But in those of the popular character, in Governments purely elective, it is a spirit not to be encouraged. From their natural tendency, it is certain there will always be enough of that spirit for every salutary purpose. And, there being constant danger of excess, the effort ought to be, by force of public opinion, to mitigate and assuage it. A fire not to be quenched, it demands a uniform vigilance to prevent its bursting into a flame, lest, instead of warming, it should consume.

It is important, likewise, that the habits of thinking in a free country should inspire caution, in those entrusted with its administration, to confine themselves within their respective constitutional spheres, avoiding in the exercise of the powers of one department to encroach upon another. The spirit of encroachment tends to consolidate the powers of all the departments in one, and thus to create, whatever the form of government, a real despotism. A just estimate of that love of power, and proneness to abuse it, which predominates in the human heart, is sufficient to satisfy us of the truth of this position. The necessity of reciprocal checks in the exercise of political power, by dividing and distributing it into different depositories, and constituting each the Guardian of the Public Weal against invasions by the others, has been evinced by experiments ancient and modern; some of them in our country and under our own eyes. To preserve them must be as necessary as to institute them. If, in the opinion of the people, the distribution or modification of the constitutional powers be in any particular wrong, let it be corrected by an amendment in the way, which the constitution designates. But let there be no change by usurpation; for, though this, in one instance, may be the instrument of good, it is the customary weapon by which free governments are destroyed. The precedent must always greatly overbalance in permanent evil any partial or transient benefit, which the use can at any time yield.

Of all the dispositions and habits, which lead to political prosperity, Religion and Morality are indispensable supports. In vain would that man claim the tribute of Patriotism, who should labor to subvert these great pillars of human happiness, these firmest props of the duties of Men and Citizens. The mere Politician, equally with the pious man, ought to respect and to cherish them. A volume could not trace all their connections with private and public felicity. Let it simply be asked, Where is the security for property, for reputation, for life, if the sense of religious obligation desert the oaths, which are the instruments of investigation in Courts of Justice? And let us with caution indulge the supposition, that morality can be maintained without religion. Whatever may be conceded to the influence of refined education on minds of peculiar structure, reason and experience both forbid us to expect, that national morality can prevail in exclusion of religious principle.

It is substantially true, that virtue or morality is a necessary spring of popular government. The rule, indeed, extends with more or less force to every species of free government. Who, that is a sincere friend to it, can look with indifference upon attempts to shake the foundation of the fabric ?

Promote, then, as an object of primary importance, institutions for the general diffusion of knowledge. In proportion as the structure of a government gives force to public opinion, it is essential that public opinion should be enlightened.

As a very important source of strength and security, cherish public credit. One method of preserving it is, to use it as sparingly as possible; avoiding occasions of expense by cultivating peace, but remembering also that timely disbursements to prepare for danger frequently prevent much greater disbursements to repel it; avoiding likewise the accumulation of debt, not only by shunning occasions of expense, but by vigorous exertions in time of peace to discharge the debts, which unavoidable wars may have occasioned, not ungenerously throwing upon posterity the burthen, which we ourselves ought to bear. The execution of these maxims belongs to your representatives, but it is necessary that public opinion should cooperate. To facilitate to them the performance of their duty, it is essential that you should practically bear in mind, that towards the payment of debts there must be Revenue; that to have Revenue there must be taxes; that no taxes can be devised, which are not more or less inconvenient and unpleasant; that the intrinsic embarrassment, inseparable from the selection of the proper objects (which is always a choice of difficulties), ought to be a decisive motive for a candid construction of the conduct of the government in making it, and for a spirit of acquiescence in the measures for obtaining revenue, which the public exigencies may at any time dictate.

Observe good faith and justice towards all Nations; cultivate peace and harmony with all. Religion and Morality enjoin this conduct; and can it be, that good policy does not equally enjoin it? It will be worthy of a free, enlightened, and, at no distant period, a great Nation, to give to mankind the magnanimous and too novel example of a people always guided by an exalted justice and benevolence. Who can doubt, that, in the course of time and things, the fruits of such a plan would richly repay any temporary advantages, which might be lost by a steady adherence to it ? Can it be, that Providence has not connected the permanent felicity of a Nation with its Virtue? The experiment, at least, is recommended by every sentiment which ennobles human nature. Alas! is it rendered impossible by its vices ?

In the execution of such a plan, nothing is more essential, than that permanent, inveterate antipathies against particular Nations, and passionate attachments for others, should be excluded; and that, in place of them, just and amicable feelings towards all should be cultivated. The Nation, which indulges towards another an habitual hatred, or an habitual fondness, is in some degree a slave. It is a slave to its animosity or to its affection, either of which is sufficient to lead it astray from its duty and its interest. Antipathy in one nation against another disposes each more readily to offer insult and injury, to lay hold of slight causes of umbrage, and to be haughty and intractable, when accidental or trifling occasions of dispute occur. Hence frequent collisions, obstinate, envenomed, and bloody contests. The Nation, prompted by ill-will and resentment, sometimes impels to war the Government, contrary to the best calculations of policy. The Government sometimes participates in the national propensity, and adopts through passion what reason would reject; at other times, it makes the animosity of the nation subservient to projects of hostility instigated by pride, ambition, and other sinister and pernicious motives. The peace often, sometimes perhaps the liberty, of Nations has been the victim.

So likewise, a passionate attachment of one Nation for another produces a variety of evils. Sympathy for the favorite Nation, facilitating the illusion of an imaginary common interest, in cases where no real common interest exists, and infusing into one the enmities of the other, betrays the former into a participation in the quarrels and wars of the latter, without adequate inducement or justification. It leads also to concessions to the favorite Nation of privileges denied to others, which is apt doubly to injure the Nation making the concessions; by unnecessarily parting with what ought to have been retained; and by exciting jealousy, ill-will, and a disposition to retaliate, in the parties from whom equal privileges are withheld. And it gives to ambitious, corrupted, or deluded citizens, (who devote themselves to the favorite nation,) facility to betray or sacrifice the interests of their own country, without odium, sometimes even with popularity; gilding, with the appearances of a virtuous sense of obligation, a commendable deference for public opinion, or a laudable zeal for public good, the base or foolish compliances of ambition, corruption, or infatuation.

As avenues to foreign influence in innumerable ways, such attachments are particularly alarming to the truly enlightened and independent Patriot. How many opportunities do they afford to tamper with domestic factions, to practice the arts of seduction, to mislead public opinion, to influence or awe the Public Councils! Such an attachment of a small or weak, towards a great and powerful nation, dooms the former to be the satellite of the latter.

Against the insidious wiles of foreign influence (I conjure you to believe me, fellow-citizens,) the jealousy of a free people ought to be constantly awake; since history and experience prove, that foreign influence is one of the most baneful foes of Republican Government. But that jealousy, to be useful, must be impartial; else it becomes the instrument of the very influence to be avoided, instead of a defense against it. Excessive partiality for one foreign nation, and excessive dislike of another, cause those whom they actuate to see danger only on one side, and serve to veil and even second the arts of influence on the other. Real patriots, who may resist the intrigues of the favorite, are liable to become suspected and odious; while its tools and dupes usurp the applause and confidence of the people, to surrender their interests.

The great rule of conduct for us, in regard to foreign nations, is, in extending our commercial relations, to have with them as little political connection as possible. So far as we have already formed engagements, let them be fulfilled with perfect good faith. Here let us stop.

Europe has a set of primary interests, which to us have none, or a very remote relation. Hence she must be engaged in frequent controversies, the causes of which are essentially foreign to our concerns. Hence, therefore, it must be unwise in us to implicate ourselves, by artificial ties, in the ordinary vicissitudes of her politics, or the ordinary combinations and collisions of her friendships or enmities.

Our detached and distant situation invites and enables us to pursue a different course. If we remain one people, under an efficient government, the period is not far off, when we may defy material injury from external annoyance; when we may take such an attitude as will cause the neutrality, we may at any time resolve upon, to be scrupulously respected; when belligerent nations, under the impossibility of making acquisitions upon us, will not lightly hazard the giving us provocation; when we may choose peace or war, as our interest, guided by justice, shall counsel.

Why forego the advantages of so peculiar a situation? Why quit our own to stand upon foreign ground? Why, by interweaving our destiny with that of any part of Europe, entangle our peace and prosperity in the toils of European ambition, rivalship, interest, humor, or caprice?

It is our true policy to steer clear of permanent alliances with any portion of the foreign world; so far, I mean, as we are now at liberty to do it; for let me not be understood as capable of patronizing infidelity to existing engagements. I hold the maxim no less applicable to public than to private affairs, that honesty is always the best policy. I repeat it, therefore, let those engagements be observed in their genuine sense. But, in my opinion, it is unnecessary and would be unwise to extend them.

Taking care always to keep ourselves, by suitable establishments, on a respectable defensive posture, we may safely trust to temporary alliances for extraordinary emergencies.

Harmony, liberal intercourse with all nations, are recommended by policy, humanity, and interest. But even our commercial policy should hold an equal and impartial hand; neither seeking nor granting exclusive favors or preferences; consulting the natural course of things; diffusing and diversifying by gentle means the streams of commerce, but forcing nothing; establishing, with powers so disposed, in order to give trade a stable course, to define the rights of our merchants, and to enable the government to support them, conventional rules of intercourse, the best that present circumstances and mutual opinion will permit, but temporary, and liable to be from time to time abandoned or varied, as experience and circumstances shall dictate; constantly keeping in view, that it is folly in one nation to look for disinterested favors from another; that it must pay with a portion of its independence for whatever it may accept under that character; that, by such acceptance, it may place itself in the condition of having given equivalents for nominal favors, and yet of being reproached with ingratitude for not giving more. There can be no greater error than to expect or calculate upon real favors from nation to nation. It is an illusion, which experience must cure, which a just pride ought to discard.

In offering to you, my countrymen, these counsels of an old and affectionate friend, I dare not hope they will make the strong and lasting impression I could wish; that they will control the usual current of the passions, or prevent our nation from running the course, which has hitherto marked the destiny of nations. But, if I may even flatter myself, that they may be productive of some partial benefit, some occasional good; that they may now and then recur to moderate the fury of party spirit, to warn against the mischiefs of foreign intrigue, to guard against the impostures of pretended patriotism; this hope will be a full recompense for the solicitude for your welfare, by which they have been dictated.

How far in the discharge of my official duties, I have been guided by the principles which have been delineated, the public records and other evidences of my conduct must witness to you and to the world. To myself, the assurance of my own conscience is, that I have at least believed myself to be guided by them.

In relation to the still subsisting war in Europe, my Proclamation of the 22d of April 1793, is the index to my Plan. Sanctioned by your approving voice, and by that of your Representatives in both Houses of Congress, the spirit of that measure has continually governed me, uninfluenced by any attempts to deter or divert me from it.

After deliberate examination, with the aid of the best lights I could obtain, I was well satisfied that our country, under all the circumstances of the case, had a right to take, and was bound in duty and interest to take, a neutral position. Having taken it, I determined, as far as should depend upon me, to maintain it, with moderation, perseverance, and firmness.

The considerations, which respect the right to hold this conduct, it is not necessary on this occasion to detail. I will only observe, that, according to my understanding of the matter, that right, so far from being denied by any of the Belligerent Powers, has been virtually admitted by all.

The duty of holding a neutral conduct may be inferred, without any thing more, from the obligation which justice and humanity impose on every nation, in cases in which it is free to act, to maintain inviolate the relations of peace and amity towards other nations.

The inducements of interest for observing that conduct will best be referred to your own reflections and experience. With me, a predominant motive has been to endeavour to gain time to our country to settle and mature its yet recent institutions, and to progress without interruption to that degree of strength and consistency, which is necessary to give it, humanly speaking, the command of its own fortunes.

Though, in reviewing the incidents of my administration, I am unconscious of intentional error, I am nevertheless too sensible of my defects not to think it probable that I may have committed many errors. Whatever they may be, I fervently beseech the Almighty to avert or mitigate the evils to which they may tend. I shall also carry with me the hope, that my Country will never cease to view them with indulgence; and that, after forty-five years of my life dedicated to its service with an upright zeal, the faults of incompetent abilities will be consigned to oblivion, as myself must soon be to the mansions of rest.

Relying on its kindness in this as in other things, and actuated by that fervent love towards it, which is so natural to a man, who views it in the native soil of himself and his progenitors for several generations; I anticipate with pleasing expectation that retreat, in which I promise myself to realize, without alloy, the sweet enjoyment of partaking, in the midst of my fellow-citizens, the benign influence of good laws under a free government, the ever favorite object of my heart, and the happy reward, as I trust, of our mutual cares, labors, and dangers.

George Washington
United States—September 17, 1796

THE GETTYSBURG ADDRESS

Fourscore and seven years ago our fathers brought forth on this continent a new nation, conceived in liberty and dedicated to the proposition that all men are created equal.

Now we are engaged in a great civil war, testing whether that nation or any nation so conceived and so dedicated can long endure. We are met on a great battlefield of that war. We have come to dedicate a portion of that field as a final resting place for those who here gave their lives that that nation might live. It is altogether fitting and proper that we should do this.

But in a larger sense, we cannot dedicate, we cannot consecrate, we cannot hallow this ground. The brave men, living and dead, who struggled here have consecrated it far above our poor power to add or detract. The world will little note nor long remember what we say here, but it can never forget what they did here. It is for us the living rather to be dedicated here to the unfinished work which they who fought here have thus far so nobly advanced. It is rather for us to be here dedicated to the great task remaining before us—that from these honored dead we take increased devotion to that cause for which they gave the last full measure of devotion—that we here highly resolve that these dead shall not have died in vain, that this nation under God shall have a new birth of freedom, and that government of the people, by the people, for the people shall not perish from the earth.

ABRAHAM LINCOLN'S 2ND INAUGURAL ADDRESS

Fellow Countrymen,

At this second appearing to take the oath of the Presidential office there is less occasion for an extended address than there was at the first. Then a statement somewhat in detail of a course to be pursued seemed fitting and proper. Now, at the expiration of four years, during which public declarations have been constantly called forth on every point and phase of the great contest which still absorbs the attention and engrosses the energies of the nation, little that is new could be presented. The progress of our arms, upon which all else chiefly depends, is as well known to the public as to myself, and it is, I trust, reasonably satisfactory and encouraging to all. With high hope for the future, no prediction in regard to it is ventured.

On the occasion corresponding to this four years ago all thoughts were anxiously directed to an impending civil war. All dreaded it, all sought to avert it. While the inaugural address was being delivered from this place, devoted altogether to saving the Union without war, insurgent agents were in the city seeking to destroy it without war—seeking to dissolve the Union and divide effects by negotiation. Both parties deprecated war, but one of them would make war rather than let the nation survive, and the other would accept war rather than let it perish, and the war came.

One-eighth of the whole population were colored slaves, not distributed generally over the Union, but localized in the southern part of it. These slaves constituted a peculiar and powerful interest. All knew that this interest was somehow the cause of the war. To strengthen, perpetuate, and extend this interest was the object for which the insurgents would rend the Union even by war, while the Government claimed no right to do more than to restrict the territorial enlargement of it. Neither party expected for the war the magnitude or the duration which it has already attained. Neither anticipated that the cause of the conflict might cease with or even before the conflict itself should cease. Each looked for an easier triumph, and a result less fundamental and astounding. Both read the same Bible and pray to the same God, and each invokes His aid against the other. It may seem strange that any men should dare to ask a just God's assistance in wringing their bread from the sweat of other men's faces, but let us judge not, that we be not judged. The prayers of both could not be answered. That of neither has been answered fully. The Almighty has His own purposes. 'Woe unto the world because of offenses; for it must needs be that offenses come, but woe to that man by whom the offense cometh.' If we shall suppose that American slavery is one of those offenses which, in the providence of God, must needs come, but which, having continued through His appointed time, He now wills to remove, and that He gives to both North and South this terrible war as the woe due to those by whom the offense came, shall we discern therein any departure from those divine attributes which the believers in a living God always ascribe to Him? Fondly do we hope, fervently do we pray, that this mighty scourge of war may speedily pass away. Yet, if God wills that it continue until all the wealth piled by the bondsman's two hundred and fifty years of unrequited toil shall be sunk, and until every drop of blood drawn with the lash shall be paid by another drawn with the sword, as was said three thousand years ago, so still it must be said 'the judgments of the Lord are true and righteous altogether.

With malice toward none, with charity for all, with firmness in the right as God gives us to see the right let us strive on to finish the work we are in, to bind up the nation's wounds, to care for him who shall have borne the battle and for his widow and his orphan, to do all which may achieve and cherish a just and lasting peace among ourselves and with all nations.

"I Have a Dream"

I am happy to join with you today in what will go down in history as the greatest demonstration for freedom in the history of our nation.

Five score years ago, a great American, in whose symbolic shadow we stand today, signed the Emancipation Proclamation. This momentous decree came as a great beacon light of hope to millions of Negro slaves who had been seared in the flames of withering injustice. It came as a joyous daybreak to end the long night of their captivity.

But one hundred years later, the Negro still is not free. One hundred years later, the life of the Negro is still sadly crippled by the manacles of segregation and the chains of discrimination. One hundred years later, the Negro lives on a lonely island of poverty in the midst of a vast ocean of material prosperity. One hundred years later, the Negro is still languishing in the corners of American society and finds himself an exile in his own land. So we have come here today to dramatize a shameful condition.

In a sense we have come to our nation's capital to cash a check. When the architects of our republic wrote the magnificent words of the Constitution and the Declaration of Independence, they were signing a promissory note to which every American was to fall heir. This note was a promise that all men, yes, black men as well as white men, would be guaranteed the unalienable rights of life, liberty, and the pursuit of happiness.

It is obvious today that America has defaulted on this promissory note insofar as her citizens of color are concerned. Instead of honoring this sacred obligation, America has given the Negro people a bad check, a check which has come back marked "insufficient funds." But we refuse to believe that the bank of justice is bankrupt. We refuse to believe that there are insufficient funds in the great vaults of opportunity of this nation. So we have come to cash this check — a check that will give us upon demand the riches of freedom and the security of justice. We have also come to this hallowed spot to remind America of the fierce urgency of now. This is no time to engage in the luxury of cooling off or to take the tranquilizing drug of gradualism. Now is the time to make real the promises of democracy. Now is the time to rise from the dark and desolate valley of segregation to the sunlit path of racial justice. Now is the time to lift our nation from the quick sands of racial injustice to the solid rock of brotherhood. Now is the time to make justice a reality for all of God's children.

It would be fatal for the nation to overlook the urgency of the moment. This sweltering summer of the Negro's legitimate discontent will not pass until there is an invigorating autumn of freedom and equality. Nineteen sixty-three is not an end, but a beginning. Those who hope that the Negro needed to blow off steam and will now be content will have a rude awakening if the nation returns to business as usual. There will be neither rest nor tranquility in America until the Negro is granted his citizenship rights. The whirlwinds of revolt will continue to shake the foundations of our nation until the bright day of justice emerges.

But there is something that I must say to my people who stand on the warm threshold which leads into the palace of justice. In the process of gaining our rightful place we must not be guilty of wrongful deeds. Let us not seek to satisfy our thirst for freedom by drinking from the cup of bitterness and hatred.

We must forever conduct our struggle on the high plane of dignity and discipline. We must not allow our creative protest to degenerate into physical violence. Again and again we must rise to the majestic heights of meeting physical force with soul force. The marvelous new militancy which has engulfed the Negro community must not lead us to distrust of all white people, for many of our white brothers, as evidenced by their presence here today, have come to realize that their destiny is tied up with our destiny and their freedom is inextricably bound to our freedom. We cannot walk alone.

As we walk, we must make the pledge that we shall march ahead. We cannot turn back. There are those who are asking the devotees of civil rights, "When will you be satisfied?" We can never be satisfied as long as the Negro is the victim of the unspeakable horrors of police brutality. We can never be satisfied, as long as our bodies, heavy with the fatigue of travel, cannot gain lodging in the motels of the highways and the hotels of the cities. We can never be satisfied as long as a Negro in Mississippi cannot vote and a Negro in New York believes he has nothing for which to vote. No, no, we are not satisfied, and we will not be satisfied until justice rolls down like waters and righteousness like a mighty stream.

I am not unmindful that some of you have come here out of great trials and tribulations. Some of you have come fresh from narrow jail cells. Some of you have come from areas where your quest for freedom left you battered by the storms of persecution and staggered by the winds of police brutality. You have been the veterans of creative suffering. Continue to work with the faith that unearned suffering is redemptive.

Go back to Mississippi, go back to Alabama, go back to South Carolina, go back to Georgia, go back to Louisiana, go back to the slums and ghettos of our northern cities, knowing that somehow this situation can and will be changed. Let us not wallow in the valley of despair.

I say to you today, my friends, so even though we face the difficulties of today and tomorrow, I still have a dream. It is a dream deeply rooted in the American dream.

I have a dream that one day this nation will rise up and live out the true meaning of its creed: "We hold these truths to be self-evident: that all men are created equal."

I have a dream that one day on the red hills of Georgia the sons of former slaves and the sons of former slave owners will be able to sit down together at the table of brotherhood.

I have a dream that one day even the state of Mississippi, a state sweltering with the heat of injustice, sweltering with the heat of oppression, will be transformed into an oasis of freedom and justice.

I have a dream that my four little children will one day live in a nation where they will not be judged by the color of their skin but by the content of their character.

I have a dream today.

I have a dream that one day, down in Alabama, with its vicious racists, with its governor having his lips dripping with the words of interposition and nullification; one day right there in Alabama, little black boys and black girls will be able to join hands with little white boys and white girls as sisters and brothers.

I have a dream today.

I have a dream that one day every valley shall be exalted, every hill and mountain shall be made low, the rough places will be made plain, and the crooked places will be made straight, and the glory of the Lord shall be revealed, and all flesh shall see it together.

This is our hope. This is the faith that I go back to the South with. With this faith we will be able to hew out of the mountain of despair a stone of hope. With this faith we will be able to transform the jangling discords of our nation into a beautiful symphony of brotherhood. With this faith we will be able to work together, to pray together, to struggle together, to go to jail together, to stand up for freedom together, knowing that we will be free one day.

This will be the day when all of God's children will be able to sing with a new meaning, "My country, 'tis of thee, sweet land of liberty, of thee I sing. Land where my fathers died, land of the pilgrim's pride, from every mountainside, let freedom ring."

And if America is to be a great nation this must become true. So let freedom ring from the prodigious hilltops of New Hampshire. Let freedom ring from the mighty mountains of New York. Let freedom ring from the heightening Alleghenies of Pennsylvania!

Let freedom ring from the snowcapped Rockies of Colorado!

Let freedom ring from the curvaceous slopes of California!

But not only that; let freedom ring from Stone Mountain of Georgia!

Let freedom ring from Lookout Mountain of Tennessee!

Let freedom ring from every hill and molehill of Mississippi. From every mountainside, let freedom ring.

And when this happens, When we allow freedom to ring, when we let it ring from every village and every hamlet, from every state and every city, we will be able to speed up that day when all of God's children, black men and white men, Jews and Gentiles, Protestants and Catholics, will be able to join hands and sing in the words of the old Negro spiritual, "Free at last! free at last! thank God Almighty, we are free at last!"

Martin Luther King, Jr.
August 28, 1963
Delivered from the steps of the Lincoln Memorial
during the March on Washington for Jobs and Freedom

Glossary

– A –

Adams, John (1735–1826) Founding Father and proponent of a bicameral legislature, Adams defended the British soldiers involved in the Boston Massacre and later served as 2nd President of the United States, where he made his "midnight appointments" that were the basis of the *Marbury v. Madison* case.

Adams, John Quincy (1767–1848) Sixth President of the United States, Adams is known for formulating the Monroe Doctrine.

Adams, Samuel (1722–1803) Second cousin to John Adams, he was a Massachusetts statesman and organizer of the Boston Tea Party. Adams served in the Continental Congress and signed the Declaration of Independence, but was opposed to a strong federal government.

Alger Jr., Horatio (1832–1899) Author of rags-to-riches dime novels extolling the virtues of hard work, determination, courage, and concern for others and helped to shape the modern idea of the American dream.

Alien and Sedition Acts Laws passed by Congress in 1798 to try and stifle the "seditious" writings of French propagandists against the neutrality of the United States with regards to the French and British War.

American Exceptionalism The idea that the American experience was different or unique from others, and therefore America had a unique or special role in the world, such as a "city upon a hill."

Anarchy Lack of authority from a failure to agree on a common course of action; part of the human predicament cycle.

Antietam A severe Civil War battle that took place on September 17, 1862. It was the bloodiest day in American history. After the battle Abraham Lincoln issued the Emancipation Proclamation.

Anti-federalists Political group that was against the ratification of the Constitution.

Areté Greek term for human virtue, the backbone of republican morality. Striving for excellence.

Aristocracy Rule based on distinguished or wise ancestors and heritage.

Articles of Confederation Document outlining an alliance of sovereign, equal states in which there was a weak central governing Continental Congress.

Autocracy One of the four alternative forms of government; sees people as children in need of a carefully controlled environment provided by government.

Auxiliary precautions Structure in the government to make it more difficult for power to become concentrated in any one groups hands, seen by the founders as a backup system to virtue.

– B –

Bank run When most depositors try to withdraw their funds simultaneously from a bank.

Beard, Charles A. (1874–1948) A leader of the "Progressive School" of historiography who attacked the Founders as being motivated by economic self-interest.

Bell, John (1797–1869) A wealthy slaveowner from Tennessee who served in both the House and the Senate, Bell ran for U.S. President against Lincoln, Breckinridge, and Douglas in 1860 with the Constitutional Union Party on a moderate pro-slavery platform.

Bicameral legislature A legislature in which there are two separate divisions or houses.

Big stick Part of the Theodore Roosevelt phrase: "Speak softly and carry a big stick," which represented the military might of the United States.

Bill of Rights First ten amendments to the Constitution regarding basic protections of rights from the government, passed in response to the Anti-Federalist argument against the initial Constitution.

Bonaparte, Napoleon (1769–1821 French Emperor and European conqueror who sold France's North American holdings to the United States as the Louisiana Purchase.

Boosterism Promoting one's town or city, sometimes in an excessive or exaggerated manner, in order to increase both its quality and its public perception.

"Boss" Tweed See *Tweed, William Marcy.*

Boston Tea Party On December 16, 1773, American colonists protested the British tax on tea by dumping 342 crates of British tea into Boston harbor.

Breckinridge, John (1821–1875) A Senator from Kentucky and the fourteenth Vice President of the United States, Breckinridge ran against Lincoln, Bell, and Douglas in the 1860 Presidential election on an extreme pro-slavery platform.

Broad construction Constitutional clauses that are written to be interpreted in a more broad or general manner.

Brown, John (1800–1859) A controversial abolitionist who tried to start a slave rebellion and used sometimes violent guerrilla tactics in fighting against the institution of slavery.

Brown, Robert (1550–1630) Writer and proponent of the Separatist movement that demanded separation from the Church of England. His writings inspired groups such as the Pilgrims to emigrate to America for religious freedom.

Bryan, William Jennings (1860–1925) A lawyer, statesman, and popular speaker, Bryan ran for President on the Democratic ticket three different times. He was a prominent leader in the Progressive movement and served as Secretary of State to Woodrow Wilson. He may be most well known as one of the lawyers in the famous Scopes Trial about teaching evolution in schools.

– C –

Calvin, John (1509–1564) A French theologian during the Protestant Reformation who greatly influenced Puritan beliefs. He taught that the Bible was the final authority for matters of faith and that salvation came through grace only (not works). He also taught the doctrine of predestination.

Capitalism The philosophy of a free market economy in which the government serves only to create an acceptable environment in which to make exchanges; see also ***market economy.***

Cardinal virtues See ***greek cardinal virtues***.

Carnegie, Andrew (1835–1919) A well-known robber baron who was owner of Carnegie Steel. In his later years, he donated most of his money to establish schools, libraries, and universities around the world.

Cartel A group of firms that act together like a monopoly.

Checks and balances Bridging the separation of powers between branches of government by placing part of each power within two separate branches.

Christian calling, the From the theology of John Calvin—people should pursue a "calling" in some sort of worldly work where they are to rise early in the morning, work hard, save their money, and invest it wisely. Prosperity indicates God's approval.

Christian virtues Ideals of morality based on Christian principles (meekness, compassion, love for one's neighbor, etc.) that people should lead lives of common decency and public uprightness.

Cicero (106–43 B.C.) An orator, statesman, political theorist, lawyer, and philosopher of Ancient Rome.

City on a hill Biblical ideal, invoked by John Winthrop, of a society governed by civil liberty (where people did only that which was just and good) that would be an example to the world.

Civil liberty According to John Winthrop, "Where men were free to do only that which is good, just, and honest."

Civil rights Rights defined using narrow, concrete language, full of specific terms and qualifiers.

Civil Rights Movement Movement by African-Americans citizens in the 1960s to gain equal civil rights and to end racial discrimination and segregation.

Classical republicanism One of the four alternative forms of government; sees people (and government) as mostly good but corruptible and so government should have restricted power and try to encourage a good moral climate.

Clay, Henry (1777–1852) American statesman and congressman who founded the Whig party.

Coincidence of wants When two parties each possess something desired by the other, promoting an exchange.

Cold War The armed stalemate of the United States and the Soviet Union during the latter half of the 20th century; it was portrayed as a war of freedom versus tyranny, of democracy versus totalitarianism, of capitalism versus communism.

Collusion When sellers are conspiring to maintain a high price and avoid competing with one another; see also ***monopoly***.

Columbus, Christopher (1451–1506) Genoese mariner who discovered the Americas while searching for a new trade route to India.

Command system An economy that is heavily controlled by government and does not readily allow free exchange.

Commerce Clause Constitutional clause that gives Congress the power to regulate certain types of trade, also a justification for the Civil Rights Act of 1964.

Committees of correspondence Groups organized by local colonial governments for the purposes of coordinating written communication with the other colonies. They disseminated the colonial interpretation of British actions among the colonies and to foreign governments. The network of committees would later provide the basis for formal political union among the colonies.

Common Law Law that is considered to be from natural law principles but that is framed in a form that can be interpreted more concretely.

Common Sense A political tract written by Thomas Paine that helped convince colonists about the necessity to fight against Britain and to become independent.

Commonwealth ideology The idea that the "Country party" had the best strategy and opportunity to preserve liberty against the "Court party."

Competing groups Groups that, in a state of anarchy, fight for supreme power and control; part of the human predicament cycle.

Competition When there are sufficient buyers and sellers in the market so that no single seller or buyer has a significant influence on price.

Confederacy Alliance of southern states that seceded from the Union over slavery.

Confederation Defensive alliance among sovereign equals.

Consent Rule of law principle that states laws must be generally acceptable to those who must live by them.

Constitutional drift When power in the government does not remain where it was originally placed.

Constitutional mechanism Parts of the Constitution that help organize and control power.

Constitutional structure The nature and arrangement of mechanisms in a constitution that organize the government.

Continental Congress A body of representatives from the British North American colonies who met to respond to England's Intolerable Acts. They declared independence in July 1776 and later drafted the Articles of Confederation.

Cooper, James Fenimore (1789–1851) A prolific and popular American writer, Cooper is particularly known for his 1826 novel, *The Last of the Mohicans*.

Corporate communities Colonial settlements established for economic or financial purposes by various companies. Although usually chartered by the Crown, their remote circumstances helped foster the idea and practice of self-governance.

Country party English opposition to the "Court party" that consisted of commonwealth men (everyday citizens). The Court party was considered morally independent with pure motives.

Court party English royal court and the the center of British political power; known also as the "Tories" and characterized by corruption and subversion.

Covenant communities Settlements based on religious or moral values, mostly interested in being an example to Europe or to living according to their own moral liberty.

– D –

Darwin, Charles (1809–1882) English naturalist known for writing *The Origin of Species*, in which he proposed the idea of natural selection as the primary means of species diversity.

De Tocqueville, Alexis (1805–1859) Frenchman who wrote *Democracy in America* (1835), in which he explored the uniqueness of American character and its sources.

Debs, Eugene V. (1855–1926) Candidate for U.S. President 5 times as a socialist, Debs was also a founder of International Workers of the World.

Declaration of Independence 1776 document expressing the desire and intention of the American colonies to break ties with Britain due to the injustices perpetrated by King George III.

Deep change Fundamental alteration in the way life is lived.

Demand See ***Law of Demand***.

Demigod Being half human and half godlike, a trait sometimes wrongly attributed to the Founders.

Democratic-Republican Party Political party led by Jefferson and Madison that championed a society of self-reliant individuals to protect rights, a smaller federal government, and a narrow and strict interpretation of the Constitution.

Democratic revolution Change in political power by the voting of the people.

Despot A ruler exercising absolute power.

Divine right of kings Political theory that royal lines are established by God and that kings rule by divine decree.

Division of labor See ***specialization***.

Douglas, Stephen A. (1813–1861) An Illinois statesman who ran against Lincoln, Bell, and Breckinridge in the 1860 Presidential election on a popular sovereignty platform for slavery, Douglas also authored the Kansas-Nebraska Act, which repealed the Missouri Compromise and heightened the slavery debate.

Due process Rule of law principle that states laws must be administered impartially.

– E –

Economic System A society's structure for making and distributing goods and services.

Efficient Economy An economy in which all benefits are maximized to the point that one party cannot increase their benefits without decreasing the benefits of others.

Efficiency A goal for economic systems to produce the most goods for the most people.

Einstein, Albert (1879–1955) German-born theoretical physicist who is most known for his Special and General Theories of Relativity and the formula for mass-energy equivalence, $E=mc^2$.

Electoral college The group of electors selected by the people who are responsible for the selection of the president.

Emancipation Proclamation Presidential order issued by Abraham Lincoln on January 1, 1863 that freed slaves in the areas of insurrection.

Enumeration The written listing of the powers of government.

Equilibrium price The price at which the amount demanded is equal to the amount supplied.

Equity A goal for economic systems to distribute goods and rewards fairly.

European Enlightenment 18th century philosophical movement that proposed individual self-interest, rather than Greek virtue or Christian humility, as the motivating factor in human behavior.

Exchange Trade between two parties.

– F –

Faction A group of individuals who share the same specific political agenda.

Factionalism When a city-state or nation has multiple factions that compete against each other. Madison felt that an extended republic would prevent factionalism from leading to tyranny because no faction could be large enough to dominate.

Federal Reserve System A quasi-governmental organization formed to regulate the money supply and help keep the economy stable.

Federalism The dividing of powers between the national and state governments.

Federalist Party Political party founded by Hamilton and John Adams that envisioned a great Western empire with a strong federal government, and a broad interpretation of Constitutional powers.

Federalist, The Series of essays published in New York newspapers under the pseudonym Publius for the express purpose of gaining support for ratification of the Constitution. Written by James Madison, Alexander Hamilton, and John Jay.

Federalists A political group that was for the ratification of the Constitution, later used to describe members of the Federalist Party.

Fifteenth Amendment All male citizens are granted

the right to vote regardless of race, color, or previous condition of servitude.

Filtered consent When the selection of government officials is distanced from direct election by the people in order to protect against mob rule and public whim. Filters include indirect election, time between elections, and size of representative regions.

Frick, Henry Clay (1849–1919) Partner of Andrew Carnegie who later helped form the United States Steel Corporation.

Ford, Henry (1863–1947) American automobile manufacturer who pioneered the assembly line as a means of mass production.

Founding A conscious, deliberate act of creating a system of government that benefits the people.

Fourteen Points Moralistic ideals of Woodrow Wilson that were to be implemented after World War I in an attempt to have a lasting peace.

Fourteenth Amendment Defined citizenship and overturned the three-fifths compromise for slaves when determining representation, repudiated Confederate debts, and prohibited Confederate leaders from holding public office.

Franklin, Benjamin (1706–1790) One of the most well-known Founders, Franklin was also a leading printer, scientist, inventor, and diplomat. He helped secure France as an ally during the Revolutionary War.

Free rider program A good or service provided by one individual or company that benefits or is consumed at not charge by another individual or company.

Freedom A goal for economic systems to preserve the liberty of the people.

French and Indian War 1754–1763 conflict between the French and British/Americans and their respective Indian allies in which the French forces were defeated.

Freud, Sigmund (1856–1939) Considered the father of psychoanalytical psychology, Freud's theories were based on the idea that people were influenced in their behavior by subconscious and external factors beyond their control.

– G –

Generality Rule of law principle that states laws must apply to broad categories of people and must not single out individuals or groups for special treatment.

Gerry, Elbridge (1744–1814) A Massachusetts delegate to the Continental Congress and a signer of the Declaration of Independence, Gerry was one of three men who refused to sign the Constitution because it did not contain a bill of rights.

Get out the vote activity Aspect of party politics in which voters are systematically rounded up and helped to get to the polling place.

***Gibbons v. Ogden* (1824)** Supreme Court case in which the power of the federal government was expanded by broad interpretation of the commerce clause.

Glorious Revolution 1688 bloodless English revolution against the King, making the King subject to Parliament; considered a true founding of government.

God's Elect From John Calvin's predestination theology, the doctrine that God has already chosen those who will be saved. These elect people are to build a holy community as an example.

Good Society Reasonably stable and prosperous society without an oppressive tyranny. Usually includes peace, respect, vibrant culture, and personal freedom to live the way one chooses.

Gould, Jay (1836–1892) Robber baron and skilled business man involved with Tammany Hall and Boss Tweed and later had controlling interest in 15% of the country's railroad tracks.

Great Compromise, The Proposed by Roger Sherman, it brought together the New Jersey and Virginia Plans by having the upper congressional house representation equal by state and the lower house representation proportional by population.

Great Depression, The Extended recession in the 1930s that led to widespread unemployment, bank failure, and a general downturn in the economy until World War II.

Great Oughts, The Natural rights that don't proclaim an "is" so much as an "ought" about the world—the way things "should" be.

Greek cardinal virtues Elements of Greek *areté*: Temperance, Courage, Wisdom, Justice.

Greek freedom The privilege of taking part in the political process and observing society's rules.

Growth of government The steady drift of power from the states to the federal government, with increasing involvement of the federal government in American life.

Growth of personal rights The broadening judicial interpretation of personal rights that were construed fairly narrowly in the past.

Growth of privacy A broadening of the toleration that ought to be extended to an array of lifestyles, behaviors, choices, and value systems as well as a decrease of government prescription in individuals' lives.

– H –

Hamilton, Alexander (1755–1804) Hamilton served as the first Secretary of the Treasury under Washington and founded the Federalist Party. He also co-wrote *The Federalist* and championed a strong central government.

Hamiltonians See ***Federalist Party***.

Henry, Patrick (1736–1799) Best known for his famous "Give me liberty, or give me death" speech in the Virginia House of Burgesses, Henry was an Anti-Federalist who pushed for a bill of rights to be added to the Constitution after its ratification.

Heritage The traditions, beliefs, principles, events, etc. that we inherit (or choose to inherit) from the past.

Hoover, Herbert (1874–1964) 31st President of the United States, Hoover lost the 1932 election for a second term (to Franklin D. Roosevelt) when his responses to the Stock Market Crash of 1929 failed to end the economic recession and the nation slid into the Great Depression.

House of Burgesses An assembly of representatives elected by the common people of the Virginia colony, similar to the House of Commons.

Human nature The fundamental disposition of humans that determines their behavior.

Human predicament The cycle from tyranny to anarchy, to which sovereign power and its ill effects give rise.

Hume's filter/indirect election When the people select the most virtuous representatives, who in turn select even more virtuous government officials.

Huntington, Collis P. (1821–1900) One of the Big Four with Leland Stanford, Huntington was involved in both railroads and shipping. He founded Newport News Shipping, the largest privately owned shipyard in the United States.

– I –

Implied rights The doctrine that the Constitution protects rights that are not explicitly stated or enumerated therein.

Indentured servants See ***indentured servitude***.

Indentured servitude Land owners would pay the passage of those willing to come to the colonies in exchange for an agreed-upon term of service, after which the indentured servant was released from his obligation and was then free to seek his own fortune.

Initiative Progressive reform in which citizens could put propositions directly on the ballot through petition and have them become laws by garnering a majority vote.

Invisible hand, the Adam Smith's term for the natural self-regulation of a market economy driven by self-interest and efficiency.

Isolationism Political ideology that favored not becoming entangled with European powers, either as friend or foe, but did favor robust trade relations with a variety of partners.

– J –

Jackson, Andrew (1767–1845) The seventh President of the United States, Jackson championed the U.S. as a democracy, pushing for more political involvement by the common man. He also vetoed the U.S. Bank's charter and made other reforms to keep the federal government small.

Jay, John (1745–1829) A Founding Father, Jay served as a President of the Continental Congress, co-wrote *The Federalist* with Hamilton and Madison, and served as the first Chief Justice of the United States Supreme Court.

Jefferson, Thomas (1743–1826) Third President of the United States, Jefferson was the principal author of the Declaration of Independence and an influential Founding Father of the United States. He founded the Democratic-Republican Party and promoted the idea of a small federal government.

Jeffersonians See ***Democratic-Republican Party***.

Jingoism An aggressive style of bragging by European governments that helped bring about World War I.

Judicial activism When the Supreme court uses judicial review in order to achieve social goals.

Judicial legislation When courts do not feel bound by the letter of the law nor by their own precedents, and instead appropriate the legislative function of making laws in resolving issues.

Judicial review Political power of the Supreme Court to rule on the constitutionality of laws.

Judiciary Act of 1789 Congressional act passed in 1789 to form the federal court system and to authorize writs of mandamus.

Jungle, The Upton Sinclair's muckraker book that exposed the practices of Chicago meat-packing plants.

– K –

Kesler, Charles A senior fellow of The Claremont Institute and editor of the *Claremont Review of Books*. He received his Ph.D. in Government from Harvard University and is currently Director of the

Henry Salvatori Center at Claremont McKenna College. A Constitutional scholar, Kesler asserts there is a moral text in the Constitution.

Keynes, John Maynard (1883–1946) A British economist whose ideas would influence Roosevelt's New Deal intervention for the U.S. economy.

Keynesian economics Economic theory in which the economy would regulate itself, but in the case of extreme depression the government would be needed to artificially stimulate demand by increasing spending or cutting taxes.

King Jr, Martin Luther (1929–1968) A Baptist minister and political activist, King was also a leader of the civil rights movement in the 1960s. His most famous speech was delivered on the steps of the Lincoln Memorial, entitled "I Have a Dream."

–L–

Laissez-faire Policy in which there is little or no interference with exchange, trade, or market prices by the government.

Land Ordinance of 1785 Called for the systematic survey of the Northwest Territory and division into mile-square plots and organization into townships.

Law of comparative advantage Economic principle that because of varying opportunity costs between producers, it is more beneficial for producers to specialize and exchange than to try to produce everything individually.

Law of Demand As the price of a particular good or service rises, individuals will buy less of that good or service.

Law of Supply As the price of a particular good or service rises, suppliers will produce more of that good or service.

League of Nations One of Woodrow Wilson's Fourteen Points at the end of WWI; it called for the creation of a group of nations to help ensure peace. The U.S. never joined because of a veto by Congress. After WWII, the United Nations was formed with similar goals.

Lee, Robert E. (1807–1870) Confederate general and commander of the Army of Northern Virginia during the Civil War. After surrendering at Appomattox on April 9, 1865, Lee urged reconciliation with the North.

Liberalism One of the four alternative forms of government; sees people in the most favorable light, but institutions or other influences can corrupt them, so government is necessary to protect them from such corruption.

Libertarianism One of the four alternative forms of government; sees the most important value as individual freedom and holds that government should only protect that freedom and nothing more.

Liberty An ideal of freedom from oppression, tyranny, and government, allowing individuals to pursue happiness through positive action.

Lincoln, Abraham (1809–1865) 16th President of the United States, Lincoln sought to end slavery and preserve the Union. He signed the Emancipation Proclamation and delivered his famous "Gettysburg Address."

Locke, John (1632–1704) English philosopher whose *Treatises of Government* espousing natural rights, consent of the governed, and social compacts greatly influenced the Founding Fathers.

Louisiana Purchase Land purchased by Thomas Jefferson from France. Consists of much of the midwest United States.

Loyal opposition When losers in the political game continue to support the system, even when it is against their ideology.

–M–

Madison, James (1751–1836) Fourth President of the United States and Founding Father, Madison is often called the "Father of the Constitution." He co-authored *The Federalist* with Hamilton and Jay, and helped Jefferson create the Democratic-Republican Party.

Majority The candidate who receives more than 50% of total votes wins.

Manifest Destiny The belief that American expansion of an "Empire of Liberty" through the Western Hemisphere was justified by the benefits of bringing the American way of life to acquired territories.

Marbury, William A "Midnight Appointment" by John Adams, Marbury sued Secretary of State James Madison for delivery of his commission, which was being withheld by order of President Jefferson.

Marbury v. Madison Supreme court case in which judicial review was established.

Market economy An economic model proposed by Adam Smith in which the government serves only to create an acceptable environment in which to make exchanges; see also ***capitalism***.

Markets Divisions of the economy that specialize in certain goods or services.

Marshall, John (1755–1835) Fourth Chief Justice of the United States, Marshall ruled that writs of mandamus were unconstitutional in the case *Marbury v. Madison*, thereby establishing a precedent for judicial review.

Marx, Karl (1818–1883) German philosopher who wrote the *Communist Manifesto*, championing communism and socialism and attacking market economies.

Mason, George (1725–1792) Virginia delegate to the Constitutional Convention, Mason refused to sign the Constitution because it did not contain a declaration of rights.

McClellan, George B. (1826–1885) Union General during the Civil War. Although he helped raise and train the Union Army as general-in-chief, McClellan failed to press his advantage at the Battle of Antietam, and was later relieved of his command by President Lincoln.

***McCullough v. Maryland* (1819)** Supreme Court case in which greater federal power was established by maintaining the national bank.

Mercantilism An economic theory that emphasized the importance of gold and silver to the economic power of a nation. Mercantilists regulated the economy by encouraging exports and restricting imports.

Midnight appointments Judiciary appointments made by Federalist John Adams soon before leaving the presidency in response to the Democratic-Republican victory in the Congress and Presidency.

Missouri Compromise 1820 agreement between slavery and anti-slavery factions in the United States that regulated slavery in western territories, prohibiting slavery above the border of Arkansas (except Missouri) and permitting it south of that border.

Modernism A cultural movement embracing human empowerment and rejecting traditionalism as outdated. Rationality, industry, and technology were cornerstones of progress and human achievement.

Monetarists Supporters of Friedman's economic theory that the economy be controlled by regulating the money supply.

Monopoly When one person or group is the sole seller in a market with many buyers; the lack of competition in a market. See also ***collusion.***

Monroe Doctrine Ideology of James Monroe in which any aggressive move by a European power in the Western Hemisphere would be regarded as a challenge and affront to the United States.

Montesquieu, Charles-Louis de Secondat, Baron de La Brède et de (1689–1755) A French political thinker who favored the British system of rule and lauded the idea of separation of powers.

Moral consensus A general agreement on standards of right and wrong that was more prevalent in early America than it is today.

Moral self-governance Puritan ideal that all must live a righteous life largely on their own, with each man being responsible for his own actions and those of his family—with an eye on his neighbor as well.

Morality A system of conduct based on beliefs of right and wrong.

Morgan, J. P. (1837–1913) Banker and financier whose firm was one of the most powerful banking houses in the world. It financed the formation of the United States Steel Corporation, the world's first billion-dollar corporation.

Morris, Gouverneur (1752–1816) Pennsylvania representative at the Constitutional Convention, Morris is credited with authoring large sections of the Constitution, including the Preamble.

Muckraker Journalists that portrayed the leaders of corporations and the actions of their companies in unfavorable circumstances, writing "yellow journalism."

–N–

NAACP National Association for the Advancement of Colored People; civil rights organization on behalf of African Americans to protect their rights.

Narrow construction Constitutional clauses that were written to be interpreted in a more narrow or direct manner.

National self-interest An approach to foreign policy that places the value and interests of one's own country above that of others. Unlike individual self-interest that promotes cooperation and exchange in a market, national self-interest tends to be aggressive and destructive.

Natural law Law that classical Greeks believed resided in the human heart and reflected our innate sense of right and wrong.

Natural liberty Where men are free to do what they please, without regard for the moral value of their actions.

Natural rights Fundamental rights granted by nature that government could not abrogate and which government was bound to protect.

Navigation Acts Economic regulations passed by British Parliament to enforce trade regulations in the colonies: all trade had to go through British or colonial merchants and be shipped in British or colonial ships with the end goal to generate large exports from England, with few imports, so that gold and silver would flow into the motherland.

New Deal Plan by Franklin D. Roosevelt involving

the creation of various government agencies and programs designed to stimulate the economy and help the U.S. escape the Great Depression.

New Jersey Plan Plan presented during the Constitutional Convention in which each state would have equal representation in the Congress.

Northwest Ordinance of 1787 Called for the governmental development of the west based on creating self-governing republics that would be systematically added to the Union.

Northwest Territory Lands north of the Ohio River.

– O –

Octopus, The Frank Norris's novel that recounted the depredations of California railroads.

Oligarchy A form of government where most or all political power effectively rests with a small segment of society, typically the most powerful, whether by wealth, family, military strength, ruthlessness, or political influence.

Opportunity cost The value of the best alternative not chosen.

Ordinance of 1784 Plan of Thomas Jefferson to organize the national domain into discrete territories along with a three-stage development of government institutions.

Original consent Giving consent to a provision or law the first time, such as the ratification of the Constitution.

– P –

Paine, Thomas (1737–1809) Author of *Common Sense* and the *American Crisis* papers, which helped convince many Americans of the need for independence.

Parks, Rosa (1913–2005) African American seamstress and civil rights activist who refused to give up her seat to a white passenger while riding a Montgomery, Alabama, bus on December 1, 1955. Her subsequent arrest became the basis for challenging the legality of segregation laws.

Party newspaper A journal used by a political party for disseminating party information to and encouraging more active participation among the grass roots voters.

Patria Latin for "fatherland," from the Greek *patris*, also the root for the word *patriot*. A sense of homeland.

Paterson, William (1745–1806) New Jersey representative at the Constitutional Convention who presented the New Jersey Plan, which gave equal representation to states regardless of size or population.

Perfect competition When buyers and sellers have no influence on price and terms of exchange.

Periodic consent Giving continuing consent at certain intervals to a provision or law to which original consent has already been given.

Pinckney, Charles (1757–1824) A South Carolina representative to the Constitutional Convention, Pinckney was a strong promoter of Federalism and helped persuade ratification of the Constitution in South Carolina.

Pilgrims Small congregation of separatists seeking to distance themselves, physically and spiritually, from the Church of England by emigrating to New England.

Plato (c. 427–347 B.C.) Greek philosopher and author of the *The Republic*, which extolled civic virtue and the necessity of areté.

Plurality Recieving the largest percentage of the votes.

Poleis (plural of Polis) City or city-state, often self-governed by its citizens as the ancient Greek city-states were.

Political convention Large meeting of party delegates for the purpose of nominating candidates, often held with much pomp and ballyhoo.

Political legitimacy Ruling by a sanction higher than stark necessity; sanction may stem from religion, history, consent, etc.

Political machine Group of party loyalists organized to deliver the vote on election day. Historically they often used questionable or illegal means such as buying votes or intimidation at the polls.

Popular campaigning Promoting candidates as being from (and therefore representing) the common masses, rather than as elite gentlemen-politicians.

Popular government "Government of the people, by the people, for the people."

Popular sovereignty The idea that power is created by and subject to the will of the people. It was the basis for Madison's proportional representation in Congress and a justification by the South for the continuance of slavery.

Populism 1880s' political movement favoring nationalizing banks and railroads to protect farms and rural towns from the private power and corruption of big corporations.

Progressivism Post-populist, urban-based political movement against private power and corporate corruption that looked hopefully towards the future, emphasizing the benefits of science and technology.

Postmodernism A skeptical paradigm that critiques

the ideals of modernism such as materialism and rational purpose and questions the true objectivity of viewpoints.

Proportional representation When representation is determined by population.

Prospectivity Rule of law principle that states laws must apply to future action and not past action.

Public togetherness Aspect of party politics in which groups of political party members would gather together in order to have more solidarity and support.

Publicity Rule of law principle that states laws must be known and certain, such that everyone knows of their existence and their enforcement is reasonably reliable.

Public goods A good or service that, if consumed by one individual, does not diminish the amount of that good or service that remains available to others.

Puritans British religious emigrants who wanted to reform the Church of England rather than sever all ties with it; their beliefs of God's Elect, the Christian Calling, and Moral Self-Governance would help shape the Founding and American national character.

–R–

Randolph, Edmund (1753–1813) Governor of Virginia and delegate to the Constitutional Convention, Randolph refused to sign the Constitution in Philadelphia, but later was instrumental in persuading Virginia leadership to ratify it.

Recall Progressive reform in which citizens can call a special election by petition to recall an elected official; a majority vote removes the person from office.

Referendum Progressive reform in which laws passed by legislatures can be directly submitted to the people for a vote; a majority vote against the law removes it from the books.

Republic From the Latin *res publica*, the "public thing," when citizens of the political state govern themselves rather than submit to a despot or an oligarchy.

Republican Party Political party that stems from the controversy over slavery. It was dedicated to keeping future territories and states free from slavery.

Republican problem The question of how the benefits of self-government can be enjoyed without incurring its inherent problems.

Revolution A means of removing tyranny from power; part of the human predicament cycle.

Robber baron Muckraker term used for leaders of large corporations and trusts to reflect their power and unscrupulous natures.

Rockefeller, John D. (1839–1937) Founder of Standard Oil company. Known for his practice of buying out his competitors, Rockefeller was a favorite target of muckrakers. He was also a generous philanthropist. His name has beecome synonymous with massive wealth.

Roe, Jane The anonymous pseudonym used for Norma Leah McCorvey in the landmark case *Roe v. Wade*, which legalized abortion. McCorvey initially claimed to have been pregnant by rape, but in the 1980s confessed that she had fabricated the rape story. She also became a pro-life activist and lobbied the Supreme Court to reverse its decision.

***Roe v. Wade* (1973)** Case in which the Supreme Court decided that abortion was protected by the Bill of Rights.

Role of money Money facilitates exchange by eliminating the necessity for a "coincidence of wants," functioning as a generally acceptable medium for exchange.

Role of prices In a market economy, prices determine the quantity of goods supplied.

Role of profits In a market economy, as profits increase, the number of suppliers and resources for making that good will increase.

Rolfe, John (c. 1585–1622) Virginia colonist who pioneered the cultivation of tobacco as a profitable agricultural enterprise. Rolfe also married Pocahontas in 1614.

Roosevelt, Franklin D. (1882–1945) The 32nd President of the United States, Roosevelt served four terms, the only U.S. President to serve more than two terms. His exuberant public personality helped bolster the nation's confidence as it struggled through the Depression and then entered World War II.

Roosevelt, Theodore (1858–1919) The 26th President of the United States, Roosevelt was known for his boisterous personality. He was known for trust-busting, championing environmental causes, and his "big stick" foreign policy that called for American policing of the Western Hemisphere to protect its economic interests.

Rule of law A set of metalegal principles developed by the English legal system as a way of distinguishing whether a particular law supported freedom or not.

– S –

Scarcity Economic condition where resources and supply cannot meet all demands, and therefore decisions must be made about which demands to meet.

Secession Formal withdrawl of states or regions from a nation.

Second Treatise of Government John Locke's work arguing that true political authority comes not from God or precedent but from the people.

Sectionalism Factionalism on a larger, more regional scale, with fewer but larger factions. Sectionalism during the 1800s over the slavery issue nullified the benefits of Madison's extended republic and led to the Civil War.

Separation of powers Dividing powers of government between the separate branches.

Scott, Dred (c. 1795–1858) Slave who sued unsuccessfully for his freedom in 1857 because he had lived with his owner in several states where slavery was illegal. The ruling of *Dred Scott v. Sandford* determined that slaves were property and could not be freed by state laws. The ruling essentially nullified the Missouri Compromise and was a major factor contributing to the Civil War.

Sherman, Roger (1721–1793) Connecticut delegate to the Constitutional Convention, Sherman proposed the great compromise of one legislative house having proportional representation while the other had equal representation.

Shortage When the amount demanded is greater than the amount supplied.

Single representative districts Representational structure where each geographical region elects its one representative independent of outcomes in other regions.

Smith, Adam (1723–1790) Scottish philosopher and economist who wrote *The Wealth of Nations*. He is considered the father of modern economics.

Social compact The social concept of a group of autonomous individuals without government making a common agreement about the sort of political world they want to live in.

Social Darwinism Belief that society, like everything else, is in a state of constant change and development, evolving into ever higher and more complex forms.

Socialism Where government plans and regulates the economy.

Sovereignty Ultimate political power—having the final say.

Specialization The economic practice of focusing resources on production of one or a few goods.

Spencer, Herbert (1820–1903) Considered the father of Social Darwinism, Spencer coined the phrase "survival of the fittest" in his 1864 book *Principles of Biology*.

Stanford, Leland (1824–1893) 8th governor of California and President of the Central Pacific Railroad. He hammered in the famous golden spike on May 10, 1869.

State of nature Hypothetical condition assumed to exist in the absence of government where human beings live in perfect freedom and general equality.

State sovereignty When ultimate political power resides in the state rather than the federal government.

Structure Rules and restrictions designed to better harness virtue.

Supply See *Law of Supply*.

Surplus When the amount supplied is greater than the amount demanded.

– T –

Tabula rasa Latin for *clean slate* or *blank slate*. Puritans felt that the new world was a *tabula rasa* on which mankind could begin the human story anew.

Taney, Roger B. (1777–1864) Fifth Chief Justice of the Supreme Court, Taney ruled in *Dred Scott v. Sandford* that the Missouri Compromise was unconstitutional.

Taxation without Representation Rallying cry of the colonists during the Revolutionary period because of the taxes placed on them by a Parliament in which they had no representation.

Tea Act Legislation passed by the British government in 1773 designed to give the British East India Company a monopoly on tea in the colonies, the Act led to the infamous Boston Tea Party.

Theocracy Divinely inspired rule, or rule by religion.

Thirteenth Amendment Abolished slavery in the United States.

Three-fifths compromise Part of the Slavery Compromise, where 3 out of every 5 slaves were counted as part of state population for taxation and representation.

Tories English royal court and the the center of British political power, known also as the "Court party" and characterized by corruption and subversion.

Trust A business entity created with the intent to form a monopoly and dominate a market.

Turner, Frederick Jackson (1861–1932) American

historian who studied and wrote about the American experience and what made it unique.

Tweed, William Marcy (1823–1878) "Boss" Tweed was head of the Tammany Hall political machine in New York City. He was convicted and eventually imprisoned for stealing millions of dollars from the city through graft.

Tyranny Absolute power centralized in one person (or small group); part of the human predicament cycle.

–V–

Vanderbilt, Cornelius (1794–1877) A robber baron and ruthless businessman who made his fortune in shipping and then later in railroads.

Virginia Plan Plan presented during the Constitutional Convention in which each state would have proportional representation in the Congress.

–W–

Warren, Earl (1891–1974) Fourteenth Chief Justice of the Supreme Court, Warren ruled in *Brown v. Board of Education* that segregation was unequal and therefore unconstitutional.

Washington, George (1732–1799) Known as the "Father of His Country," General Washington led the Continental Army to victory during the Revolutionary War, presided over the Constitutional Convention, and was elected as the First President of the United States.

Wealth of Nations, The Book written by Scottish economist Adam Smith that criticized mercantilism and proposed a free market economy in which the "invisible hand" determined prices.

Weddington, Sarah (b. 1945) Attorney with Linda Coffee in the *Roe v. Wade* case.

Whig Party England's first political party, organized in political opposition to the King; Americans later formed their own Whig party during the Jacksonian democracy era, but the two parties did not hold the same ideology.

William of Orange (1650–1702) Husband of Mary (daughter of King James II), who acceded the throne with his wife in 1689, and became William III of England.

Wilson, James (1742–1798) A primary framer of the Constitution, Wilson proposed the three-fifths compromise for slave representation and election of the president by the people. He was also key in Pennsylvania's ratification of the Constitution.

Wilson, Woodrow (1856–1924) 28th President of the United States, Wilson helped frame the Treaty of Versailles ending WWI and proposed Fourteen Points that included the formation of the League of Nations.

Winthrop, John (1587–1689) Governor of the Massachusetts Bay Colony, Winthrop is known for his sermon "A Model of Christian Charity," in which he stated that the Puritan colony would be "a city upon a hill."

Writ of mandamus A court document forcing an action by a certain party.

INDEX

– K –

– L –

– M –

– N –

– O –

– P –

Figures

8.2 J. Mund
8.3 public domain art
8.4 Kent Minson, Devin LuBean
8.5 Kent Minson, Jennifer Berry
8.6 Kent Minson
8.7 Kent Minson, Jennifer Berry
8.8 Kent Minson, Jennifer Berry
8.9 Kent Minson, Jennifer Berry
8.10 Courtesy National Park Service, 1960
8.11 Courtesy loc.gov
8.12 Kent Minson

Chapter 9

Splash photos.com
9.1 Devin LuBean
9.2 Devin LuBean
9.3 photos.com
9.4 public domain photo, photos.com
9.5 Courtesy loc.gov
9.6 Courtesy loc.gov
9.7 Courtesy loc.gov
9.8 Devin LuBean
9.9 Devin LuBean
9.10 Currier & Ives, 1863
9.11 photos.com
9.12 photos.com
9.13 Devin LuBean
9.14 Currier & Ives, 1876
9.15 public domain photo, courtesy of railways-atlas.tapor.ualberta.ca
9.16 courtesy of egyptiansoccer.com
9.17 Devin LuBean
9.18 Kent Minson
9.19 Courtesy of David Stott
9.20 Kent Minson
9.21 photos.com
9.22 Devin LuBean
9.23 © held by Wal-mart, Microsoft
9.24 public domain, wikipedia.org
9.25 public domain, wikipedia.org

Public domain portraits:
Charles Darwin
Karl Marx
Sigmund Freud
Albert Einstein
Andrew Canegie
John D. Rockefeller

Chapter 10

Splash Photos.com
10.1 Kent Minson
10.2 Newell Covers Wyeth, 1919
10.3 Gaylord Watson, 1881
10.4 Hunter & Co., 1870
10.5 Kent Minson
10.6 ©2006 DigitalGlobe, TerraMetrics
10.7 Kent Minson, Devin LuBean
10.8 Currier & Ives, 1876
10.9 Herbert Morton Stoops, 1815
10.10 public domain cartoon, courtesy of wikipedia.org
10.11 Courtesy loc.gov
10.12 public domain cartoon, courtesy of wikipedia.org
10.13 Grant Wood, 1931
10.14 public domain photos, courtesy of wikipedia.org
10.15 courtesy of state.ca.gov

Public domain portraits:
Thomas Jefferson
Andrew Jackson
John Quincy Adams
Daniel Webster
Henry Clay
Frederick Jackson Turner

Chapter 11

Splash Photos.com
11.1 Alexander Gardner, 1862
11.2 Alexander Gardner, 1862
11.3 Kent Minson
11.4 courtesy loc.gov
11.5 Jennifer Berry
11.6 Timothy O'Sullivan
11.7 courtesy loc.gov, 1865
11.8 Devin LuBean
11.9 public domain, courtesy of civil-war.net
11.10 Devin LuBean
11.11 Devin LuBean
11.12 public domain art
11.13 Harper's Weekly, 1861
11.14 Matthew Brady, 1863
11.15 L. Prang & Co., 1887
11.16 Kent Minson
11.17 public domain cartoon, colorized by Devin LuBean

Public domain portraits:
Stephen A. Douglas
John Brown
Dred Scott
Abraham Lincoln
Stephen Douglas
John Bell
John Brekinridge
Jefferson Davis
Robert E. Lee
Ulysses S. Grant

Chapter 12

Splash Courtesy loc.gov
12.1 public domain art
12.2 public domain, wikipedia.org
12.3 public domain, wikipedia.org
12.4 public domain, wikipedia.org
12.5 public domain, wikipedia.org
12.6 public domain, wikipedia.org
12.7 courtesy financialhistory.org
12.8 public domain photo
12.9 public domain,wikipedia.org
12.10 courtesy loc.gov
12.11 courtesy loc.gov
12.12 public domain cartoon, courtesy of nevadaobserver.com
12.13 public domain cartoon, courtesy of cartoons.osu.edu
12.14 courtesy loc.gov
12.15 public domain cartoon, courtesy of elections.harpweek.com
12.16 public domain cartoon, courtesy of wikipedia.org
12.17 public domain, wikipedia.org
12.18 Devin LuBean
12.19 courtesy loc.gov
12.20 courtesy loc.gov
12.21 courtesy loc.gov
12.22 public domain photo, courtesy of teachpol.tcnj.edu
12.23 courtesy of memory.loc.gov
12.24 public domain photo, courtesy of teachpol.tcnj.edu
12.25 public domain image, courtesy of wikipedia.org
12.26 Devin LuBean

Public domain portraits:
Jay Gould
Leland Standford
Cornelius Vanderbilt
Collis Huntington
Henry Clay Frick
John D. Rockefeller
Andrew Carnegie
J.P. Morgan
Eugene V. Debs
William Marcy Tweed
Herbert Spencer
Franklin Delano Roosevelt
John Maynard Keynes
Charles A. Beard
Thomas Woodrow Wilson

Chapter 13

Splash mil.gov, wikipedia.org
13.1 H.S. Sands
13.2 William Hubbard, 1832
13.3 Kent Minson
13.4 Devin LuBean
13.5 John Gast, 1872
13.6 Kent Minson
13.7 Kent Minson
13.8 public domain, courtesy wikipedia.org
13.9 public domain, courtesy wikipedia.org
13.10 public domain, courtesy wikipedia.org
13.11 New York Times, 1918
13.12 courtesy history.navy.mil
13.13 public domain photos, courtesy of wikipedia.org
13.14 courtesy Mell Roy Peterson family
13.15 public domain, 2incolor.com
13.16 public domain, mbe.doe.gov
13.17 public domain, wikipedia.org
13.18 public domain, archives.gov
13.19 public domain, loc.gov
13.20 public domain, wikipedia.org
13.21 public domain, loc.gov
13.22 public domain, wikipedia.org
13.23 public domain, wikipedia.org
13.24 used by permission

Chapter 14

Splash photos.com
14.1 public domain, courtesy wikipedia.org
14.2 public domain, courtesy wikipedia.org
14.3 Asher Brown Durand, 1886
14.4 photos.com, Devin LuBean
14.5 public domain, wikipedia.org
14.6 public domain, loc.gov
14.7 public domain, loc.gov
14.8 public domain, wikipedia.org
14.9 public domain, loc.gov
14.10 public domain, loc.gov
14.11 public domain, loc.gov
14.12 Devin LuBean
14.13 public domain, wikipedia.org
14.14 photos.com
14.15 public domain art

Public domain portraits:
Rosa Parks
Earl Warren

Chapter 15

Splash Photos.com
15.1 public domain art
15.2 photos.com
15.3 Kent Minson
15.4 photos.com
15.5 public domain, wikipedia.org
15.6 photos.com
15.7 courtesy nola.com
15.8 public domain art
15.9 public domain, wikipedia.org
15.10 public domain, loc.org
15.11 public domain, wikipedia.org
15.12 public domain, loc.org
15.13 photos.com, modified by Kent Minson

Appendix A

Splash Photos.com
A.1 public domain art
A.2 Kent Minson
A.3 Devin LuBean
A.4 Devin LuBean
A.5 Devin LuBean
A.6 Kent Minson

Appendix B

Splash Photos.com
B.1 Devin LuBean
B.2 Devin LuBean
B.3 Kent Minson
B.4 Kent Minson
B.5 Devin LuBean
B.6 Kent Minson

Cover

Kent Minson
Original images courtesy, Photos.com, loc.gov

Splash Page Design

Kent Minson